HAZARDOUS GASES

HAZARDOUS GASES

Risk Assessment on the Environment and Human Health

Edited by

JASPAL SINGH
Assistant Professor, Department of Chemistry, Gurukula Kangri (Deemed to be University), Haridwar, Uttarakhand, India

R. D. KAUSHIK
Head, Department of Chemistry, Gurukula Kangri (Deemed to be University), Haridwar, Uttarakhand, India

MALVIKA CHAWLA
Scholar, Department of Chemistry, Gurukula Kangri (Deemed to be University), Haridwar, Uttarakhand, India

ELSEVIER

ACADEMIC PRESS
An imprint of Elsevier

Academic Press is an imprint of Elsevier
125 London Wall, London EC2Y 5AS, United Kingdom
525 B Street, Suite 1650, San Diego, CA 92101, United States
50 Hampshire Street, 5th Floor, Cambridge, MA 02139, United States
The Boulevard, Langford Lane, Kidlington, Oxford OX5 1GB, United Kingdom

Library of Congress Cataloging-in-Publication Data
A catalog record for this book is available from the Library of Congress

British Library Cataloguing-in-Publication Data
A catalogue record for this book is available from the British Library

ISBN: 978-0-323-89857-7

For information on all Academic Press publications
visit our website at https://www.elsevier.com/books-and-journals

Publisher: Andre Gerhard Wolff
Acquisitions Editor: Kattie Washington
Editorial Project Manager: Barbara Makinster
Production Project Manager: Swapna Srinivasan
Cover Designer: Miles Hitchen

Typeset by SPi Global, India

Contents

5. Bromine: Risk assessment, environmental, and health hazard

Manis Kumar Jha and Sushma

6. Bromine chloride: Risk assessment, environmental, and health hazard

Sushma

7. Carbon dioxide: Risk assessment, environmental, and health hazard

Malvika Chawla

8. Carbon monoxide: Risk assessment, environmental, and health hazard

Raj Nath Yadava and Vipul Bhatt

11. Cyanogen chloride: Risk assessment, environmental, and health hazard

Rajeev Jain and Sonu Kumar

12. Diazomethane: Risk assessment, environmental, and health hazard

Sonu Kumar

13. Fluorine: Risk assessment, environmental, and health hazard

Preeti Das and Pooja

14. Formaldehyde: Risk assessment, environmental, and health hazard

Santosh Kumar Dubey and Preeti Das

15. Hydrogen cyanide: Risk assessment, environmental, and health hazard

Manila and Payal Devi

16. Hydrogen selenide: Risk assessment, environmental, and health hazard

C. S. Mathela and Payal Devi

17. Hydrogen sulfide: Risk assessment, environmental, and health hazard

Ali Mohammed and Payal Devi

18. Methane: Risk assessment, environmental, and health hazard

Nitish Kumar and Himanshu Gupta

19. Methyl bromide: Risk assessment, environmental, and health hazard

R. D. Kaushik

20. Methyl isocyanate: Risk assessment, environmental, and health hazard

Manila

21. Nitric oxide: Risk assessment, environmental and health hazard

Krishan Chandra Singh and Kavita Rawat

28. Silane: Risk assessment, environmental, and health hazard

Shuvasish Choudhury, Ajendra Kumar, and Neeraj Kumar

29. Styrene: Risk assessment, environmental, and health hazard

Ritu Arora and Manila

30. Sulfur dioxide: Risk assessment, environmental, and health hazard

Mayank Kumar Malik and Sonu Kumar

Contributors

Ritu Arora Department of Mathematics and Statistics, Gurukula Kangri (Deemed to be University), Haridwar, Uttarakhand, India

Vipul Bhatt Gurukula Kangri (Deemed to be University), Haridwar, Uttarakhand, India

Malvika Chawla Department of Chemistry, Gurukula Kangri (Deemed to be University), Haridwar, Uttarakhand, India

Shuvasish Choudhury Central Instrumentation Laboratory, Assam University, Silchar, Assam, India

Preeti Das Department of Chemistry, Gurukula Kangri (Deemed to be University), Haridwar, Uttarakhand, India

Payal Devi Department of Chemistry, Gurukula Kangri (Deemed to be University), Haridwar, Uttarakhand, India

Santosh Kumar Dubey Department of Chemistry, Kurukshetra University, Kurukshetra, Haryana, India

Himanshu Gupta Department of Physics, Gurukula Kangri (Deemed to be University), Haridwar, Uttarakhand, India

Rajeev Jain Pondicherry University, Puducherry, India

Manis Kumar Jha Metal Extraction and Forming Division, National Metallurgical Laboratory (CSIR), Jamshedpur, Jharkhand, India

R. D. Kaushik Department of Chemistry, Gurukula Kangri (Deemed to be University), Haridwar, Uttarakhand, India

Nishant Khandelwal Department of Chemistry, Gurukula Kangri (Deemed to be University), Haridwar, Uttarakhand, India

Ajendra Kumar Department of Mathematics and Statistics, Gurukula Kangri (Deemed to be University), Haridwar, Uttarakhand, India

Ashwani Kumar Department of Chemistry, Kurukshetra University, Thanesar, Haryana, India

Neeraj Kumar Building Energy Efficiency Group, CSIR-Central Building Research Institute, Roorkee, Uttarakhand, India

Nitish Kumar Department of Chemistry, Gurukula Kangri (Deemed to be University), Haridwar, Uttarakhand, India

Ravendra Kumar Department of Chemistry, G.B. Pant University of Agriculture and Technology, Pantnagar, Uttarakhand, India

Sonu Kumar Department of Chemistry, Gurukula Kangri (Deemed to be University), Haridwar, Uttarakhand, India

Sanjeev Maken Department of Chemistry, Deenbandhu Chhotu Ram University of Science and Technology, Murthal, Haryana, India

Mayank Kumar Malik Department of Pharmaceutical Sciences; Department of Chemistry, Gurukula Kangri (Deemed to be University), Haridwar, Uttarakhand, India

Manila Department of Chemistry, Gurukula Kangri (Deemed to be University), Haridwar, Uttarakhand, India

C. S. Mathela Former Professor and Head, Department of Chemistry, Kumaun University, Nainital, Uttarakhand, India

Gagan Matta Department of Zoology & Environmental Science, Gurukula Kangri (Deemed to be University), Haridwar, Uttarakhand, India

Ali Mohammed Department of Applied Chemistry, Faculty of Engineering and Technology, Aligarh Muslim University, Aligarh, Uttar Pradesh, India

Sushil K. Pandey Department of Chemistry, University of Jammu, Jammu, India

Pooja Department of Chemistry, Gurukula Kangri (Deemed to be University), Haridwar, Uttarakhand, India

Lakshami Prashad Purohit Department of Physics, Gurukula Kangri (Deemed to be University), Haridwar, Uttarakhand, India

Kavita Rawat Department of Chemistry, H.N.B. Garhwal University, Badshahithaul Campus, Tehri Garhwal, Uttarakhand, India

Karuna Saini Department of Chemistry, Gurukula Kangri (Deemed to be University), Haridwar, Uttarakhand, India

Ashok Kumar Sharma Department of Chemistry, Deen Bandhu Chhotu Ram University of Science and Technology, Murthal, Haryana, India

Jaspal Singh Department of Chemistry, Gurukula Kangri (Deemed to be University), Haridwar, Uttarakhand, India

Krishan Chandra Singh Department of Chemistry, Maharishi Dayanand University, Rohtak, Haryana, India

Om Singh Department of Chemistry, College of Engineering Roorkee, Roorkee, Uttarakhand, India

Rakesh Kumar Soni Department of Chemistry, C.C.S University, Meerut, Uttar Pradesh, India

Sushma Department of Chemistry, Gurukula Kangri (Deemed to be University), Haridwar, Uttarakhand, India

Raj Nath Yadava Purnea University, Purnia, Bihar, India

Preface

The current increasingly rapid development of technological media, the rapid growth of industry, the thousands of cars that circulate every day in the streets of major cities, and the use of pesticides and other poisonous substances for the amplified production of products have increased atmospheric pollution, resulting in the daily exposure of humans to small concentrations of harmful gaseous substances, such as carbon monoxide, carbon dioxide and likewise, which have serious harmful effects on human health as well as the environment. It is therefore vital that those in government, industry, academia and research be informed of the associated risks and of possible means to manage these risks for the most vulnerable, as well as for the environment.

Hazardous Gases: Risk Assessment on the Environment and Human Health is aimed at providing readers with an in-depth knowledge of some of the hazardous gases that are potentially harmful to human health and the environment. It is an attempt to produce an effective one-stop reference for academia, students, researchers, and others engaged in an evaluation of chemicals; governmental agencies in healthcare; nonhealthcare agencies whose programs affect health; and pharmaceutical companies and private research consultancies, as well as others with an interest in this crucial topic.

The book is divided into 30 chapters incorporating information about the effect of poisonous gases on the environment and human health hazards. Some of the hazardous gases included in this edition of the book are arsenic pentafluoride, arsine, boron trifluoride, bromine, bromine chloride, carbon dioxide, carbon monoxide, chlorine, cyanogen, cyanogen chloride, diazomethane, fluorine, formaldehyde, hydrogen cyanide, hydrogen sulfide, hydrogen selenide, methane, methyl bromide, nitric oxide, nitrogen dioxide, oxygen, ozone, phosgene, phosphine, styrene, silane, and sulfur dioxide. The chapters are alphabetically arranged to make the content easy to access. Each chapter covers important definitions, theoretical explanations, pictorial and tabular data, historical background, physicochemical properties, physiological role discussing the mechanism of toxicity, the effects on humans as well as the environment, emergency guidelines, hazard control measures, and handling and storage of hazardous gases. In addition, the related risks of each gas as well as management of the risks have been included, which are in line with the accident case studies related, that turned into disasters. Also, this book examines recent research trends for specific toxic gases.

We hope this book will prove valuable to academia and students, government agencies in healthcare, pharmaceutical companies, and private research consultancies, as it gives immediate access to nearly 30 common hazardous gases.

We would like to extend our sincere gratitude and appreciation for all of the hard work and dedication provided by our authors. Without the endless hours they spent working on this book project, with their experience, efforts, and support, this book would not have been possible.

Having an idea and turning it into a book is as difficult as it sounds. The experience is both internally challenging and rewarding. We especially want to thank all those individuals who helped to make this happen.

Finally, without the experience and support of the Elsevier team, this book would not exist. The team has shown their support throughout the process and have given us the opportunity to lead a great group of individuals. Thank you to Barbara Makinster, Kattie Washington, Swapna Srinivasan, and team. We could not have done it without your help.

Jaspal Singh

R. D. Kaushik

Malvika Chawla

CHAPTER

1

Arsenic pentafluoride: Risk assessment, environmental, and health hazard

Om Singh

Department of Chemistry, College of Engineering Roorkee, Roorkee, Uttarakhand, India

Abbreviations

AAS atomic absorption spectrometry
AHRA arsenic health risk assessment
ATP adenosine triphosphate
CC cloud computing
DMSA dimercaptosuccinic acid
DNA deoxyribonucleic acid
EDTA ethylenediaminetetraacetic acid
GIC general informative criteria
Iips International Institute for Population Sciences
Mips microprocessor without interlocked pipelined stages
NAD+ nicotinamide adenine dinucleotide
SDS safety data sheet

1.1 Introduction

Arsenic pentafluoride (AsF_5) is a colorless, noncombustible gas that is extremely toxic and corrosive. Arsenic and hydrogen fluoride are formed when arsenic pentafluoride is hydrolyzed in the presence of water and a humid environment. This gas is hazardous to the aquatic environment (Greenwood & Earnshaw, 1997). It is heavier than air and is also a potential carcinogen as listed by the International Agency for Research on Cancer. Highly toxic, acute, or chronic health hazards result from this gas if gasped, swallowed, or absorbed by the skin (Cotton, Wilkinson,

Hazardous Gases
https://doi.org/10.1016/B978-0-323-89857-7.00007-4

Murillo, & Bochmann, 1999). Arsenic pentafluoride destroys red blood cells and gives off irritation vapors on inhalation in moist air causing widespread systemic injury.

1.2 Occurrence

Arsenic is present everywhere in the environment containing various sulfur-containing minerals, such as arsenopyrite. This depends basically on the geologic participations from mineral-enduring procedures, whereas volumes in inherent floras and faunas consider race diversity. The naturally high quantity of arsenic is restrained by some species of aquatic plants, such as algae and seaweed, and aquatic organisms, such as shellfish (lobster, crab, barnacle, etc.). However, synthetic sources include lead, zinc, and gold that are result of the smelting of nonferrous metal minerals, mostly copper (Paul, Chakraborty, Ali, & Ray, 2015). Arsenic pentafluoride is formed by the reaction of As and F and other reactions as shown from Eqs. (1.1)–(1.4), respectively:

$$2As + 5F_2 \rightarrow 2AsF_5 \quad (1.1)$$

$$AsF_3 + F_2 \rightarrow AsF_5 \quad (1.2)$$

$$2As_2O_5 + 10F_2 \rightarrow 4AsF_5 + 5O_2 \quad (1.3)$$

$$2As_2O_3 + 10F_2 \rightarrow 4AsF_5 + 3O_2 \quad (1.4)$$

It is a great acceptor as revealed in reaction (1.5). As with sulfur tetrafluoride develops an ionic complex-forming halide complex (Azeem, Brownstein, & Gillespie, 1969).

$$AsF_5 + SF_4 \rightarrow SF_3^{+} + AsF_6^{-} \quad (1.5)$$

1.3 Physicochemical properties

Arsenic pentafluoride has a trigonal bipyramidal structure with five bonding pairs and zero lone pairs (Greenwood & Earnshaw, 1997) (Fig. 1.1). In the solid state, the axial and equatorial As—F bond lengths are 171.9 and 166.8 pm, respectively. At room temperature, it is a colorless gas, and at −53°C, a condensed yellow liquid. It is also formed by the reaction of fluorine with arsenic metal or arsenic trifluoride. Fluoride works as a strong Lewis acid; it is also used for the formation of ionic complexes; and in combination with Bronsted acid forms conjugate superacid. Vapor density measurements show some grade of association, but it is a monomeric covalent compound with a high degree of coordinating ability. Due

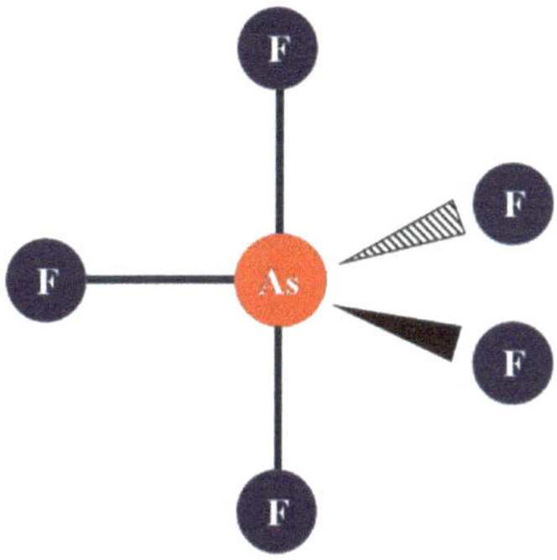

FIG. 1.1 Structure of arsenic pentafluoride. *Credit: Om Singh.*

to its potential high toxicity, safety must be taken in the management of any arsenic compound (Singh, 2017).

It is also prepared with graphite and constant intercalation compounds, which show electric conduction similar to Ag.

Some of the physical properties of arsenic pentafluoride are shown in Table 1.1 (Fischer & Thompson, 1978).

1.4 Physiological role

Metabolic processes in organisms can be influenced by hydrogen fluoride or fluoride ions released after hydrolysis of arsenic pentafluoride. After absorption, As(V) reacts with As(III), which is subsequently

TABLE 1.1 Physical properties of arsenic pentafluoride.

Synonyms	Arsenic pentafluoride Pentafluoroarsorane Arsorane, pentafluoro-
Molecular weight	169.91361 g/mol
Color	Colorless
Physical form	Solid branded carbon beads and poisonous gas confined within a wrapped stainless-steel cylinder
Molecular formula	AsF_5
Boiling point	−63 °F (−53°C)
Freezing point	−112 °F (−80.0°C)
Vapor pressure	11,790 mmHg @ 20°C
Mass	7.71 g/L
Solvent solubility	Alcohol, ether, benzene

methylated and eliminated via the urine. However, the corrosiveness of this gas does not induce the intake of doses augmenting the physiological elimination of arsenic and fluoride. The gas reacts with macromolecules in tissues, metabolized via some pathways after the delay. The whole mechanism is still a topic of research (Jaishankar, Tseten, Anbalagan, Mathew, & Beeregowda, 2014).

1.5 Uses

Arsenic pentafluoride is used to make electroconductive polymers, semiconductor fabrication, and general synthetic chemicals (Vogel, 1977). Intercalation compounds of graphite such as "graphite/AsF_5" have been suggested for use both as extremely electronically conducting constituents (Fischer & Thompson, 1978) and chemical catalysts (Foley, Zeller, Falardeau, & Vogel, 1977; Olah, Kaspi, & Bukala, 1977; Selig & Gani, 1975). There is documentation showing that arsenic pentafluoride oxidizes graphite but with an uncertainty of the extent of oxidation and amount of AsF_3 (Ebert, Mills, Scanlon, & Selig, 1981). The compounds formed are considered among the most highly anisotropic of any general informative criteria (McRae, Mareche, Lelaurain, & Furdin, 1987).

1.6 Risk assessment

A number of health effects, including neurologic and neurobehavioral effects, heart disease, pulmonary or hepatic fibrosis, hematological effects, etc., can result postexposure to this gas. Therefore, some important points in risk assessment should be kept in mind (Singh, 2017):

- Cloud computing (CC) might fill the gaps between radio communication and environmental or health interference.
- The intangible outline of a cloud-based pronouncement support structure for arsenic encompasses substance health risk analysis (CC-AHRA).
- Risk assessment clarifies how the CC-AHRA may support uncovered societies in considering the danger of cancer due to arsenic contact from drinking water.
- The CC-AHRA may support water organization and system, investigator, and private initiative production with well-versed judgments for mitigating health hazards due to arsenic ingestion.

1.7 Hazardousness

Arsenic pentafluoride is a hazardous gas that destroys red blood cells often affecting the lungs, skin, kidneys, and liver. When the poisoning becomes acute, symptoms lead to diarrhea, nausea, hematuria, cramping muscles, alopecia, colic pain, convulsions, headache, delirium, vomiting, body ache, angina, and pain in the abdomen or loins.

While arsenic pentafluoride has not been recognized as a congenital abnormality or a multiplicative hazard, arsenic and definite compounds are teratogens and might lead to reproductive damage, such as productiveness and interference with menses (Kim & Kim, 2015).

With the help of certain mechanisms, arsenic and its metabolites interrupt adenosine triphosphate (ATP) production. Arsenic also prevents pyruvate dehydrogenase by opposing phosphate, which separates oxidative phosphorylation at the level of the citric acid cycle, and hence prevents energy-linked reduction of NAD^+, mitochondrial exhalation, and ATP synthesis. Arsenic can alter DNA-mending enzyme activity, and interfere with DNA methylation designs and cell production (Casarett, Klaassen, & Watkins, 2003; Kitchin & Wallace, 2008).

Arsenic is absorbed by inhalation or ingestion, but when it comes in contact with the skin it is absorbed in a smaller amount. After inhalation or ingestion, it is distributed all over the body and reduces to arsenite. Arsenic or its metabolites are eliminated in urine. Arsenic binds with metal and forms protein metallothionein, which reduces the noxious effects of arsenic by acting as an antioxidant (Pokhrel, Bhandari, & Viraraghavan, 2009).

1.8 Treatment

In concern with how to handle AsF_5 toxicity, treatment of arsenic poisoning can be done with the help of chelation therapy. Some chelating agents are dimercaprol, ethylenediaminetetraacetic acid, or dimercaptosuccinic acid. For less severe cases, charcoal tablets and a high sulfur diet can be used (Pokhrel et al., 2009).

1.9 Risk management

- Due to probable adverse effects in exposed persons, many rules and plans have been identified for numerous inorganic and organic procedures for arsenic by international, federal, and state organizations.
- Treatment approaches as well as chemical precipitation (clotting procedures), ion exchange, reverse osmosis/electrodialysis, use of activated carbon or alumina, and oxidation have been suggested for arsenic elimination in contaminated water.
- CC, an internet-based, manageable, economical, and on-demand calculating facility, has appeared as a key constituent in real-time administrative progressions. It permits user to access, examine, and assess data anywhere (Mondal, Majumder, & Mohanty, 2006).

1.10 Current research

The following current research or experimental work has been recognized in different science areas (Ebert & Selig, 1977):

- AsF_5 is a poisonous gas that can be used in the making of electroconductive polymers. Also, the AsF_5 molecule can compromise various health problems, for example, nervous system syndrome. Therefore, the assessment of AsF_5 gas is essential but there are concerns about public health and environmental pollution. Recently, scientific research reported a novel Kagome phosphorene nanosheet (Kagome-PNS) for the study of the performance of AsF_5 molecules using the density-functional theory method. The Kagome-PNS can be skillfully used as a chemical sensor to identify the molecule of AsF_5 gas (Nagarajan & Chandiramouli, 2020).
- Current progress in the assessment of arsenic by arsine generation and atomic absorption spectrometry with argon–hydrogen combustions has provided very sensitive and fast analyses of trace quantities.
- The technique is useful to investigate environmental pollution (H_2O, soil, and wastewater), steel, and silicate rocks. For rock examination, the sample is decayed by H_2SO_4, HNO_3, HF acids, and $KMNO_4$ solution; frequent fuming is required to eliminate fluoride, but the process is time consuming (Wahed et al., 2006).
- The process can be practically functional for the assessment of arsenic in rocks, soils, deposits, and numerous reserves. All researches can be admitted to the cloud/Internet component of the CC-AHRA through a web portal.
- Various nanomaterial-based trace element sample pretreatment procedures have been established in the past for simplifying actions and growing the analytical presentation of the approaches (Mumuni & Peacock, 2001).
- Solid-phase abstraction, particularly that use microprocessor without interlocked pipelined stages/International Institute for Population Sciences, provides the important involvement of element preconcentration and speciation examination in biological, food, environmental examination, etc. (Horwitz, 1979).

1.11 Safety measures

The following are some of the safety measures that should be followed (Kim & Kim, 2015):

- For eye protection, immediately wash the eye and internal surface of the eyelid with water for a few minutes and keep the eye open if required.

- Preferably wear a fire prevent lab coat, pants, and closed-toe shoes.
- Contact with skin, eyes, and clothing should be avoided.
- Wash the hands immediately after handling the product.
- Move immediately to fresh air in case of accidental inhalation of the gas.

1.12 Handling and storage (Jaishankar et al., 2014)

Handling

- Gas cylinders must be handled in sealed material.
- The cylinder should not be heated in any way to enhance the rate of the product.
- In a mandate to protect subatmospheric (<760 Torr) packages, the temperature of the cylinder must not be more than 24°C.
- If the temperature is greater than 24°C, the container may be pressurized, which advances the chance of seepage or contact with hazardous vapors (Mondal et al., 2006).

Storage

- Arsenic pentafluoride should be stored according to all recent guidelines and ethics (Saunders & Wheeler, 1991).
- Keep the gas in a dry, cool, and ventilated area with a noncombustible infrastructure.
- Protect the gas container from contact with sunlight, mechanical losses, and temperatures below 55°C.
- Gas containers should be stored away from oxygen or other extremely oxidized or combustible materials.
- Gas cylinders should have exit plugs and valves with protective caps in the vessel.

1.13 Precautions

There are different precautions for handling arsenic pentafluoride (Horwitz, 1979):

- Wash hands immediately after handling.
- Wear protective gloves, clothes, eyeglasses, and face covers.
- Do not drink or smoke.
- Choose outdoor or well-ventilated areas.
- Be familiar with all relevant information.
- Use respiratory protection.

References

Azeem, M., Brownstein, M., & Gillespie, R. J. (1969). An investigation of the structures of the adducts of SF4 with BF3, PF5, AsF5, and SbF5 in the solid state and in solution in HF. *Canadian Journal of Chemistry, 47*(22), 4159–4167.

Casarett, L. J., Klaassen, C. D., & Watkins, J. B. (2003). *Casarett and doull's essentials of toxicology*: (p. 25). New York: McGraw-Hill/Medical Pub. Div. Chapter 1.

Cotton, F. A., Wilkinson, G., Murillo, C. A., & Bochmann, M. (1999). *Advanced inorganic chemistry*: (p. 869). New York: John Willey & Sons Inc.

Ebert, L. B., Mills, D. R., Scanlon, J. C., & Selig, H. (1981). Magnetic resonance of intercalation compounds of graphite: Questions of ionicity and mobility of inserted species. *Materials Research Bulletin, 16*(7), 831–840.

Ebert, L. B., & Selig, H. (1977). Determination by magnetic resonance of the oxidation state of species intercalated in graphite. *Materials Science and Engineering, 31*, 177–181.

Fischer, J. E., & Thompson, T. E. (1978). Graphite intercalation compounds. *Physics Today, 31* (7), 36–45.

Foley, G. T., Zeller, C., Falardeau, E. R., & Vogel, F. L. (1977). Room temperature electrical conductivity of a highly two dimensional synthetic metal: AsF_5-graphite. *Solid State Communications, 24*(5), 371–375.

Greenwood, N. N., & Earnshaw, A. (1997). *Chemistry of the elements*. Butterworth.

Horwitz, W. (1979). *Arsenic and tin in foods: Reviews of commonly used methods of analysis*. (Vol. 9). Food & Agriculture Org.

Jaishankar, M., Tseten, T., Anbalagan, N., Mathew, B. B., & Beeregowda, K. N. (2014). Toxicity, mechanism and health effects of some heavy metals. *Interdisciplinary Toxicology, 7*(2), 60–72.

Kim, Y. J., & Kim, J. M. (2015). Arsenic toxicity in male reproduction and development. *Development and Reproduction, 19*(4), 167.

Kitchin, K. T., & Wallace, K. (2008). The role of protein binding of trivalent arsenicals in arsenic carcinogenesis and toxicity. *Journal of Inorganic Biochemistry, 102*(3), 532–539.

McRae, E., Mareche, J. F., Lelaurain, M., & Furdin, G. (1987). Conductivity anisotropy in AsF_5—Intercalated graphite. *Journal of Physics and Chemistry of Solids, 48*(10), 957–963.

Mondal, P., Majumder, C. B., & Mohanty, B. (2006). Laboratory based approaches for arsenic remediation from contaminated water: Recent developments. *Journal of Hazardous Materials, 137*(1), 464–479.

Mumuni, A. R., & Peacock, C. J. (2001). Development of a simple system for the determination of arsenic after hydride generation atomic absorption spectrophotometry. *Journal of Applied Sciences and Environmental Management, 5*(1).

Nagarajan, V., & Chandiramouli, R. (2020). Arsenic pentafluoride surface adsorption studies on Kagome-phosphorene–a DFT outlook. *Physics Letters A*, 126552.

Olah, G. A., Kaspi, J., & Bukala, J. (1977). Heterogeneous catalysis by solid superacids. 3. Alkylation of benzene and transalkylation of alkylbenzenes over graphite-intercalated Lewis acid halide and perfluorinated resin sulfonic acid (Nafion-H) catalysts. *The* Journal of Organic Chemistry, *42*(26), 4187–4191.

Paul, S., Chakraborty, S., Ali, N., & Ray, D. P. (2015). Arsenic distribution in environment and its bioremediation: A review. *International Journal of Agriculture, Environment and Biotechnology, 8*(1), 189–204.

Pokhrel, D., Bhandari, B. S., & Viraraghavan, T. (2009). Arsenic contamination of groundwater in the Terai region of Nepal: An overview of health concerns and treatment options. *Environment International, 35*(1), 157–161.

Saunders, R., & Wheeler, T. (1991). *Handbook of safety management*. Pitman.

Selig, H., & Gani, O. (1975). Graphite intercalation compounds with xenon oxide tetrafluoride and iodine pentafluoride. *Inorganic and Nuclear Chemistry Letters, 11*(1), 75–77.

Singh, S. K. (2017). Conceptual framework of a cloud-based decision support system for arsenic health risk assessment. *Environment Systems and Decisions*, *37*(4), 435–450.

Vogel, F. L. (1977). The electrical conductivity of graphite intercalated with superacid fluorides: Experiments with antimony pentafluoride. *Journal of Materials Science*, *12*(5), 982–986.

Wahed, M. A., Chowdhury, D., Nermell, B., Khan, S. I., Ilias, M., Rahman, M. , et al. (2006). A modified routine analysis of arsenic content in drinking-water in Bangladesh by hydride generation-atomic absorption spectrophotometry. *Journal of Health, Population and Nutrition*, 36–41.

CHAPTER

2

Arsine: Risk assessment, environmental, and health hazard

Rakesh Kumar Soni[a] *and Om Singh*[b]

[a]Department of Chemistry, C.C.S University, Meerut, Uttar Pradesh, India
[b]Department of Chemistry, College of Engineering Roorkee, Roorkee, Uttarakhand, India

Abbreviations

ALS advanced life support
ATD average total dose
CBC complete blood count
CDI chronic daily intake
CGA PL Compressed Gas Association, Pamphlet
CR cancer risk
GAP good agricultural practices
HACCP hazard analysis critical control points
Hb hemoglobin
HI hazard index
HQ hazard quotient
MOE margin of exposure
PRA probabilistic risk assessment
PVC poly vinyl chloride
USEPA United States Environmental Protection Agency
WHO World Health Organization

2.1 Introduction

Arsine is a nonirritating, toxic, transparent, and flammable gas with a mild garlic or fish-like odor, which is detectable at levels sufficient to cause poisoning (Plantamura, Dorandeu, Burnat, & Renard, 2011). The general formula of an organoarsenic compound is $AsH_3 - xRx$, (R is aryl or alkyl

https://doi.org/10.1016/B978-0-323-89857-7.00028-1

group), as shown in Fig. 2.1. Arsenic on reaction with acid forms arsine, which is also known as stibine (smells like rotten eggs).

2.2 Occurrence

Arsine does not occur in nature and was first obtained in 1775 by Scheele. Commercially, it is prepared by electrochemical reduction of arsenic compounds in a solution of Zn_3As_2 and acid (Eq. 2.1). It can also be synthesized by reacting arsenic trioxide and sodium borohydride catalyzed by zinc in acid (Sengupta, Sawalha, Ohira, Idowu, & Dasgupta, 2010).

$$3H_2 \rightarrow HAsO_2 \rightarrow AsH_3 + 2H_2O \tag{2.1}$$

From Eq. (2.1), it can be seen that the nascent hydrogen is responsible for the mitigation of arsenicals to arsine. Despite this known perspective, until now there is no evidence available supporting the essential role of nascent hydrogen in the preparation of arsine.

Another synthetic approach that more sufficiently explains arsine preparation is the thermodynamically favored reaction given in Eq. 2.2:

$$6H^+ + 2AI(O) + HAsO_2 \rightarrow AsH_3 + 2H_2O + 2AI_3(E = 1.48\ \text{V}) \tag{2.2}$$

2.3 Historical background

Arsine gas was recognized in 1775 by Scheele. In 1815, Cehlen, a German chemist, was the first victim reported to die from accidental exposure to arsine (Elkins & Fahy, 1967), which created havoc in the vicinity.

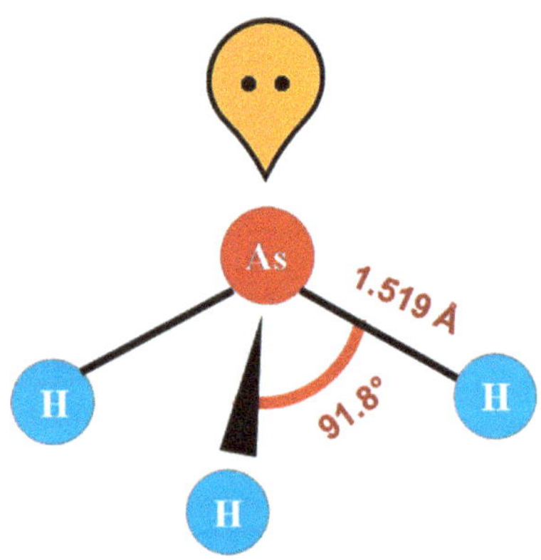

FIG. 2.1 Structure of arsine. *Credit: Om Singh.*

2.4 Physicochemical properties

Arsine is generally prepared from inorganic arsenic (such as arsenite or arsenate) under definite chemical conditions (Winski, Barber, Rael, & Carter, 1997). Arsine smells like a clove of mild garlic or fish when present at above 0.5ppm; otherwise it is a colorless, odorless, and flammable gas. It is soluble in water but slightly soluble at 20°C. This gas is 2.7 times denser than air. Arsine is a stable compound but decomposes at room temperature slowly; this decomposition rate is dependent on factors like humidity, presence of light, specific catalysts, etc.

Arsine can react hazardously with halogens like fluorine and chlorine, which can even result in an explosion. It can react with strong oxidizing agents like permanganate and nitric acid and easily oxidizes with concentrated or dilute O_2 engagement, represented by reaction (2.3):

$$2AsH_3 \rightarrow 3O_2 \rightarrow As_2O_3 + 3H_2O \quad (2.3)$$

2.5 Physiological role

There are reports of toxicity of arsine in humans showing a dependency on oxidative pressure (Blair et al., 1990; Hatlelid, 1996; Hatlelid, Brailsford, & Carter, 1995; Hatlelid & Carter, 1997; Pernis & Magistretti, 1960) or reaction with sulfhydryl groups (Levinsky, Smalley, Hillyer, & Shindler, 1970; Winski et al., 1997). The following reaction (Eq. 2.4) depicts the hemolytic activity of arsine as postulated by Hatlelid and Carter (1997).

$$H_2As - H + HbO_2 \rightarrow H_2As^* + Hb - O_2 - H \quad (2.4)$$

On the addition of H_2As-H_b or H_2As–heme, the reaction further leads to the formation of methemoglobin, arsine peroxide, hydrogen peroxide, or arsenic, as shown by Eq. (2.5):

$$H_2As^* + H_b - O_2 - H \rightarrow \text{adduct} + H_2O_2 \quad (2.5)$$

Moreover, the formed adduct may damage hemoglobin (Hatlelid & Carter, 1997), which can cause restructuring of the protein membrane as well as alter membrane rigidity. This initiates the oxidation of heme iron for the formation of hemin with the help of protein sulfhydryl groups present in the membrane (Hatlelid, 1996). Eventually the overriding number of Heinz bodies and hemin sources rupture the red blood cell (RBC) membrane, hence causing cell fragmentation (Weed & Reed, 1966).

Some studies also suggest the hindrance of hemoglobin oxidation, which is responsible for erythrocyte structure by sulfhydryl groups of glutathione (Blair et al., 1990; Pernis & Magistretti, 1960). The imbalanced sodium–potassium pump causes red cell inflammation and bleeding when the sulfhydryl group of Na^+K^+-ATPase reacts (Fig. 2.2).

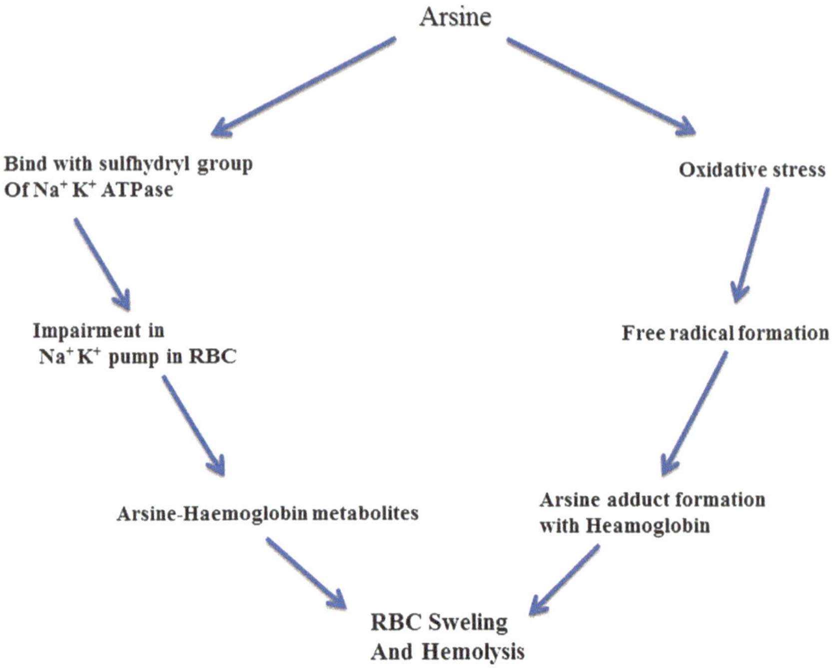

FIG. 2.2 Schematic showing the physiological role of arsine gas. *Credit: Om Singh.*

Distortion in the cellular membrane and RBC due to the imbalance in the sodium and potassium channel and higher than normal hematocrit has been observed with the help of ultrastructure (Winski et al., 1997). However, no significant effect of arsine was observed on ATP levels and ATPase.

2.6 Uses

Due to the colorless and odorless nature of arsine, in earlier times it was used as a chemical weapon and was considered ideal for a covering effect. Some of the chief uses of arsine are discussed in a later section (Pakulska & Czerczak, 2006). Arsine is frequently utilized in industries analogous to the production of semiconductors and dopants, and for soldering, galvanizing, burnishing, lead plating, and as a computer chip.

The GaAs semiconductor is prepared by chemical vapor deposition at 700–900°C (Eq. 2.6):

$$Ga(CH_3)_3 + AsH_3 \rightarrow GaAs + 3CH_4 \tag{2.6}$$

As—H bonds are often bound due to acidic properties. Therefore, it can be deprotonated as given in Eq. (2.7):

$$AsH_3 \rightarrow NaNH_2 \rightarrow NaAsH_2 + NH_3 \quad (2.7)$$

2.7 Sources of contaminants

The source of arsine exposure appears mostly from industrial procedures used in various chemical and metallurgical industries (Pakulska & Czerczak, 2006), such as casting and purifying of metals, galvanization, rubbing, darkening of zinc/aluminum, printing, plating, coal transformation to gas, overcharging of lead batteries, and other operations like the production of nascent hydrogen and reaction with acids and unpolished metals containing arsenic in the form of adulteration. Other sources of exposure are detailed in Table 2.1 (Chou & World Health Organization, 2003; National Research Council, 2003).

2.8 Risk assessment

While dealing with the risk analysis of arsine, several approaches have been employed (Hong, Song, & Chung, 2014), e.g., the United States Environmental Protection Agency recognized a technique to assess cancer-causing and noncausing risk associated with consistent ingesting of substances containing arsenic. Alternatively, four significant constraints, i.e., average total dose, chronic daily intake, cancer risk, and hazard quotient (HQ), essential to evaluate health hazards have also been taken advantage of. Personnel handling arsine should undergo measurable approaches with the purpose of risk assessment, e.g., probabilistic risk assessment, margin of exposure, and hazard index.

2.9 Hazardousness

Arsine is a noxious form of arsenic (Wald & Becker, 1986), which generally acts as a hematotoxin responsible for causing hemoglobinuria, jaundice, and hemolytic anemia (Blair et al., 1990). The hazardousness of the arsine gas increases osmotic brittleness of RBCs, fragment erythrocytes, and consequently affects the formation of blood cellular components. Arsine absorption occurs in humans and animals through the mucous surface of the lungs and respiratory tract. It is degraded by metabolic activity to trivalent arsenic (As(III)), (As(V)), and methylated to monomethylarsonate(V) and dimethylarsinate(V) (Pakulska & Czerczak, 2006). However, the dispersal of gas in body tissues is comparatively slow and correspondingly shows symptoms after long-time exposure only. Acute arsine

TABLE 2.1 Different sources of exposure to arsine.

Place of introduction	Development of exposure
Chemical industry	Workers are vulnerable while handling arsenic, nascent hydrogen, and aluminum.
Copper smelting and refinery	The reaction of sulfuric acid with a galvanized pail can be caused by the accidental release of arsine gas.
Bronzing process	The presence of zinc in the technical process of bronze plating containing arsenic and ferric chloride in concentrated HCl can lead to the formation of arsine.
Transmission repair shop	Some pesticides deposited on the aluminum transmission cover of trucks in contact with acidic detergents can release arsine gas.
Paint workshop	The reaction of nascent hydrogen and arsenic, present on the surface of painting material, can react to form arsine.
Fe^{+2} metal foundry	Arsine is released when hot waste holding an arsenical impurity is covered with water.
Polishing of metallic elements	In an acidic environment, arsenic changes into arsine in the presence of nascent hydrogen.
Darkening of zinc/aluminum alloy parts	Zinc/aluminum alloy in an acid solution forms arsine.
Soil-confined arsenic compounds	With the help of biotransformation by fungi or bacteria (*Alcaligenes* and *Pseudomonas*), arsenic converts into arsine.
Transistor industry	Arsine can drip while it is distributed to a semiconductor unit.
Microelectronic industry	The production of semiconductors may release arsine.
Lead–acid battery manufacturing	Arsine may be produced when battery acid reacts with lead–arsenic alloy.
Moist houses	In moist houses, mold growth on paper leads to the formation of compounds containing trimethylarsine and possibly arsine itself.
Home conditions, cot mattresses	Polyvinyl chloride-covered cot mattresses may also release arsine gas.

poisoning may cause sleepiness, difficulty in breathing, anorexia, nausea, diarrhea, immobility, and even death in some cases. On the other hand, chronic arsine poisoning can cause headaches, vomiting, low temperature, numbness, hematolysis, loss of blood, liver and kidney function disturbance, and peculiar Mees' lines (Ayala-Fierro, Baldwin, Wilson, Valeski, & Carter, 2000; Risk & Fuortes, 1991).

2.10 Treatment

Though there is no specific available antidote for the management of arsine, early chelation therapy can be effectively employed postexposure. In serious conditions, symptomatic treatment such as fluid resuscitations can be given to control blood pressure and for cardiac monitoring (Kuivenhoven & Mason, 2019). Treatment should be prioritized by examination of blood tests or other specific tests like kidney function tests (creatinine, bilirubin levels).

If hypotension or sign of shock is detected, then fluid management should be preferred. If adult systolic pressure is lower than 80 mmHg, then lactated Ringer's solution or intravenous saline may be suitable to avoid fluid imbalance (Reigart & Roberts, 1999). Diuretics such as Lasix and furosemide and other drugs like British anti-Lewisite, dimercaptopropanesulfonic acid, and succimer(2,3-dimercaptosuccinic acid) can also be used as per the symptomatic condition. Likewise, other treatments such as magnesium sulfate, defibrillation lidocaine, amiodarone in ventricular tachycardia and benzodiazepine, and general anesthesia in the case of annexations (Kuivenhoven & Mason, 2019) can also be considered.

2.11 Risk management

Implementation of standards by supervisory bodies must be given priority keeping in view the level of risk (Hong et al., 2014). The practice of chemical involvement and biotic control of As, Cd, Hg, and Pb in food should be thoroughly studied. The investigation reports suggest use of good agricultural practices, hazard analysis critical control points, or other production-specific practices to avoid adulteration in food manufacturing and processing (International Agency for Research on Cancer, 2012). As stated by the World Health Organization, a toxico-vigilance center should be available to assist as a chain linking instructors, controllers, and medical officers with actual individuals in native societies.

2.12 Current research

There are reports of biomarkers for the detection of toxic arsenic exposure, genomic vulnerability, and genotoxicity (Khan, Chakraborty, Bundschuh, Bhattacharya, & Parvez, 2020). The studies recognize genomic variations of arsenic methyltransferase that could govern vulnerability to arsenic-associated health consequences. Arsenic stimulating DNA destruction and variation of gene and protein appearance have also been studied. Studies concerning the techniques responsible for the improvement of sustainable magnetic chitosan biosorbent beads for kinetic reduction or removal of arsenic-contaminated water (Ayub, Raza, Majeed, Tariq, & Irfan, 2020) have also been done.

2.13 Arsine toxicity: Case studies

The first arsine toxicity incident took place in 1815 when Scheele, a German chemist, died after breathing in a small quantity of arsine in his laboratory.

On January 21, 2005, four employees were admitted to hospital who were working on an electrocoagulation system for removing arsenic from process water. After an investigation by doctors, an excessive level of arsine was found in their bodies.

Some reports are available for arsine acute exposure (Hesdorffer, Milne, Terblanche, & Meyers, 1986; Morse & Setterlind, 1950; Pinto, Petronella, Johns, & Arnold, 1950) but these reports are based on signs and symptoms such as intestinal and muscular pain, vomiting and diarrhea, and blood in urine, and are dependent on the level of exposure (Levinsky et al., 1970).

2.14 Safety measures

It is recommended to follow the following safety measures (Flora, Flora, & Saxena, 2009):

- Do not attempt to lift an arsine cylinder without a protection cap.
- Do not pull or drop cylinders because they may be physically damaged. Use a cart designed to transport cylinders even for a short distance.
- Make sure that all pipelines and apparatus are properly connected.
- Ensure that a fume hood or walled system has been installed.
- Always wear protective gloves made from neoprene, butyl rubber, polyvinyl chloride, polyethylene, or Teflon.
- Contact lenses should be removed and not reused after use. Eyeglasses should be washed thoroughly.
- After use, personal protective equipment should be placed in a polybag and disposed of (Czerczak et al., 2002).

2.15 Handling

- Due to high reactivity, keep arsine away from heat, open flames, and hot surfaces.
- Ventilated areas should be preferred.

- Equipment in use in the vicinity of arsine must be free from sparking.
- Aluminum, carbon steel, and stainless steel are the best materials for arsine handling.
- Avoid using brass.
- Gasket materials such as Kel-F and Teflon or Viton and nylon are preferred.
- Use a reverse flow protective device in piping.
- An adjustable strap wrench should be used for removing overtight or rusted caps.
- After each use, close the valve.
- Do not use a wrench, screwdriver, pry bar, etc. in valve cap openings. This may cause leakage.
- Only well-trained and correctly educated individuals must handle arsine.
- Arsine must be handled using good industrial hygiene and safety procedures.
- Cylinders should be kept with valve protection caps in place.

2.16 Storage (Hong et al., 2014)

- Store away from combustible, nonventilated, moist, and heavily-trafficked areas.
- Storage temperature must be maintained below 52°C/125 °F.
- Filled and unfilled cylinders must be stored separately.
- Always keep and hold pressured gas cylinders in agreement with the Compressed Gas Association pamphlet, "Safe Handling of Compressed Gases in Containers."

2.17 Precautions

- Follow all compacted gas universal handling protections.
- Nonsparking or explosion-proof electrical apparatus must be used.

References

Ayala-Fierro, F., Baldwin, A. L., Wilson, L. M., Valeski, J. E., & Carter, D. E. (2000). Structural alterations in the rat kidney after acute arsine exposure. *Laboratory Investigation*, *80*(1), 87–97.

Ayub, A., Raza, Z. A., Majeed, M. I., Tariq, M. R., & Irfan, A. (2020). Development of sustainable magnetic chitosan biosorbent beads for kinetic remediation of arsenic contaminated water. *International Journal of Biological Macromolecules*, *163*, 603–617.

Blair, P. C., Thompson, M. B., Bechtold, M., Wilson, R. E., Moorman, M. P., & Fowler, B. A. (1990). Evidence for oxidative damage to red blood cells in mice induced by arsine gas. *Toxicology, 63*(1), 25–34.

Chou, C. H., & World Health Organization (2003). *Hydrogen sulfide: Human health aspects.* World Health Organization.

Czerczak, S., Fishbein, L., & World Health Organization (2002). *Arsine: Human health aspects.* World Health Organization.

Elkins, H. B., & Fahy, J. P. (1967). Arsine poisoning from aluminium tank cleaning. *Indian Medical Surgery, 36*(11), 747–749.

Flora, S. J., Flora, G., & Saxena, G. (2009). Arsenicals: Toxicity, their use as chemical warfare agents, and possible remedial measures. In *Handbook of Toxicology of Chemical Warfare Agents* (pp. 109–133). Academic Press.

Hatlelid, K. M. (1996). Reactions of arsine with hemoglobin. *Journal of Toxicology and Environmental Health Part A, 47*(2), 145–157.

Hatlelid, K. M., Brailsford, C., & Carter, D. E. (1995). An in vitro model for arsine toxicity using isolated red blood cells. *Fundamental and Applied Toxicology, 25*(2), 302–306.

Hatlelid, K. M., & Carter, D. E. (1997). Reactive oxygen species do not cause arsine-induced hemoglobin damage. *Journal of Toxicology and Environmental Health, 50*(5), 463–474.

Hesdorffer, C. S., Milne, F. J., Terblanche, J., & Meyers, A. M. (1986). Arsine gas poisoning: The importance of exchange transfusions in severe cases. *British Journal of Industrial Medicine, 43*(5), 353.

Hong, Y. S., Song, K. H., & Chung, J. Y. (2014). Health effects of chronic arsenic exposure. *Journal of Preventive Medicine and Public Health, 47*(5), 245.

International Agency for Research on Cancer (2012). *Effectiveness of tax and price policies for tobacco control.* (Vol. 14). World Health Organization.

Khan, K. M., Chakraborty, R., Bundschuh, J., Bhattacharya, P., & Parvez, F. (2020). Health effects of arsenic exposure in Latin America: An overview of the past eight years of research. *Science of the Total Environment, 710,* 136071. https://doi.org/10.1016/j.scitotenv.2019.136071.

Kuivenhoven, M., & Mason, K. (2019). Arsenic (arsine) toxicity. In *StatPearls [internet].* StatPearls Publishing (NCBI).

Levinsky, W. J., Smalley, R. V., Hillyer, P. N., & Shindler, R. L. (1970). Arsine hemolysis. *Archives of Environmental Health: An International Journal, 20*(3), 436–440.

Morse, K. M., & Setterlind, A. N. (1950). Arsine poisoning in the smelting and refining industry. *Achieves of Industrial Hygiene and Occupational Medicine,* 2(2), 148–169.

National Research Council (2003). *Acute exposure guideline levels for selected airborne chemicals.* (Vol. 3). National Academies Press.

Pakulska, D., & Czerczak, S. (2006). Hazardous effects of arsine: A short review. *International Journal of Occupational Medicine and Environmental Health, 19*(1), 36–44.

Pernis, B., & Magistretti, M. (1960). A study of the mechanism of acute hemolytic anemia from arsine. *Medicina del Lavoro, 51*(1), 37–41.

Pinto, S. S., Petronella, S. J., Johns, D. R., & Arnold, M. F. (1950). Arsine poisoning. A study of thirteen cases. *Achieves of Industrial Hygiene and Occupational Medicine,* 1(4), 437–451.

Plantamura, J., Dorandeu, F., Burnat, P., & Renard, C. (2011, July). Arsine: An unknown industrial chemical toxic. *Annales Pharmaceutiques Francaises, 69*(4), 196–200.

Reigart, J. R., & Roberts, J. R. (1999). *Recognition and management of pesticide poisonings.* US Environmental Protection Agency.

Risk, M., & Fuortes, L. (1991). Chronic arsenicalism suspected from arsine exposure: A case report and literature review. *Veterinary and Human Toxicology, 33*(6), 590–595.

Sengupta, M. K., Sawalha, M. F., Ohira, S. I., Idowu, A. D., & Dasgupta, P. K. (2010). Green analyzer for the measurement of total arsenic in drinking water: Electrochemical reduction of arsenate to arsine and gas phase chemiluminescence with ozone. *Analytical Chemistry, 82*(9), 3467–3473.

Wald, P. H., & Becker, C. E. (1986). Toxic gases used in the microelectronics industry. *Occupational Medicine (Philadelphia, Pa)*, *1*(1), 105–117.

Weed, R. I., & Reed, C. F. (1966). Membrane alterations leading to red cell destruction. *The American Journal of Medicine*, *41*(5), 681–698.

Winski, S. L., Barber, D. S., Rael, L. T., & Carter, D. E. (1997). Sequence of toxic events in arsine-induced hemolysis in vitro: Implications for the mechanism of toxicity in human erythrocytes. *Toxicological Sciences*, *38*(2), 123–128.

CHAPTER

3

Boron trichloride: Risk assessment, environmental, and health hazard

Ravendra Kumar[a] *and Nishant Khandelwal*[b]

[a]Department of Chemistry, G.B. Pant University of Agriculture and Technology, Pantnagar, Uttarakhand, India

[b]Department of Chemistry, Gurukula Kangri (Deemed to be University), Haridwar, Uttarakhand, India

Abbreviations

DNEL	derived no-effect level
PNEC	predicted no-effect concentration
PPE	personal protective equipment
STP	standard temperature and pressure

3.1 Introduction

Boron trichloride is an inorganic compound with the formula BCl_3. It is a colorless gas (under pressure) or liquid that is normally used as a reagent in organic synthesis. It is extremely reactive toward water and is a Lewis acid generally used for the demethylation of methyl ethers (ATSDR, 2007). BCl_3 in gaseous form has been used as soldering fluxes for alloys of aluminum, iron, zinc, tungsten, and Monel; in the plasma etching of aluminum alloys, tungsten, and tungsten silicide; to remove nitrides, carbides, and oxides in the refining of aluminum, magnesium, zinc, and copper alloys; to treat the melt in the casting of aluminum; as a source of boron in high energy fuels and rocket propellants; to douse fires in heat-treating ovens that contain magnesium products; as reagents in the synthesis of drug candidates for a range of diseases; and in the fiber industry as unrefined materials needed to produce pyrolytic boron nitride and boron fiber, components of high-tech composite structures.

https://doi.org/10.1016/B978-0-323-89857-7.00023-2

Boron trichloride is a corrosive and toxic substance that can cause injury to the skin, eyes, and respiratory tract. This rapidly evaporating liquid can cause frostbite too. High inhalation exposure of BCl_3 can cause pulmonary edema. BCl_3 reacts violently with H_2O evolving HCl and boric acid (H_3BO_3). It also forms HCl on contact with moisture present in air and soil. It reacts violently with H_2O producing a visible vapor cloud. Short-term exposure to BCl_3 causes second- and third-degree skin burns and severe eye injury. Inhalation can cause edema and severe upper respiratory tract irritation, and vapors of BCl_3 are intolerable even at low concentrations (HSDB, 2011). The boron trichloride (gas) can be collected and measured as a liquid by condensing in a cooled centrifuge tube and then transferred to the reaction system by distillation with a slow stream of nitrogen.

3.2 Occurrence

Boron is an element that presents in the environment mostly through natural processes. Boron occurs naturally in the environment due to the discharge into air, soil, and water during weathering. It may also occur in groundwater in very little amounts. Humans add boron to the environment by manufacturing glass, combusting coal, melting copper, and through the addition of fertilizers. The concentrations of boron that are added by humans are smaller than the naturally added concentrations through natural weathering.

3.3 Physicochemical properties

Boron trichloride hydrolyzes easily in H_2O, moisture containing air, or ethanol to boric acid and HCl (ATSDR, 2007), as represented by the following equation

$$BCl_3 + 3\,H_2O \rightarrow H_3BO_3 + 3\,HCl$$

The pH of a saturated solution of BCl_3 is equivalent to that of concentrated HCl (boron trichloride rapidly and completely decomposes in water). A concentrated solution of hydrochloric acid has an HCl concentration of 12.338 mol/L (Lide, 2002). This corresponds to a pH of 1.0–1.3. This would require a mass of 481 g of boron trichloride to be decomposed into HCl and dissolved into the water for every liter of solution. However, if boron trichloride reacts with H_2O in an environment that is open to the atmosphere, the HCl gas produced from decomposition will not fully dissolve into the solution. The quantity of a gas that dissolves into a liquid solvent is directly proportional to its partial pressure above the liquid.

TABLE 3.1 Physicochemical properties of BCl_3.

Property name	Property value
Molecular formula	BCl_3
Molecular weight	117.17 g/mol
Physical state/appearance	Colorless fuming liquid at low temperatures
Odor	Pungent, suffocating odor
Boiling point (°C)	12.5
Melting point (°C)	−107
Vapor density (g/m^3)	4.03 (air = 1)
Vapor pressure (kPa at 27°C)	166
Water solubility	Hydrolyzes upon contact with water into H_3BO_3 and HCl
Other solubilities	Decomposes in alcohol to give HBr and HCl
Flammability in air	Nonflammable
Density (g/cm^3)	1.3728 at 0°C
Synonyms	Trichloroborane, trichloroboron

This is demonstrated by Henry's Law given below (Zumdahl & Zumdahl, 2010):

$$P = kC$$

where

P = partial pressure of the gas above the solution.
k = Henry's law constant for the given solute/solvent combination.
C = concentration of the gas dissolved.

Following the entry into the upper GI tract, boron trichloride hydrolyzes. Some of the physicochemical properties are mentioned below in Table 3.1.

3.4 Physiological role

BCl_3 decomposes quickly in the presence of moisture to boric acid and HCl (U.S. EPA, 2004). Thus, oral, inhalation, or dermal exposure results in the production of these decomposition products. Toxic kinetic studies of BCl_3 have not been reported in either humans or animals. Discussion of the toxico kinetics of BCl_3 is based on its two hydrolysis products, boric acid and HCl, as follows (Eq. 3.1):

$$BCl_3 + 3\,H_2O \rightarrow H_3BO_3 + 3\,HCl \tag{3.1}$$

3.5 Uses

It has established its major usage as a reagent in various synthetic reactions. Some of its uses are detailed in this segment.

- Used in metal refining by removing oxides, nitrides, and carbides of Al, Mg, and Zn.
- Used as a flux for soldering alloys of Al, Fe, Zn, and W.
- Used for improving the casting of Al.
- Used for making electrical resistors.
- Used as a boron source, a high energy fuel for the development of rocket propellants.

3.6 Source of contaminants

Boron disclosure through air and drinking water is not very likely to occur, but the risk of exposure to borate dust in the working areas does exist. Boron exposure may also occur from consumer products such as cosmetics and laundry.

3.7 Risk assessment

A risk assessment should be conducted and documented in a working place to evaluate the risks related to the use of BCl_3 and to select the PPE that matches the relevant risk. The following recommendations should be measured for risk assessment:

- Preferably use permanent leak-tight connections.
- Systems under pressure should be frequently checked for leakages.
- Gas detectors should be used when a toxic amount releases.
- Give sufficient general or local ventilation.
- Consider the work permit system.
- Product should be handled in a closed environment and under severely restricted conditions.
- Wear personal protective equipment kit.
- Eye and face defense: Protect eyes, face, and skin from liquid splashes. It is suggested to wear safety eyewear, goggles, or face-shield to avoid

disclosure to liquid splashes. (Guideline: EN136 Respiratory defensive devices. Full face masks. Requirements, testing, marking).

- Skin protection hand protection advice: Wear cold insulating gloves (EN 511). Protective gloves against cold. Chemically resistant neoprene gloves complying with EN 374 should be worn at all times when handling chemical products if a risk assessment indicates this is necessary (EN 374-1/2/3 Protective gloves against chemicals and micro-organisms). Wear working gloves and safety shoes while handling containers.
- Body protection: Protect eyes, face, and skin from contact with the product. Maintain appropriate chemically resistant defensive clothing readily available for emergency use. Personal protective equipment (PPE) for the body should be selected based on the task being performed and the risks concerned. (Guideline: EN 943: Protective clothing against liquid and gaseous chemicals, aerosols, and solid particles. Performance necessities for gas-tight (Type 1) chemical protective suits for emergency teams).
- Other protection: Wear working gloves and safety shoes while handling containers (ISO 20345 Safety footwear).
- Respiratory protection: Keep the self-contained breathing apparatus willingly obtainable for emergency use. Use SCBA in the presence of high concentrations. The choice of the Respiratory Protective Device (RPD) must be based on known or anticipated exposure levels, the hazards of the product, and the safe working limits of the selected RPD. When allowed by a risk assessment Respiratory Protective Equipment (RPE) may be used. (Guideline: EN136 Respiratory protective devices. Full face masks. Requirements, testing, marking Respiratory protection).
- For short-term use: Material: Filter B-P3 Guideline: EN 14387: Respiratory protective devices. Gas filter(s) and combined filter(s). Requirements, testing, marking thermal hazards if there is a risk of contact with the liquid, all protective equipment should be appropriate for particularly low temperatures.
- Environmental exposure controls: Specific risk management measures are not required beyond good manufacturing hygiene and protection procedures. Refer to local regulations for restriction of emissions to the atmosphere.

3.8 Hazardousness

The exposure above the 5-ppm level may bring severe irritation in the upper respiratory tract; this may lead to the burning of the throat and choking. The long duration of exposure may lead to pulmonary edema, ulceration to the nose, and laryngeal spasm. Life-threatening results within a few minutes of exposure in the range of 1500–2000 ppm have been reported in the literature, which includes a severe stage of damage to the liver and kidney. In case of contact with eyes, it can fatally harm the eye, whereas ingestion can cause burns similar to frostbite (HSDB, 2011).

The only available data on acute inhalation of BCl_3 is lethality tests in which the LC_{50} (1-h) for male rats was found to be 2541 ppm (12,140 mg/m^3) (Vernot, MacEwen, Haun, et al., 1977) and the LC_{50} (1-h) for female rats was 21,100 mg/m^3 (HSDB, 2011).

3.9 Environmental effects

Plants absorb boron from the ground and through plant-consuming animals; boron can end up in food chains. Boron has been found in animal tissue, but it is not likely to accumulate.

When animals absorb large amounts of boron over a relatively long period of time through food or drinking water, the male reproductive organs get affected. When animals are exposed to boron during pregnancy their offspring may suffer from birth defects or overdue development. Furthermore, animals are likely to suffer from nose irritation when they breathe in boron.

3.10 Risk management

- Consider a work permit system for example, for maintenance activities.
- Make sure there is adequate air ventilation.
- Grant sufficient general and local exhaust aeration.
- Keep concentrations well below occupational exposure limits.
- Gas detectors should be used when toxic quantities may be released.
- Systems under pressure should be frequently checked for leakages.
- Product should be handled in a closed system and under strictly controlled conditions.
- Only use permanent leak-tight installations (e.g., welded pipes).
- Do not eat, drink, or smoke when using the product.

Some occupational exposure limits are given in Table 3.2.

3.11 Safety measures

- For prevention from exposure, the engineering controls must be checked accurately.

TABLE 3.2 Occupational exposure limits of Boron trichloride.

Type	Value
DNEL values	
Short term local inhalation	16 ppm
Long term local inhalation	8 mg/m^3
Long term systemic inhalation	16 mg/m^3
PNEC values	
Sewage treatment plant	39 μg/L
Freshwater	0.039 mg/L
Aquatic (marine water)	39 μg/L
Marine water	0.039 mg/L
Aquatic (intermit releases)	48 μg/L
Freshwater—intermittent	0.048 mg/L
Soil	11 μg/kg
Air	16 mg/m^3
Aquatic (freshwater)	39 μg/L
STP	0.039 mg/L

- The location with chemicals should be enclosed with exhaust aeration.
- Good work practices can help to reduce dangerous exposures as workplace controls are healthier than personal protective equipment.
- Use corrosion-proof equipment and a well-ventilated exhaust system in the worker's breathing zone.
- Use of explosion-proof, corrosion-resistant, forced-draft fume hood is preferred.
- Wear safety glasses with side or face shields when transfer filling or breaking transfers connections.
- Provision of readily accessible eyewash stations and safety showers should be there.
- Protective clothing like metatarsal shoes, work gloves should be used for cylinder handling.
- When workplace conditions warrant respirator use, follow a respiratory protection program.

3.12 Handling and storage

Handling

- Only skilled and properly instructed people should handle gas.
- The substance must be handled in accordance with good industrial cleanliness and protection procedures.
- Use only a properly specified apparatus which is suitable for this product, its supply pressure, and temperature.
- Make contact with your gas supplier if in uncertainty.
- Keep away from disclosure, obtain special directions before use.
- Do not smoke while handling the product.
- Ensure the complete gas system has been (or is frequently) checked for leaks before use.
- Flush out the system with dry inert gas (e.g., He or N_2) before the gas is introduced and when the system is placed out of service.
- Avoid suck back of water, acid, and alkalis. Refer to supplier's handling directions.
- Do not allow back feed into the container.
- Protect containers from physical damage (do not drag, roll, slide, or drop).
- When moving containers, even for short distances, use suitable equipment (trolley, hand truck, and fork truck, etc.)
- Leave valve protection caps in place until the container has been protected against either a wall or bench or placed in a container stand and is ready for use.
- If consumer experiences any complexity while operating container valve, discontinue using it and contact the supplier.
- Do not attempt to repair or modify container valves or safety relief devices.
- Damaged valves should be immediately reported to the supplier.
- Keep container valve outlets clean and free from contaminates, mainly oil and water.
- Close the container valve after each use and when it is empty (even if still connected to equipment).
- Never attempt to transfer gases from one container to another.
- Never use direct flame or electrical heating devices to raise the pressure of a container.
- Do not take out or deface labels provided by the supplier for the credentials of the container contents.

Storage

- Keep the container in a spacious place below 50°C. Cylinders should be stored in the vertical position and appropriately protected to prevent spilling.

- Secure cylinders to prevent them from falling.
- Stored containers should be occasionally checked for general conditions and escape.
- Container valve guards or caps should be in position.
- Store containers in a location free from fire risk and away from sources of heat and ignition.
- Keep away from flammable materials.
- Observe all regulations and local requirements regarding the storage of containers.
- Containers should not be stored in an environment that is likely to encourage corrosion.

3.13 Precautions

- One should be well-skilled about handling and storage of the cylinders.
- BCl_3 should be kept away from water, as it may release HCl gas and heat.
- It must be stored to avoid contact with hexa fluoroiso propylidene amino lithium, NH_3, $C_6H_5NH_2$, PH_3, N_2O_4, strong acids, and strong bases (violent reactions occur with these chemicals).
- BCl_3 is not compatible with alcohols, fats, and grease, oxidizing reagents (perchlorates, peroxides, permanganates, chlorates, nitrates, Cl, Br, and F).
- It should be stored in a tightly closed container in a cool, dry, spacious area away from metals and moisture.

References

ATSDR (Agency for Toxic Substances and Disease Registry) (2007). *Toxicological profile for boron (draft for public comment).* Atlanta, GA: U.S. Department of Health and Human Services, Public Health Service.

HSDB (Hazardous Substances Data Bank) (2011). *Boron trichloride (CAS No. 10294 34 5).* National Library of Medicine. Retrieved from: (2011). *http://toxnet.nlm.nih.gov/.*

Lide, D. R. (Ed.). (2002). *Handbook of chemistry and physics.* (83rd ed.). Boca Raton, FL: CRC Press.

U.S. EPA (Environmental Protection Agency) (2004). *Toxicological review of boron and compounds (CAS No. 7440-42-8). In Support of summary information on the integrated risk information system (IRIS)* EPA/635/04/052. Retrieved from: (2004). http://www.epa.gov/ncea/iris/toxreviews/0410tr.pdf.

Vernot, E. H., MacEwen, J. D., Haun, C. C., et al. (1977). Acute toxicity and skin corrosion data for some organic and inorganic compounds and aqueous solutions. *Toxicology and Applied Pharmacology, 42*(2), 417–423.

Zumdahl, S. S., & Zumdahl, S. A. (2010). *Chemistry* (8th ed.). Belmont, CA: Brooks/Cole.

CHAPTER

4

Boron trifluoride: Risk assessment, environmental, and health hazard

Ashwani Kumar[a] *and Nishant Khandelwal*[b]

[a]Department of Chemistry, Kurukshetra University, Thanesar, Haryana, India

[b]Department of Chemistry, Gurukula Kangri (Deemed to be University), Haridwar, Uttarakhand, India

Abbreviations

DFT density functional theory
G2YNT graphdiyne nanotube
NIOSH National Institute for Occupational Safety and Health
OSHA Occupational Safety and Health Administration

4.1 Introduction

Boron trifluoride, also well-known as trifluoroborane, is an inorganic colorless gas with a pungent odor. BF_3 is also a toxic and corrosive gas that is frequently used as a Lewis acid to catalyze numerous organic reactions such as the alkylation of aromatic rings and also as a versatile building block for the synthesis of numerous boron compounds. Harshaw Chemical patented the preparation of boron trifluoride from borax and HF in 1939. BF_3 is conveniently handled in the lab as an adduct with ethyl ether. When dissolved in water it produces hydrofluoric acid, a highly corrosive substance. The vapors of boron trifluoride are heavier than air, which on exposure to fire or heat may bring out violent rupturing.

Hazardous Gases
https://doi.org/10.1016/B978-0-323-89857-7.00017-7

4.2 Occurrence

Boron trifluoride does not exist in nature in a free state. It was synthesized for the first time in the year 1808 by Joseph Louis, Gay-Lussac, and Louis Jacques Thénard (Gay-Lussac & Thenard, 1809). Boron trifluoride is mainly prepared either by the reaction of hydrofluoric acid and boron trioxide (Eq. 4.1) or when sodium tetrafluoroborate, boron trioxide, and sulfuric acid react altogether in a definite proportion (Eq. 4.2).

$$B_2O_3 + 6\,HF \rightarrow 2\,BF_3 + 3\,H_2O \quad (4.1)$$

$$6\,NaBF_4 + B_2O_3 + 3\,H_2SO_4 \rightarrow 8\,BF_3 + 6\,NaHSO_4 + 3\,H_2O \quad (4.2)$$

Alternatively, BF_3 may be produced by the reaction of boric acid, sulfuric acid, and ammonium bifluoride.

4.3 Historical background

Joseph Louis Gay-Lussac and Louis Jacques Thénard in 1808 discovered boron trifluoride. They failed to isolate hydrofluoric acid from calcium fluoride and vitrified boric acid, as the resulting vapors formed by the above reaction failed to corrode the glass; hence, they named it fluoric gas (Gay-Lussac & Thenard, 1809).

4.4 Physicochemical properties

It is an inorganic gas that acts as a strong Lewis acid. It forms a dihydrate of itself when bought in contact with moist air. Some of the physicochemical properties of its nonhydrated form are shown in Table 4.1.

4.5 Uses

Boron trifluoride is widely used as a reagent in organic synthesis. It is most commonly used as a Lewis acid (Brotherton, Weber, Guibert, & Little, 2000). Examples include:

- Used as an initiator in the synthesis of polymers from unsaturated compounds
- Used as a catalyst in organic reactions such as isomerization, alkylation, esterification, dehydration (Sowa, Hennion, & Nieuwland, 1935), acylation (Mani, Herbert, & Manise, 1991), and condensation.

Other, less common applications for boron trifluoride include:

TABLE 4.1 Physicochemical properties of boron trifluoride.

Property name	Property value
Physical state	Gas at 20°C and 1013hPa
Color	Colorless
Odor	Pungent gas with suffocating odor.
Boiling point	−148°F
Melting point	−196°F
Solubility in water	332g/100g at 273K; Soluble in cold water, undergoes hydrolysis in hot water
Molecular weight	67.8g/mol
Density	3.08kg/m^3 at 25°C and 1013hPa
Vapor pressure	4.87MPa at −13°C
Freezing/boiling points	−126.8°C/−101°C at 1013hPa
Flammability	Nonflammable gas

- Used as a dopant in ion implantation
- p-Type dopant for epitaxially (crystal growth and material decomposition) grown silicon
- Used as a sensitive neutron detector in ionization chambers
- Used in devices for monitoring radiation levels in the atmosphere
- Used as a fumigant in pest control
- Used as a flux for making magnesium solders
- Used in the preparation of diborane

4.6 Sources of contaminants

Boron trifluoride is manufactured in closed systems in a continuous or batch process on industrial scale, to lessen the occupational exposure potential. Sometimes workers may get exposed to this during maintenance, cleaning, transfer, sampling, and analysis.

4.7 Risk assessment

The risk assessment should be conducted and recognized in each work area to measure the related risks during the use and handling of BF_3 and to select the protective personal equipment that matches the relevant

TABLE 4.2 Classification according to Regulation (EC) No. 1272/2008.

Category	Code	Statement
Hazard	H280	Contains gas under pressure; may detonate if heated
	H314	Causes rigorous skin burns and eye damage
	H330	Fatal if inhaled
	H373	May cause serious damage to organs through prolonged or repeated exposure
Prevention	P260	Do not breathe gas/vapors
	P280	Wear protective gloves/protective clothing/eye protection/face shield
	P303+P361 +P353+P315	IF ON SKIN (or hair): Take off immediately all contaminated clothing. Rinse skin with water/shower. Get immediate medical advice/attention
	P304+P340 +P315	IF INHALED: Take out a person to fresh air and keep comfortable for breathing. Get immediate medical advice/attention
	P305+P351 +P338+P315	IF IN EYES: Rinse cautiously with water for several minutes. Remove contact lenses, if present, and easy to do. Continue rinsing with water. Get immediate medical advice/attention
	P308+P313	IF exposed or concerned: Get medical advice/attention
Storage	P403	Store in a well-ventilated place
	P405	Store locked up

hazard. Therefore, some recommendations should be considered while working with BF_3. Keep the self-contained breathing device always ready with oneself for emergency use. Appropriate chemical resistant protective clothing should be readily available for emergency use. Special protective equipment for the protection of body should be selected that is based on the task being performed and the risks involved. Protect the eyes, face, and skin from contact with the product (Table 4.2). One ppm ($3\,mg/m^3$) is the safe exposure limits of BF_3 as proposed by NIOSH (National Institute for Occupational Safety and Health) and OSHA (Occupational Safety and Health Administration).

4.8 Environmental effects

It reacts violently when bought in direct contact with water however in a humid atmosphere, boron trifluoride forms a dihydrate ($BF_3 \cdot 2H_2O$). An extensive survey of the literature reveals that boron trifluoride is hydrolytically unstable. Its ecological fate and pathways are based on the properties of its hydrolysis product (consisting of boric acid). This shows that it has very low aquatic toxicity for stimulating bio-accumulation in the food chain.

4.9 Hazardousness

Routes of exposure—Inhalation, skin, eye, and through mucous membrane contact.

Lengths of exposure—It has been found lethal in animal studies for acute (4 h) and subchronic (several months) tests.

The severity of the effect—Depends on concentration and duration.

Target organs—Lungs, blood, bones, and teeth.

Type of effect—An exposure to high concentration produces pulmonary irritation. Subchronic, low concentration exposure produces generalized effects of fluoride exposure, including hypocalcemia. See Chapter 13 for more insight on fluoride-related hazardousness.

Boron trifluoride and its organic complexes are highly corrosive. They are found to be destructive to almost all tissues of the body. They react with the moisture present in the skin and tissues of the body forming hydrofluoric acid and fluoroboric acid, which causes severe burns. On further reaction, these compounds form boron trifluoride gas that causes breathing trouble and severe irritation to the skin, eyes, and mucous membrane. The extreme level of inhalation may cause respiratory failure and even death. Exposure to the eyes can cause severe burns and even lead to blindness.

4.10 Risk management

Organizational measures

1. Collect the most recent available Safety Data Sheets.
2. Put into practice good basic standards of occupational hygiene.
3. Ensure operatives are well informed of the hazards and trained to minimize exposure.
4. Handle and stock up under the indications.

Engineering controls

1. Use material of high reliability for loading and unloading.
2. The provision of local exhaust ventilation should be there on the premises.
3. The periodic monitoring of the work atmosphere should be conducted.
4. Ensure that the safety showers, respirators, and handwash stations are close to workstation locations.

Fire fighting measures

1. Use suitable extinguishing media like dry powder, carbon dioxide, foam, water spray, or fog to control fire fumes.
2. Use special protective kit against both liquid and gaseous chemicals, aerosols, and solid particles, such as gas-tight chemically protective clothing equipped with self-contained breathing apparatus.

Accidental release measures

1. Use personal safety measures, protective kit, and emergency procedures.
2. Evacuate the affected area.
3. Use chemically protective clothing equipped with breathing equipment.
4. Ensure adequate air ventilation throughout the affected area.
5. Monitor the concentration of the released product.

For maintenance of activities always consider a work permit system. Provide enough general and local exhaust ventilation. Keep concentrations well below set exposure limits. In case of release in toxic quantities, gas detectors should be used. Systems under pressure should be checked for leakages on a regular basis. Product should be handled in a loose system and under strictly guarded conditions. Only use permanent leak-tight installations (e.g., welded pipes). Do not eat, drink, or smoke in the working area when using the product (GmbH, Linde Gas, & Poisoning Information, 2017).

4.11 Current research

Boron trifluoride is hygroscopic, hence it is speculated to react with moisture to form stable dehydrate, thus resulting in cost of respiratory irritation. Recently, an expedient method demonstrated the detection of toxic BF_3 gas molecules using graphdiyne nanotube (G2YNT) (Maria, Nagarajan, & Chandiramouli, 2020). This methodology investigated the structural characteristics and electronic attributes of G2YNT using density-functional theory (DFT) as well as the adsorption behavior of neutral BF_3 gas molecules on neutral G2YNT.

4.12 Safety measures

Emergency eye wash fountains and safety showers should be available near any possible exposure site. Always wear leather safety gloves and safety shoes during the handling of cylinders. Protect cylinders from any physical damage; do not roll, drag, slide, or drop. While transporting the cylinder, always keep in place a removable valve cover. Never try to lift a cylinder by its cap; the cap is intended solely to protect the valve. When moving cylinders, even for short distances, use a cart (trolley, hand truck, etc.) designed to transport cylinders. Never insert any entity (e.g., wrench, screwdriver) into cap openings; doing so may damage the valve and may causes a leak. Always use an adjustable strap wrench to remove overtight or rusted caps. Always try to open the valve slowly. If the valve is hard to open, discontinue use and immediate contact your supplier. Close the valve of the container after each use; keep closed even when empty. High temperatures may damage the container and could cause the pressure relief device to fail impulsively, venting the container contents.

4.13 Handling and storage

- A sealed and purged system should be used for handling.
- Preferably the doubly contained piping; diaphragm or bellows sealed, soft seat valves; back flow prevention devices; flash arrestors; and flow monitoring or limiting devices should be employed.
- Provisions of a suitable exhaust system with appropriate gas cabinets should be there.
- Insertion of any foreign substance into a gas cylinder is strictly prohibited.
- Relevant regulations should be followed for storing cylinders for the guidance given in CGA P-1, Safe Handling Containers containing compressed gases.
- Storage of materials should be done after segregating the hazardous substance.
- Cylinders should be protected from direct sunlight, precipitation, and mechanical damage.
- Storage temperature should not exceed 55°C.

4.14 Precautions

- Do not breathe air contaminated with the boron trifluoride gas/vapors.
- Avoid contact with water or moist air.
- Never apply flame or localized heat directly to any part of the boron trifluoride container.

References

Brotherton, R. J., Weber, C. J., Guibert, C. R., & Little, J. L. (2000). Boron compounds. In *Ullmann's encyclopedia of industrial chemistry*. Wiley-VCH.

Gay-Lussac, J. L., & Thenard, J. A. (1809). Des propriétés de l'acidefluoriqueet surtout de son action sur le métal de la potasse. *Mémoires de Physique et de Chimie de la Société d'Arcueil*, *2*, 317–331.

GmbH, Linde Gas, & Poisoning Information (2017). *SECTION 1: Identification of the substance/mixture and the company/undertaking product name: Chemical name: SECTION 2: Hazards identification.* .

Mani, R. I., Herbert, L., & Manise, D. (1991). Boron trifluoride in the synthesis of plant phenolics: Synthesis of phenolic ketones and phenylstyryl ketones. *Journal of the Tennessee Academy of Science*, *66*, 1–8.

Maria, J. P., Nagarajan, V., & Chandiramouli, R. (2020). Boron trifluoride interaction studies on graphdiyne nanotubes—A first-principles insight. *Chemical Physics Letters*, *738*, 136841.

Sowa, F. J., Hennion, G. F., & Nieuwland, J. A. (1935). Organic reactions with boron fluoride. IX. The alkylation of phenol with alcohols. *Journal of the American Chemical Society*, *57*(4), 709–711.

CHAPTER

5

Bromine: Risk assessment, environmental, and health hazard

Manis Kumar Jha[a] *and Sushma*[b]

[a]Metal Extraction and Forming Division, National Metallurgical Laboratory (CSIR), Jamshedpur, Jharkhand, India

[b]Department of Chemistry, Gurukula Kangri (Deemed to be University), Haridwar, Uttarakhand, India

Abbreviations

ACGIH	American Conference of Governmental Industrial Hygienists
ANSI	American National Standards Institute
BIF	bromine incorporation factor
BSF	bromine substitution factor
CSF	cancer slope factor
DBP	disinfection by-product
DSC	Department Safety Coordinator
EPA	Environmental Protection Agency
IDLH	immediately dangerous to life and health
LADD	life average daily dose
NOAA	National Oceanic and Atmospheric Administration

5.1 Introduction

Bromine is the third lightest naturally occurring halogen and a fuming brownish-red color liquid at room temperature that tends to volatilize with identical color fumes. It is likely found to be more soluble in water than chlorine (O'Neil et al., 2001; Teitelbaum, 2001), and acts as a good oxidizing agent in presence of water (Gosselin, Smith, & Hodge, 1984). Specifically, it has an offensive and suffocating bleach-like odor with an odor threshold ranging between 0.05 and 3.5 ppm (Kasilo & Edelman, 1999;

Hazardous Gases
https://doi.org/10.1016/B978-0-323-89857-7.00021-9

Makarovsky et al., 2007). Bromine is denser than air; hence, it is mainly found at ground level. In nature, bromine is extensively perceived from the soluble inorganic bromides, whereas some extent of it can also be found in the minerals. The vapor pressure and IDLH values (promptly hazardous to life and wellbeing) define its irritant behavior representing an inhalation risk. However, it is devoid of flammability but has the ability to initiate the fire in presence of combustibles (MCA, 1968).

Bromine is mainly used in manufacturing processes of fumigants, water cleansing agents, flameproofing agents, colors, sanitizers, synthetic organic chemicals, some medicines, etc. Moreover, in some places of southern Israel, it is also utilized for the purification of pools.

Bromine tends to instantly diffuse out of solution producing a vapor, causing damage to some major organs, such as the liver, stomach, kidneys, and lungs, if inhaled. However, considering its dark red color, any related spills can be easily detected, even at concentrations below 0.1 ppm. Mostly the laborers involved in the industry dealing with bromine or scientists are comparatively more vulnerable to exposure (Kasilo & Edelman, 1999; Makarovsky et al., 2007).

5.2 Occurrence

Bromine exists naturally as a mixture of two stable isotopes that is Bromine-79 and Bromine-81. The paramount of bromine is extensively found in the inland seas, seawater, salt lakes, and brine wells. Around 65 ppm has been recorded to be found in seawater, where about 2500–10,000 ppm concentration of bromine comes from inland seas and brine wells. The maximal occurrence of bromine comes from salt brines of the United States and China, the Dead Sea in Israel and Jordan, and ocean water in Wales and Japan. The higher levels of bromine in water are a major point of concern for researchers, as it not only adversely affects human beings but also harms marine flora and fauna (Heeb, Criquet, Zimmermann-Steffens, & Von Gunten, 2014; Luong, Peters, Young, & Perry, 1980; Singer, 1994; Westerhoff, Chao, & Mash, 2004).

The indigenous origins of bromine are rocks and Earth's crust; organobromine compounds produced by various types of marine organisms; brominated and mixed bromochloro-disinfection by-products (DBPs) (U.S. EPA, 1998; Villanueva, Cordier, Font-Ribera, Salas, & Levallois, 2015); and bromide discharges from oil and gas facilities and coal-fired power plants (Greune, 2014; Landis et al., 2016; States et al., 2013; Weaver, Xu, & Mravik, 2016; Wilson & Van Briesen, 2013).

5.3 Historical background

In 1826, a French chemist, Antoine-Jerome Balard accidentally discovered, studied, and determined properties of bromine while studying the salt marsh flora of Mediterranean waters, Montpellier, France. He successfully discovered the only liquid nonmetallic element. This was done by first crystallizing sodium chloride and sodium sulfate from the seawater, then saturating the obtained residue with chlorine, which was finally distilled to give bromine. In the same year of this great achievement, he even communicated his work to the French Academy's Journal (Wisniak, 2002). Originally, a year earlier to his work, a student at Heidelberg, Carl Löwig successfully extracted the bromine gas from waters of a natural spring at Kreuznach, but before he came to any conclusion, Ballard already got recognition for the same.

5.4 Physicochemical properties

At room temperature, bromine exists as a reddish-brown color liquid with substantial vapor pressure. It exhibits violent behavior on reaction with alkali metals and metals such as phosphorus, arsenic, aluminum, and antimony; however, the degree of behavior may vary. The 3.41 g of bromine deliquesces in 100 mL of water to give bromine water, which releases iodine from iodine-containing solutions and sulfur from H_2S. Bromine water oxidizes sulfurous acid to sulfuric acid and decomposes in the presence of sunlight, as shown in Eq. (5.1).

$$H_2O + Br_2 \rightarrow 2HBr + ½O_2 \tag{5.1}$$

A hydrate, that is clathrate incorporating 172 water molecules and 20 cavities for accommodating bromine can be isolated from bromine water. Based on temperature, bromine produces bromides, hypobromites, or bromates when dissolved in an aqueous alkali hydroxide solution. Due to its solubility in chloroform and carbon disulfide, it can be easily extracted from water. It also offers an orange color solution in presence of organic solvents.

Bromine acts as a less effective oxidizing agent in comparison to chlorine, due to its weak hydration affinity than chloride ion. Likewise, a metal-chlorine bond is comparatively stronger than a metal-bromine bond. Correlating effect has been observed in the case of organic bromo and chloro derivatives. As in these circumstances, bromo derivatives have been found to be dense and less volatile, combustible, and stable.

Bromine is present in nature with a stable oxidation state of −1; however, other oxidation states are also known, such as 0 (elemental bromine, Br_2), +1 (hypobromite, BrO^-), +3 (bromite, BrO_2^-), +5 (bromate, BrO_3^-), and +7 (perbromate, BrO_4^-). The ligands, mainly oxygen and fluorine, generally form stable compounds with bromine in positive oxidation states. While covalent bonds are formed with the compounds having oxidation numbers +1, +3, +4, +5, and +7. Some properties of bromine are given below in Table 5.1 (National Center for Biotechnology Information, 2020).

TABLE 5.1 Physical and chemical properties of bromine.

Atomic number	35
Molecular weight	159.81 g/mol
Electronic shell	[Ar] $3d^{10}$ $4s^2$ $4p^5$
Color	Dark reddish-brown
Odor	Bleachy and penetrating
Taste	Taste threshold: 0.17–0.23 mg/L
Melting point	−7.2°C
Boiling point	58.8°C
Vapor density	7.139 g/L at 0°C
Vapor pressure	212 mmHg at 25°C
Specific gravity (water = 1)	3.1 at 20°C (68°F)
Density	3.1023 at 25°C
Standard potential	1.08 V
Heat of vaporization	187 J/g at 50°C
Energy of first ionization	1142.7 kJ mol^{-1}
Surface tension	40.9 mN/m at 25°C
Ionization potential	10.55 eV
Solubility	Sparingly soluble in water; soluble in alcohol, ether, carbon tetrachloride, carbon, disulfide, aqueous solutions of bromides, chloroform

5.5 Physiological role

The most pertinent way of bromine exposure is by respiration (IPCS, 1999), following which it gets absorbed in the lungs. While, in case of ingestion of bromine liquid, it quickly gets absorbed passively from the intestine, with a half-life of somewhat between 12 and 30 days. The attributes of bromine determine the depth as well as the site of lung infiltration, hence affecting the rate of absorption. Primarily its deposition in the lung lies proportionally to its solubility in water. However, respiration of bromine in high concentrations may markedly irritate the lower airway anatomy (IPCS, 1999). Moreover, there are no reported studies directing bromine absorption via the dermal pathway.

Bromine enters the circulatory system in a neutral form, from where it is distributed to various tissues, and chiefly to the extracellular fluid (Health Council of the Netherlands, 1996). By far there are no available specifics clarifying the metabolism of bromine succeeding inhalation; however, it has been reported that bromine instantly converts into bromide in living tissues (IPCS, 1999). Subsequently, bromide gets partitioned in the body as well as replaces chloride. Sometimes, this ion also acts as a CNS depressant and overdosage of which may result in severe adverse effects, with a restricted time of exposure due to its irritant nature (IPCS, 1999).

On reaction with water, bromine forms hypobromous acid, which slowly decomposes to hydrogen bromide and O_2 (U.S. EPA, 2009). The accountability of bromine-induced tissue damage is reported to be due to the released nascent oxygen or oxygen-free radicals from the water in mucous membrane that is a potent oxidizer (HSDB, 2007; IPCS, 1999).

5.6 Uses

Bromine is commonly used in the disinfection of water, bleaching of fibers and silk, and preparation of some medicines and dyes (O'Neil et al., 2001), while some other uses are enlisted below:

- **Flame retardant:** One of the secondary uses of bromine has been recognized as a flame-retardant material. There are studies concluding the high flammability of materials generally utilized in the making of clothes, carpets, curtains, and drapes. Keeping this in mind, the researchers came to the conclusion of either coating the fabric with bromine or chemically incorporating bromine at the time of fabric preparation to make it fire-resistant. For example, tris(dibromopropyl) phosphate ($(Br_2C_3H_5O)_3PO$) that is now restricted as flame retardants due to its carcinogenic nature.

- **Drilling wells:** The process of drilling wells generally utilize calcium bromide ($CaBr_2$), sodium bromide (NaBr), or zinc bromide ($ZnBr_2$) in order to enhance the efficiency of the process.
- **Pesticides:** Bromine plays a vital role in the preparation of pesticides. For instance, one of the earliest synthesized pesticides, methyl bromide has been utilized in the fumigation of crops. However, it has been phased out from usage due to its succeeding deteriorating effect on the ozone layer. For more details on methyl bromide refer to Chapter 19.
- **Leaded gasoline:** One of the bromide compounds that is ethylene dibromide ($C_2H_4Br_2$) has found its specific utilization in "leaded gasoline." This category of gasoline is mainly responsible for the efficient burning of fuels and keeps car engines devoid of any ignition-related sound issues that is, "knocking." Consequently, knocking reduces the efficiency of a car engine. The leaded gasoline gives off the free lead which reacts with ethylene dibromide and converts it into a safe compound. However, some of the free lead tends to escape after fuel combustion, which on inhalation may damage the nervous system. Therefore, its use has been banned by several countries such as the United States.

5.7 Sources of contaminants

The natural occurrence and anthropogenic origin are among the vital sources of bromine compounds in the atmosphere. The release of bromine into the water from oil and gas facilities and coal-fired power plants can easily contaminate drinking water. This also seeps into the farming lands, thus enters into the food chain. The release of bromine in ambient air from facilities may be one of the sources of skin exposure. Moreover, due to denser nature, bromine generally resides in low-lying areas for comparatively longer times.

5.8 Risk assessment

The bromine-related risk can be assessed based on health-based guidance value provided by statutory authorities that are specifically dependent on the recognition and calculation of risks. The main attributes involved in the hazard analysis are the estimation of time of exposure, nature of the source, exposure pathway, and likewise. Subsequently, the bromine-related risk can also be analyzed by bromination fraction computed from bromine incorporation factor (BIF) and bromine substitution factor (BSF). Moreover, statistical data can be evaluated by using the Mann-Whitney rank-sum test and Monte Carlo simulation. Researchers have also reported a statistical simulated framework using the cancer slope factor (CSF) in order to estimate the human health risk incurred due to bromide contaminated water (Wang, Small, & VanBriesen, 2017). The US Environmental Protection Agency (EPA) suggests the use of oral CSF and life average daily dose (LADD) (mg/kg/day) for the computation of cancer risk levels from the ingestion of drinking water (Theodore & Dupont, 2012). In order to access the bromide-related hazard, the WHO and Food and Agricultural Organization of the United Nations have established an acceptable daily intake of 1000 μg/kg bw/day as safe (JMPR, 1989).

5.9 Environmental effects

Generally, organic bromines are utilized as disinfecting and protecting agents at numerous places such as greenhouses, farmland, from where it can easily rinse off to surface water. Succeeding to which can adversely affect the health of daphnia, fishes, lobsters, and algae. Based on the physiology of living organisms, bromine is mainly considered to play a nonessential role and contemplated as toxic even if present in low concentrations (Tchounwou, Yedjou, Patlolla, & Sutton, 2012).

In earlier days, the invasion of bromine in the cattle food has also been reported with symptoms like liver damage, loss of sight, and depletion of growth, a decrease of immunity, decrease milk production and sterility, and malformed children. Further, this has resulted in the deaths of thousands of cows and pigs in order to prevent any human contamination.

Bromine and its compounds absorbed by mammals indirectly from their prey have proven to cause impairment in them, by damaging their nervous system and DNA. Subsequently, uptake through food, respiration, and skin also initiate similar health conditions in mammals.

The dissociation property of organic bromine compounds to yield free bromine is imitated in the atmosphere, causing ozone layer depletion. As a result, many bromine compounds have been phased out from their application by many countries.

5.10 Hazardousness

The oxidizing aspect of bromine makes burn more intensifying when it comes in contact with skin and may even cause blindness in case of contact with eyes. It can have serious consequences on the tissue of the mucous membrane and upper respiratory tract. A person who is exposed to bromine may feel: spasm, inflammation, and edema of the larynx and bronchi, chemical pneumonitis and pulmonary edema, irritation, tearing and inflammation of the eyes, blepharospasm, and photophobia.

Tissue damage may occur as a result of the reaction of bromine with tissue components like epithelium and dermal keratinocytes, which tends to release radical oxygen species from the mucous membrane. Secondarily, on reacting with tissues moisture, irritation may also occur due to the formation of hydrobromic and hypobromic acids.

The major characteristic effects associated with exposure of bromine comprises conjunctivitis, lacrimation, severe eye irritation, nose and throat irritation, cough, and dyspnea. While ingestion may cause abdominal pain and hemorrhagic gastroenteritis and contact with skin may lead to burns (Haddad & Winchester, 1983; Morabia et al., 1988) or halodermas. Bromoderma tuberosum is a dermal manifestation specific to bromine exposure (Anzai, Fujiwara, & Inuzuka, 2003; Pfeifle, Grieben, & Bork, 1992).

The concentration, route, and length of exposure; age, and preexisting medical condition of the exposed personnel; exposure happened at open space or confined space, all factors play a crucial role in determining the severity of bromine on human health. The volatile characteristic of bromine has been accounted for the major cause beneath all documented incidents.

5.11 Treatment

In case of accidental exposure, symptomatic treatment such as the use of diluents that is, water or milk (adult dose: 100–200 mL; child dose: 2 mL/kg) and blood pressure, pulse, 12-lead ECG, and oxygen level monitoring should be given preference at the primary stage of management. Having aware of that, the patient should be checked for any mucosal damage or perforation as prime caution. There are studies suggesting the positive effect of early dilution therapy with water or milk on the acute alkali injury of the esophagus. However, there are chances of induced vomiting and the risk of edema in case of administration of a large volume of fluid.

5.12 Risk management

The bromine-related risks can be governed by the application of various limits given by competent bodies, for example, the short-term exposure limit of 0.1–0.3 ppm and long-term exposure limit of 0.1 ppm (NIOSH, 1997). The values are computed on the basis of the level of risk or amount of respiratory damage caused after inhalation of bromine. The management of brominated contamination from discharges should also be considered and evaluated periodically. Additionally, regression analysis of bromination factors can also be utilized as a fruitful method for the risk management of human health.

A threshold value limit (TLV) of 0.1 ppm (0.7 mg/m^3) as a TWA for a normal 8-h workday and a 40-h work week and a STEL of 0.3 ppm (2 mg/m^3) for periods not exceeding 15 min has been assigned to bromine by ACGIH (American Conference of Governmental Industrial Hygienists) (ACGIH, 1991) calculated on the severity of the risk.

5.13 Current research

In recent years bromine has attracted the attention of researchers due to its ozone-depleting characteristics. The communications circumventing the damage caused to ozone molecules on getting exposed to bromine has been conducted by the National Oceanic and Atmospheric Administration (NOAA). In their studies, they have detailed the three-reaction cycle approach. The framework specifies that the primary reaction occurs between chlorine or chlorine monoxide and ozone, which forms monoatomic (O) or diatomic oxygen (O_2). Similarly, in the second cycle also diatomic oxygen is formed. And, in the final step bromine reacts with ozone to form diatomic oxygen. For the occurrence of all three steps, sunlight is needed. As a result, ozone depletion is comparatively higher during summers than winters. For more specifics on Ozone refer to Chapter 24: Ozone: Risk assessment, environmental and health hazard. Moreover, numerous related researchers are still going on (Werner et al., 2017).

5.14 Bromine toxicity: Case study

The last bromine-related accident was reported in Cardiff city of Wales, the United Kingdom in 2003. As per reports, some contractual laborers working on the air-conditioning system in British Gas building accidentally got exposed to hydrobromic acid, as they were using the same chemical for the AC treatment. Further, hydrobromic acid on reaction with some other chemicals melted the pipes, causing the release of toxic fumes. This incident resulted in the emergency evacuation of almost 1700 workers, while 17 were provided with on the spot available first aid and around 18 were hospitalized; however, eventually the incident didn't resulted in any loss of life.

Moreover, a major incident in 2002 was obstructed by statuary authorities' actions as a considerable amount of bromine was found in the headquarters of the former Palestinian leader Arafat in Ramallah. It was speculated that this quantity of bromine was intended for some big terrorist attacks. Because by this time terrorist groups were aware of the devastating effects of bromine.

Subsequently, the toxic and denser-than-air attributes of bromine make its release more disastrous for the environment. Corresponding to which an incident was reported at a chemical plant in Geneva, Switzerland, from where somewhat around the sunrise of November 8, 1984, some dense brown fumes were observed. The reason for the incident was reported to be due to the release of 550 kg of liquid bromine. The dense cloud formed after release remained low over the ground and drifted to the center of the town before reaching Lake Geneva where it dissipated. The records suggest that the time elapsed from the release to the disappearance of the cloud over the lake was approximately 5 h.

Similarly, with the help of an ozone analyzer, the Ecotoxicological Centre of the Canton of Geneva also recorded the presence of bromine between 0.2 and 0.5 ppm, which remained in the air for the undefined time (Bentur, 2001). As a result, 91 people were hospitalized in the University Hospital in Geneva. However, no severe complications were seen even after a month-long follow-up protocol. Some of the reported symptoms were eye irritation (90%), upper airways irritation (68%), cough (47%), expectoration (34%), and headache (46%).

One tragic incident happened with the UC Berkeley graduate students while cleaning the chemical inventory inside a laboratory refrigerator. The students observed the appearance of condensate on the walls of the refrigerator and assumed it to be water. But succeeding cleaning guided them to the leaking bottle of 99% bromine, causing havoc among them. The refrigerator was immediately sealed, and information was notified to the Department Safety Coordinator (DSC) for cleanup assistance. Simultaneously, the students involved in the cleaning were hospitalized postobservation of signs of lightheadedness, nausea, and tightness in the chest.

5.15 Safety measures

1. The ANSI (American National Standards Institute) approved goggles, full face shield, 100% nitrile rubber gloves, coveralls and jacket, leather or other nonwoven ANSI approved steel-toed shoes, chemical resistant suit should be worn by the personnel handling bromine.
2. There should be provision of eyewash fountains in the premises dealing with bromine in order to irrigate eyes in case of accidental exposure.
3. The shoes and clothes contaminated with bromine should be discarded or washed on a priority basis.
4. For extra protection, a set of rubber boots should be worn over shoes.

5.16 Handling and storage

- The presence of water inside steel bromine chambers should be restricted.
- The reused bromine should never be returned to the bromine drums or chambers.
- The bromine should be stored in a dry, well-ventilated, and moisture-free region.
- The standard compartments should be preferred while storing bromine and the resembling packaging must be avoided.
- The contaminated, defective, and leaking valves should be thoroughly inspected before carrying out any bromine-related processes or handling.
- The equipment meant to be used while handling bromine should be shielded from strong alkalis, reducing agents, other metals, wood, fabric, paper, grease, and oils.
- The bromine drums either empty or full should be stored in an upright position at a cool (above −7°C/20°F) and dry place.
- In order to avoid any explosion, the drums should be housed in moisture and heat-devoid regions.
- The drums kept for a longer period of time should be timely inspected.
- The bromine should be kept away from natural acid amines, aliphatic ketones phenols, and hydrocarbons.
- The dry form of bromine should not be kept in the proximity of metals, such as titanium, potassium, aluminum.

5.17 Precautions

1. Wear self-contained breathing apparatus around bromine locus.
2. Keep oxidizers like aluminum, titanium, mercury, sodium, and potassium away.
3. Handle bromine with extreme care.
4. Keep aside from combustibles.

References

ACGIH. (1991). *Documentation of the threshold limit values and biological exposure indices* (6th ed). Cincinnati, OH: American Conference of Governmental Industrial Hygienists.

Anzai, S., Fujiwara, S., & Inuzuka, M. (2003). Bromoderma. *International Journal of Dermatology, 42*(5), 370–371.

Bentur, Y. (2001). *Therapy guidelines for hospitals during a mass toxicological event.* Haifa: Emergency & Disaster Management Division, Ministry of Health.

Gosselin, R. E., Smith, R. P., & Hodge, H. C. (1984). *Clinical toxicology of commercial products* (5th ed.). Philadelphia: Williams & Wilkins.

Greune, A. C. (2014). *Bromide occurrence in North Carolina and the relationship between bromide concentration and trihalomethane formation.* Graduate Faculty of North Carolina State University.

Haddad, L. M., & Winchester, J. F. (1983). *Clinical management of poisoning and drug overdose.* WB Saunders Company.

Hazardous Substances Data Base (HSDB) (2007). *Bromine.* Hazardous substances data base. Retrieved from:(2007). *https://semspub.epa.gov/work/03/2247167.pdf.*

Health Council of the Netherlands (1996). *Bromine—Health-based reassessment of administrative occupational exposure limits.* The Hague.

Heeb, M. B., Criquet, J., Zimmermann-Steffens, S. G., & Von Gunten, U. (2014). Oxidative treatment of bromide-containing waters: Formation of bromine and its reactions with inorganic and organic compounds—A critical review. *Water Research, 48,* 15–42.

International Programme on Chemical Safety (IPCS) (1999). *Bromine. Poisons information monograph. PIM 080.* Geneva: WHO.

JMPR (1989). Joint Food and Agricultural Organization of the United Nations (FAO) and World Health Organization Meeting on Pesticide Residues. Bromide ion. 9. In *Pesticide residues in food—1988. Evaluations. Part II. Toxicology. FAO plant production and protection paper 93.* Rome, Italy: FAO.

Kasilo, O. M. J., & Edelman, P. A. (1999). *Bromine, poisons information monographs 080.* IPCS INCHEM.

Landis, M. S., Kamal, A. S., Kovalcik, K. D., Croghan, C., Norris, G. A., & Bergdale, A. (2016). The impact of commercially treated oil and gas produced water discharges on bromide concentrations and modeled brominated trihalomethane disinfection byproducts at two downstream municipal drinking water plants in the upper Allegheny River, Pennsylvania, USA. *Science of the Total Environment, 542,* 505–520.

Luong, T., Peters, C. J., Young, R. J., & Perry, R. (1980). Bromide and trihalomethanes in water supplies. *Environmental Technology, 1*(6), 299–310.

Makarovsky, I., Markel, G., Hoffman, A., Schein, O., Brosh-Nissimov, T. M., Finkelstien, A. , et al. (2007). Bromine—The red cloud approaching. *IMAJ, 9,* 677–679.

MCA (1968). *Chemical safety data sheet SD-49: Properties and essential information for safe handling and use of bromine.* Washington, DC: Manufacturing Chemists Association.

Morabia, A., Selleger, C., Landry, J. C., Conne, P., Urban, P., & Fabre, J. (1988). Accidental bromine exposure in an urban population: an acute epidemiological assessment. *International Journal of Epidemiology, 17*(1), 148–152.

National Center for Biotechnology Information (2020). *PubChem compound summary for CID 24408, bromine. (2020). https://pubchem.ncbi.nlm.nih.gov/compound/Bromine. (Accessed 30 October 2020).*

NIOSH (1997). *National Institute for Occupational Safety and Health (NIOSH) pocket guide to chemical hazards.* DHHS (NIOSH) Publication No. 94-170, Washington, DC.

O'Neil, M. J., Smith, A., Heckelman, P. E., Obenchain, J. R., Gallipeau, J. R., & D'Arecca, M. A. (Eds.). (2001). *The Merck index: An encyclopedia of chemicals, drugs, and biologicals.* (13th ed.). Whitehouse Station, NJ: Merck & Co., Inc..

Pfeifle, J., Grieben, U., & Bork, K. (1992). Bromoderma tuberosum caused by anticonvulsive treatment with potassium bromide. *Der Hautarzt; Zeitschrift fur Dermatologie, Venerologie, und verwandteGebiete, 43*(12), 792–794.

Singer, P. C. (1994). Control of disinfection by-products in drinking water. *Journal of Environmental Engineering, 120*(4), 727–744.

States, S., Cyprych, G., Stoner, M., Wydra, F., Kuchta, J., Monnell, J. , et al. (2013). Marcellus Shale drilling and brominated THMs in Pittsburgh, Pa., drinking water. *Journal of American Water Works Association, 105*(8), E432–E448.

Tchounwou, P. B., Yedjou, C. G., Patlolla, A. K., & Sutton, D. J. (2012). Heavy metal toxicity and the environment. *Molecular, clinical and environmental toxicology, 101*, 133–164.

Teitelbaum, D. T. (2001). The halogens. In E. Bingham, B. Cohrssen, & C. H. Powell (Eds.), *Patty's toxicology* (pp. 731–826). (5th ed.). New York: John Wiley & Sons.

Theodore, L., & Dupont, R. R. (2012). *Environmental health and hazard risk assessment: Principles and calculations.* CRC Press.

U.S. EPA (1998). *National primary drinking water regulations: Disinfectants and disinfection byproducts: Final rule. 40 CFR 9, 141, and 142, Washington, DC*: (pp. 69390–69476). .

U.S. EPA. (2009). *Bromine. Interim acute exposure guidelines levels (AEGLs).* Washingon, DC: EPA.

Villanueva, C. M., Cordier, S., Font-Ribera, L., Salas, L. A., & Levallois, P. (2015). Overview of disinfection by-products and associated health effects. *Current Environmental Health Reports, 2*(1), 107–115.

Wang, Y., Small, M. J., & VanBriesen, J. M. (2017). Assessing the risk associated with increasing bromide in drinking water sources in the Monongahela River, Pennsylvania. *Journal of Environmental Engineering, 143*(3).

Weaver, J. W., Xu, J., & Mravik, S. C. (2016). Scenario analysis of the impact on drinking water intakes from bromide in the discharge of treated oil and gas wastewater. *Journal of Environmental Engineering, 142*(1)04015050.

Werner, B., Stutz, J., Spolaor, M., Scalone, L., Raecke, R., Festa, J. , et al. (2017). Probing the subtropical lowermost stratosphere and the tropical upper troposphere and tropopause layer for inorganic bromine. *Atmospheric Chemistry and Physics, 17*(2), 1161–1186.

Westerhoff, P., Chao, P., & Mash, H. (2004). Reactivity of natural organic matter with aqueous chlorine and bromine. *Water Research. 38*(6). https://doi.org/10.1016/j.watres.2003.12.014. 1502–13.

Wilson, J. M., & Van Briesen, J. M. (2013). Source water changes and energy extraction activities in the Monongahela River, 2009–2012. *Environmental Science & Technology, 47*(21), 12575–12582.

Wisniak, J. (2002). The history of bromine from discovery to commodity. *Indian Journal of Chemical Technology, 9*(3), 263–271.

CHAPTER

6

Bromine chloride: Risk assessment, environmental, and health hazard

Sushma

Department of Chemistry, Gurukula Kangri (Deemed to be University), Haridwar, Uttarakhand, India

Abbreviations

AEGL acute exposure guideline levels
ANSI American National Standards Institute
LR lactated ringer
NOAEL no-observed-adverse-effect level

6.1 Introduction

Bromine chloride also known as brominemonochloride and bromochloride (HSDB, 2011) is a reddish-yellow mobile liquid or gas with an irritating odor; often produced when an equimolar amount of bromine and chlorine react reversibly in the liquid and vapor phases (Lang, 2006). It is a strong oxidizing agent (Dagani, Barda, Benya, & Sanders, 2000). Consequently, the 60% product formed accounts for BrCl, whereas the remaining 40% dissociates in the form of bromine and chlorine.

Extensively, bromine chloride is utilized in the quantification of mercury in something and the killing of bacteria and molds in industrial facilities. It is also accustomed to being used as a water-treatment biocide and in various organic synthetic procedures. Some of the major organic pathways include the preparation of bromochloro compounds from addition across olefin double bonds and aromatic bromide and hydrogen chloride from aromatic brominations. It is also used as a brominating agent in the

https://doi.org/10.1016/B978-0-323-89857-7.00005-0

preparation of fire-retardant chemicals, pharmaceuticals, high-density brominated liquids, agricultural chemicals, dyes, and bleaching agents (Frim & Ukeles, 2011).

Howbeit, the extent of human exposure to BrCl depends on the environmental humidity and its pH and physiologic sources of moisture (e.g., sweat, moisture in the upper respiratory tract).

6.2 Historical background

In 1826 Antoine Jérôme Balard was the first to synthesize this compound by direct union of bromine and chlorine (Schmiesser & Shuster, 1966). Carl Jacob Löwig after a few years of his research successfully isolated its hydrate $BrCl{\cdot}5H_2O$ from solutions (Barratt & Stein, 1929). Though its existence was experimentally confirmed in 1929 when Gillam and Morton reported the absorption spectrum of a Br_2/Cl_2 mixture (Gillam & Morton, 1929). For the advancement in the same direction in 1952, some researchers also investigated the disproportionation equilibrium constant of BrCl that was found to be 0.145 in CCl_4, Eq. (6.1) (Wang, Kelley, Cooper, Beckwith, & Margerum, 1994).

$$2BrCl \overset{Kd}{\Leftrightarrow} Br_2 + Cl_2 \tag{6.1}$$

The researchers also confirmed that BrCl primarily exists in the form of $BrCl_6^{5-}$. Later, G.M.B. Dobson of the Clarendon Laboratory, Oxford prepared filters for ultra-violet light from mixtures of the bromine and chlorine and further found that the ultraviolet absorptions are not additive in these mixtures. Since then, this mixed halogen has been extensively studied.

6.3 Physicochemical properties

The physical properties of this mixed halogen lie somewhat intermediate between the properties of both bromine and chlorine (Frim & Ukeles, 2011; Lang, 2006); however, they differ in their visual identity and polarity. See Chapters 5 and 9, for a detailed overview of both bromine and chlorine, respectively. The mixed halogens are comparatively polar as well as reactive concerning their individual halogen molecule (Barrie, Dahlstrom, Goswami, & Kaetzel, 2012; Cotton & Wilkinson, 1980). While on the other hand, the reversible dissociation of BrCl to its elemental component makes it the least stable among all the diatomic interhalogens. Furthermore, an array of weak and strong acids viz. hydrochloric, hypochloric, hydrobromic, and hypobromous acids forms post-reaction of bromine chloride and dissociation products with water. The by-products formed may have negligible temperature dependence but are proportionally affected by the measure of hydrogen ion concentration (Liu & Margerum, 2001). The following Eqs. (6.2)–(6.5) show some of the primary reactions (Frim & Ukeles, 2011):

$$2BrCl \leftrightarrow Br_2 + Cl_2 \quad (6.2)$$

$$BrCl + H_2O \leftrightarrow HCl + HOBr \quad (6.3)$$

$$Cl_2 + H_2O \leftrightarrow HOCl + HCl \quad (6.4)$$

$$Br_2 + H_2O \leftrightarrow HOBr + HBr \quad (6.5)$$

It is evident from the above reactions that the ions such as Br^-, Cl^- also exist in equilibrium alongside the by-product (Liu & Margerum, 2001). Considerably, the vapor pressure of BrCl is roundabout four times more than dry air, as computed on the basis of its molecular weight that is, 115.36 g/mol. The vapor pressure of BrCl is 2.368 kPa (17.8 mmHg) at 25°C (IPCS, 2009).

The dissociation of BrCl molecules into bromine products and bromine and chlorine is resultant of its high reactivity and fast equilibrium. Under equilibrium conditions, this gas subsists as a fuming dark red liquid below 5°C. Hence, at low temperature, bromine chloride exists in the solid-state (Emeleus & Anderson, 1960). It indicates that the dissociation presumably prevents the precise estimation of the physical properties of BrCl. Values of melting point and boiling point can be obtained from the equilibrium temperatures for the phase changes. Additionally, since the vapor pressure of dissociated liquid bromine chloride usually requires a pressure of 760 mm at 5°C, the attainment of similar conditions of the vapor above the liquid requires a temperature of 36°C (Greenwood, 1956). It is also interesting to find that the freezing point of bromine chloride that is, −65°C is lower than that of bromine (−7°C). The advantage of this is that heating and insulation is not required during the winter months.

The solubility of bromine chloride is a function of polarity; it thus shows greater solubility in polar solvents. The solubility of BrCl in water is 2.5 times that of bromine that is, 8.5 g/100 g at 20°C. Solubility in water tends to markedly increase with the addition of chloride ion, which consequently forms the complex chlorobromate ion, $BrCl_2^-$. Moreover, at a lower temperature, this gas forms a yellow, crystalline hydrate, $BrCl_7 \cdot 34H_2O$ having a melting point of 18°C at 1 atm.

6.4 Physiological role

There is not a shred of evidence showing the metabolism or disposition of bromine chloride in humans or animals. On the other hand, there are reports describing the dissociation of the compound in bromine and chloride after entering in humans or animals. This does not preclude the fact that it might exist in an undissociated form, which is a matter of research.

6.5 Uses

For many years, bromine chloride has been used in the quantification of mercury. Broadly, it is used as a biocide, specifically as an algaecide, fungicide, and disinfectant in industrial recirculating cooling water systems. Additionally, it is used in the preparation of Li-SO batteries that consequently enhances the voltage and energy density.

BrCl has broad-spectrum biocidal activity over a wider pH range, rapidness in disinfection, effectiveness at lower residual concentrations, and lower aquatic toxicity. Hence, it is preferred as an efficient water treatment biocide over chlorine. It is also utilized in the preparation of bromochloro compounds and aromatic bromide and hydrogen chloride. The former compounds are prepared by addition across olefin double bonds, where the latter is a resultant of aromatic brominations. Bromine chloride has been found useful as a brominating agent in the preparation of fire-retardant chemicals, pharmaceuticals, high-density brominated liquids, agricultural chemicals, dyes, and bleaching agents.

Predominantly, BrCl is used as disinfectant, fungicide, algaecide, and slimicide in recirculating and once-through cooling and process water systems, heat transfer systems, air washers, and industrial scrubbing systems, containerized ponds and decorative fountains, industrial once-through cooling water systems, pulp and paper mills, wastewater systems, and likewise. Here, hypobromous acid plays a major role as a micro biocide, which is formed when bromine chloride dissolves in water.

6.6 Sources of contaminants

The dissociated equilibrium of bromine chloride can be represented as Br^- and Cl^- (Liu & Margerum, 2001). As a consequence, the chemical species formed on contact with water behave as a source of release of BrCl in the ambient air. This may have an adverse impact on human exposure to mixture of varying compositions. The extent of exposure can be positively correlated to the environmental humidity and its pH; physiologic sources of moisture (e.g., sweat, moisture in the upper respiratory tract); and products formed (National Research Council, 2014).

6.7 Risk assessment

The available information on risk evaluation specifically dealing with bromine chloride is sparse. This compound dissociates to form Br^-, Cl^-, or other by-products. Therefore, this section discusses the parameters of the related risk evaluation. To evaluate the risks, the engineering controls must be looked upon with utmost care. Predominantly three factors are mainly considered at a workplace that is level of hazardousness of the substance, quantity of release, and extent of exposure.

According to the reports, the presence of bromate in water is a matter of serious issue. Here, bromine gets oxidized to bromate in the presence of chlorine (Huang, Naiyun, & Yang, 2008), where chlorination combined with UV irradiations or catalyzed by metal oxides (CuO or NiO) results in a greater level of risks (Fang et al., 2017; Liu, Von Gunten, & Croué, 2012). Moreover, comprehensive studies provide an enhanced risk for bladder cancer associated with swimming pools disinfectants and disinfection by-products (Richardson et al., 2010). It is therefore imperative to analyze all the possible risks related to bromine chloride (Yang, Dong, Jiang, Wang, & Liu, 2019).

The level of risk analysis for the bromine chloride is based on the AEGLs, which represent the extent of exposure limit for the general public ranging from 10 min to 8 h. The three-tier system briefly elaborates the degree of severity of toxic effects. For the estimation of lethality, the benchmark concentration analysis was used to compute the no-observed-adverse-effect level (NOAEL). The 7-h BMCL05 (benchmark concentration, 95% lower confidence limit with 5% response) was recorded as 39.4 ppm. The effects of direct-acting irritants like bromine chloride are not expected to differ significantly between species or among individuals (NRC, 2001).

6.8 Environmental effects

On entering the environment, bromine chloride become lethal to aquatic organisms. It has been observed that fish exposed to this disinfectant exhibit variable mortality patterns as well as stress symptoms. Salmonids are more sensitive to chlorobrominated effluent in pertinent to chlorine. As a result, tests also reveal that less than half the amount of BrCl produces LC50 in comparison to chlorinated effluent. However, some available contradictory results show greater toxicity of chlorinated effluents (Ward & DeGraeve, 1978). There are reports supporting this converse fact of the negligible toxic effect of chlorobrominated on macro invertebrates.

Also, researchers have collected data related to the lethality of bromine chloride by performing experiments on Sprague-Dawley rats. The results obtained from measuring the mortality rate specifies the inculcation of respiratory problems during and after the exposure. However, the records found does not justify any sign of reproductive toxicity, genotoxicity, and chronic toxicity or carcinogenicity (Dow Chemical Co., 1977).

6.9 Hazardousness

The liquid BrCl causes rapid onset of irritation and burns on the skin and other tissues and may be extensively painful to the respiratory tract. The particulars on the odor threshold, lethal concentrations, developmental toxicity, reproductive toxicity, genotoxicity, or carcinogenicity of bromine chloride are sparsely accessible. Some of the acute health effects can occur after short-term exposure to bromine chloride. The likely obtained symptoms are severe irritation and burning of skin and eyes and irritation of nose and throat. Whereas, long-term exposure may result in the demise of the exposed or adversely damage the upper- and lower respiratory tract.

6.10 Treatment

If exposed to bromine chloride, the suspected person should be given immediate first aid. Adequate decontamination should be carried out at the primary level. In case of breathing issues, artificial respiration preferably with a demand valve resuscitator, bag-valve-mask device, or pocket mask or CPR should be done. Flush the eyes with an ample amount of water. Do not induce vomiting.

If required, then a patent airway that is oropharyngeal or nasopharyngeal airway should be established. The suspect should be closely observed for signs and treated accordingly. Patients must be monitored for pulmonary edema, shock, and seizures. Administer activated charcoal. The dry sterile dressing should be used for covering the burns after decontamination.

The symptomatic treatment should be given, such as beta-agonist albuterol to treat severe bronchospasm; IV administration of D5W/SRP; 0.9% saline (NS) or lactated Ringer's (LR) in case of hypovolemia; diazepam or lorazepam to treat seizures; proparacaine to assist eye irrigation (Currance, Clements, & Bronstein, 2005).

6.11 Risk management

Because of the wide applicability of BrCl as a water disinfectant its related-risk management is the need of the hour. The best way to reduce the risk includes the proper and periodic maintenance of engineering tools. The operations with even slight chance of causing toxicity must be performed in isolated areas. The facilities should have provision of ventilation at the site of chemical release.

With the purpose to determine cancer risk postingestion of BrCl or its dissociated products contaminated water, the trihalomethanes can be primarily utilized. For the removal of bromide from water, broadly three techniques can be effectively used that is membrane, electrochemical, and adsorption methods (Harrison, Le Gouellec, Cheng, & Childress, 2007; Pontie, Buisson, Diawara, & Essis-Tome, 2003; Watson, Farré, & Knight, 2012). Out of the three, adsorption is the cheapest and widely employed method. Beside these methods, the inhibition of oxidation also plays a crucial role in regulating the formation of bromate. This can be anticipated by pH depression, ammonia addition, H_2O_2 addition, and likewise inhibitory action on oxidation processes (Pinkernell & von Gunten, 2001; Yang et al., 2017).

6.12 Current research

Bromine chloride has been recognized as a promising substitute for disinfectants. Collectively, several places have replaced chlorine with BrCl due to its rapid onset of action. Recently, to make sure that it is effective against a wide array of bacteria, a chamber with flowing wastewater seeded with poliovirus has been simulated. The chamber was further treated individually with BrCl and chlorine, respectively. As a result, the former was found to be more efficient against bacteria in comparison to the latter one. The study clearly shows that engaged BrCl against viruses and bacteria is more feasible than chlorine. Hence, satisfactorily BrCl can be applied in the place of chlorine at the time of water disinfection.

6.13 Safety measures

1. There should be the provision of eyewash fountains in the vicinity of the facilities dealing with bromine chloride. In case of accidental exposure to BrCl, the eyes of the exposed person should be irrigated for a minimum time span of 30 min with a large quantity of running water. The personnel should also avoid wearing contact lenses.

2. The employees working in the BrCl facilities should take suitable precautions before entering and after leaving the workplace.
3. All protective clothing like suits, gloves, footwear, and headgear should be clean, available each day, and put on before work.
4. The individuals should wear indirect-vent impact and splash-resistant goggles along with a face shield when functioning with corrosion, highly irritating, and toxic substances.
5. The employees involved with the operating of BrCl should be well trained and acquainted with the hazardousness.

6.14 Handling and storage

In general, bromine chloride is compressed in liquid form for storage in cylinders or vessels that are provided with pipe for the uptake and release of gas under pressure of 30 psig at 25°C. It is easier to withdraw BrCl in a liquid state than in a gaseous form with accuracy and precision while taking out the desired quantities. The handler should use neoprene rubber gloves, rubber protective clothing, and respiratory protective equipment.

The building meant for the storage of BrCl should be fire-resistant, cool, dry, and well-ventilated. Otherwise, the provision of fire extinguishers should be there on the premises. BrCl should be kept away from food and feedstuffs. The storage area should be deprived of drainage and sewer systems.

6.15 Precautions

1. Wear American National Standards Institute (ANSI) approved chemical safety goggles, leather or other nonwoven ANSI approved steel-toed shoes, chemical-resistant suit, protective rubber boots overshoes, and full-face shield over eyewear.
2. Ensure wearing clothes devoid of bromine odor.
3. Do not reuse bromine/chlorine contaminated clothing.
4. Clean equipment before use.

References

Barratt, S., & Stein, C. P. (1929). On bromine chloride. *Proceedings of the Royal Society of London Series A, Containing Papers of a Mathematical and Physical Character, 122*(790), 582–588.

Barrie, M. D., Dahlstrom, D. L., Goswami, E., & Kaetzel, R. (2012). The halogens. In E. Bingham, & B. Cohrssen (Eds.), *Patty's industrial hygiene and toxicology* (pp. 1033–1108). (6th ed.). New York: Wiley.

Cotton, F. A., & Wilkinson, G. (1980). *Advanced inorganic chemistry: A comprehensive text* (4th ed.). New York: John Wiley and Sons.

Currance, P. L., Clements, B., & Bronstein, A. C. (Eds.). (2005). *Emergency care for hazardous materials exposure*. (3rd ed.). St. Louis: Elsevier Mosby.

Dagani, M. J., Barda, H. J., Benya, T. J., & Sanders, D. C. (2000). Bromine compounds. In *Ullmann's encyclopedia of industrial chemistry*. New York: John Wiley and Sons.

Dow Chemical Co (1977). *Evaluation of acute inhalation toxicity of bromine chloride in rats*. Dow report no. 77 2993. Submitted to EPA by Dow Chemical Company, Midland, MI, with cover letter dated 05/28/92. EPA document no. 88-920002267.

Emeleus, H. J., & Anderson, J. S. (1960). *Modern aspects of inorganic chemistry*: (p. 452) (3rd ed.). New York: Van Nostrand.

Fang, J., Zhao, Q., Fan, C., Shang, C., Fu, Y., & Zhang, X. (2017). Bromate formation from the oxidation of bromide in the UV/chlorine process with low pressure and medium pressure UV lamps. *Chemosphere*, *183*, 582–588.

Frim, R., & Ukeles, S. D. (2011). Bromine, inorganic compounds. In *Kirk-Othmer encyclopedia of chemical technology*. New York: John Wiley and Sons.

Gillam, A. E., & Morton, R. A. (1929). The absorption spectra of halogens and inter-halogen compounds in solution in carbon tetrachloride. *Proceedings of the Royal Society of London Series A, Containing Papers of a Mathematical and Physical Character*, *124*(795), 604–616.

Greenwood, N. N. (Ed.). (1956). *A comprehensive treatise on inorganic and theoretical chemistry* (p. 476). In Vol. 1. (p. 476). London: Longmans, Green and Co.

Harrison, C. J., Le Gouellec, Y. A., Cheng, R. C., & Childress, A. E. (2007). Bench-scale testing of nanofiltration for seawater desalination. *Journal of Environmental Engineering*, *133*(11), 1004–1014.

HSDB (Hazardous Substances Data Bank) (2011). *Bromine chloride*. Bethesda, MD: TOXNET Toxicology Data Network, U.S. National Library of Medicine. [online]. Retrieved from: (2011). *http://toxnet.nlm.nih.gov/*.

Huang, X., Naiyun, G., & Yang, D. (2008). Bromate ion formation in dark chlorination and ultraviolet/chlorination processes for bromide-containing water. *Journal of Environmental Sciences*, *20*(2), 246–251.

IPCS (International Programme on Chemical Safety) (2009). *Bromine chloride (CAS Reg. No. 13863-41-7). International Chemical Safety Card 1713*. International Programme on Chemical Safety and the Commission of the European Communities. [online]. Retrieved from: (2009). *http://www.ilo.org/dyn/icsc/showcard.display?p_lang=en&p_card_id=1713*.

Lang, J. P. (2006). Chlorine, bromine, iodine, and astatine: Inorganic chemistry. In *Encyclopedia of inorganic chemistry*. New York: John Wiley and Sons.

Liu, Q., & Margerum, D. W. (2001). Equilibrium and kinetics of bromine chloride hydrolysis. *Environmental Science and Technology*, *35*(6), 1127–1133.

Liu, C., Von Gunten, U., & Croué, J. P. (2012). Enhanced bromate formation during chlorination of bromide-containing waters in the presence of CuO: Catalytic disproportionation of hypobromous acid. *Environmental Science and Technology*, *46*(20), 11054–11061.

National Research Council (2014). Bromine chloride. *Acute exposure guideline levels for selected airborne chemicals: Vol. 18*. Washington, DC: The National Academies Press https://doi.org/10.17226/18941.

NRC (National Research Council) (2001). *Standing operating procedures for developing acute exposure guideline levels for hazardous chemicals*. Washington, DC: National Academy Press.

Pinkernell, U., & Von Gunten, U. (2001). Bromate minimization during ozonation: Mechanistic considerations. *Environmental Science and Technology*, *35*(12), 2525–2531.

Pontie, M., Buisson, H., Diawara, C. K., & Essis-Tome, H. (2003). Studies of halide ions mass transfer in nanofiltration—Application to selective defluorination of brackish drinking water. *Desalination*, *157*(1), 127–134.

Richardson, S. D., DeMarini, D. M., Kogevinas, M., Fernandez, P., Marco, E., Lourencetti, C., et al. (2010). What's in the pool? A comprehensive identification of disinfection by-products and assessment of mutagenicity of chlorinated and brominated swimming pool water. *Environmental Health Perspectives*, *118*(11), 1523–1530.

Schmiesser, M., & Shuster, E. (1966). In Z. E. Jolles (Ed.), *Bromine and its compounds*. London: E. Benn Ltd.

Wang, T. X., Kelley, M. D., Cooper, J. N., Beckwith, R. C., & Margerum, D. W. (1994). Equilibrium, kinetic, and UV-spectral characteristics of aqueous bromine chloride, bromine, and chlorine species. *Inorganic Chemistry*, *33*(25), 5872–5878.

Ward, R. W., & DeGraeve, G. M. (1978). Acute residual toxicity of several wastewater disinfectants to aquatic life. *Journal of the American Water Resources Association*, *14*(3), 696–709.

Watson, K., Farré, M. J., & Knight, N. (2012). Strategies for the removal of halides from drinking water sources, and their applicability in disinfection by-product minimisation: A critical review. *Journal of Environmental Management*, *110*, 276–298.

Yang, J., Dong, Z., Jiang, C., Wang, C., & Liu, H. (2019). An overview of bromate formation in chemical oxidation processes: Occurrence, mechanism, influencing factors, risk assessment, and control strategies. *Chemosphere*, *237*, 124521.

Yang, J., Li, J., Dong, W., Ma, J., Yang, Y., Li, J. , et al. (2017). Enhancement of bromate formation by pH depression during ozonation of bromide-containing water in the presence of hydroxylamine. *Water Research*, *109*, 135–143.

CHAPTER

7

Carbon dioxide: Risk assessment, environmental, and health hazard

Malvika Chawla
Department of Chemistry, Gurukula Kangri (Deemed to be University), Haridwar, Uttarakhand, India

Abbreviations

AHP	Analytical Hierarchy Process
AIM	Aquifer Impact Model
ALARP	As Low As Reasonably Practicable
CCS	Carbon Capture and Storage
CFD	Computational Fluid Dynamics
DNV	Det Norske Veritas
DPSIR	Drivers, Pressures, States, Impacts, and Responses
DREAM	Design for Risk Evaluation and Monitoring
EIA	Environmental Impact Assessment
ESL	Evidence Supported Logic
FEP	Features, Events, Processes
FMECA	Failure Mode, Effects and Criticality Analysis
FTA	Fault Tree Analysis technique
HAZOP	HAZard Operability
LCA	Life Cycle Assessment
NRAP-IAM-CS	NRAP-Integrated Assessment Model-Carbon Storage
NSealR	Natural Seal ROM
PHAST	Participatory Hygiene and Sanitation Transformation
QRA	Quantitative Risk Assessment
REV	Reservoir Evaluation and Visualization Wellbore Leakage Analysis Tool
STSF	Short Term Seismic Forecasting
SWIFT	Structured What IF Technique
TESLA	The Evidence Support Logic Application

Hazardous Gases
https://doi.org/10.1016/B978-0-323-89857-7.00006-2

7.1 Introduction

Carbon dioxide (CO_2) is a nonflammable, odorless, and colorless gas, undetectable to human senses, and often accumulated near the ground. It is inherently a product of ignition, fermentation, and breathing process. These peculiarities elucidate why surrounded environments are acquainted with CO_2 buildup, dislodging oxygen from the region (Zaba et al., 2011). The volcanic outgassing, burning of organic matter, exhalation, and inhalation processes of living aerobic organisms, burning of various fossil fuels for power generation, and usage for transportation are some of the major natural and man-made sources for the presence of atmospheric carbon dioxide.

The characteristic attributes of carbon dioxide prevail as comparatively more vulnerable to humans and the environment. The defined dense behavior of CO_2 that is, 1.5 times than air, proportionally accounts for an increase in the tendency of leaking from pipework or confined storage spaces, which could create hazardous situations. Likewise, colorless, tasteless, and odorless inherent property act as an add-on to the hazardous nature of the release of CO_2 (IPCC, 2005). Consequently, the enhanced atmospheric concentration of CO_2 also have phenomenal repercussions on global warming.

Moreover, the long-term exposure to airborne carbon dioxide ranging between 0.5% and 1% concentration can result in chronic effects such as metabolic acidosis, increased calcium deposits in soft tissues and some other similar effects. The further increment in the concentration above 3% can prove to be lethal to the cardiovascular system and upper respiratory tract. The leakage can adversely affect the environment as well as human health resulting in an unforeseen impact on society as a whole. This indicates the requirement to conduct risk recognition programs centered on hazard identification, mitigation, and management.

7.2 Occurrence

Carbon dioxide naturally occurs in the atmosphere as a trace gas. Due to its solubility in water, it occurs naturally in groundwater, rivers and lakes, ice caps, glaciers, and seawater. It is also present in deposits of petroleum and natural gas.

This appears to be antecedent of the times of Pliny, who in his "Naturalis Historiae" had stated, "But, by Hercules! the history of the heavens themselves would not be more difficult to relate: the abundance of metals ... the virtues of medicinal springs ... the exhalation of deadly vapors (spiritus lethales) either emitted from caverns or certain unhealthy districts; some of them fatal to birds alone as at Socrate, a district near the city; others to all animals, except to man, while others are so to man also." The interstices mentioned in this context are conventionally called vents or Charon's sewers, culpable for exhaling a deadly vapor. The written description also acknowledges a place in Asia and prophetic caves, where safe infiltrating by others is prohibited. Only the "priest of the great Mother of the Gods" is permitted to enter those caves, for predicting the future, as at Delphi. With the advent of modern times, the writings by Pliny are still unquestionable, due to his reference to the presence of carbonated water springs and CO_2 vents, such as the Grotta del Cane at Pouzzoles near Naples and likewise (Germann, 1938).

7.3 Historical background

The initial chronicle of CO_2 was described in the early 1600s by J.B. van Helmont, which was specifically delineated by Joseph Black in 1756. Furthermore, with the passage of time, the toxic attributes of carbon dioxide were evidently revealed in the 19th century.

7.4 Physicochemical properties

Under atmospheric requisite, carbon dioxide remains chemically stable and inert. It completely overcomes and restrains the combustion reactions. Carbon dioxide possesses a corrosive effect in combination with carbon steel and nonferrous metals due to the generation of carbonic acid in the presence of water. It tends to react vigorously with some definite substances particularly amines or ammonia.

CO_2 is heavier than air and therefore, expects to flow downward accumuating in pits, basements, or native depressions. In presence of even mild air movement, these pools of CO_2 can persist for many hours. The physical states of carbon dioxide exceedingly confide on pressure and temperature (Table 7.1) (National Center for Biotechnology Information, 2020).

7.5 Physiological role

The prosper amount of carbon dioxide, at the cellular level, confers the modification in buffering character, thereby shifting cellular pH. Thereupon, activating the Na^+/K^+ adenosine triphosphatase (ATPase) that subsequently causes hypokalemia. On the other hand, increased concentration of carbon dioxide with the lapse of time causes acidosis and elevates K^+ postdisintegration of the normal buffering system. In appendage, peptides,

TABLE 7.1 Physicochemical properties of carbon dioxide.

Formula	CO_2
Synonyms	Carbon dioxide Carbonic anhydride Dry ice Carbonic acid gas
Physical state	Exist as a gas or in solid form (at atmospheric pressure); the sublimes at −78.5°C
Molecular weight	44.01 g/mol
Boiling point	−78.48°C (−78.464°C (sublimes))
Melting point	−56.5°C
Solubility	1.48 mg/mL at 25°C
Vapor density	1.53 at 78.2°C (air = 1)
Vapor pressure	56.5 atm
Viscosity	21.29 μPa s at 300 K
Heat of vaporization	83.12 g cal/g
Ionization potential	13.77 eV
Refractive index	1.6630 at 24°C
Density	1.799 g/L
Heat of formation	−393.51 kJ/mol
Entropy	213.785 J/K mol
Heat capacity	37.13 J/mol K
Van der Waals constant	a = 3.658 bar sq L/sq mol; b = 0.0429 L/mol
Enthalpy of fusion	9.02 kJ/mol
Diffusion coefficient (D) in water	1.91×10^{-5} sq cm/s at 25°C
Ionization energy	13.773 eV

proteins, and amino acids are assumed to form *N*-carboxy derivatives (carbamates) in the presence of carbon dioxide, elucidating the observed neurotoxicity. The considerable perusal of literature also acquaints with some other systemic sympathetic stimulation, for instance, a rise in heart rate, cardiac output, and cardiac arrhythmias, and an increment in mean pulmonary artery pressure and pulmonary vascular resistance.

The passively absorbed carbon dioxide through the medium of lungs is primarily borne as carbonic acid and further dissociate into bicarbonate in

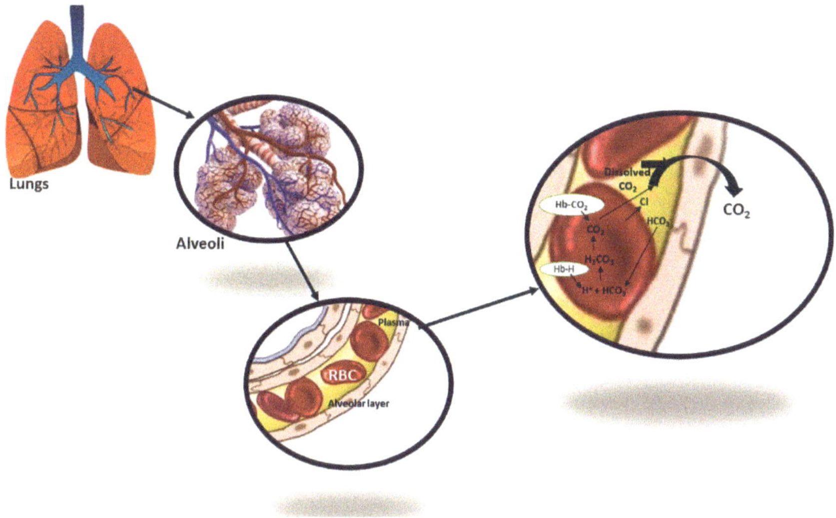

FIG. 7.1 Schematic illustration depicting carbon dioxide exchange from cells to alveoli. *Credit: Malvika Chawla.*

the blood, in solution or alliance with hemoglobin, the reaction is catalyzed with the aid of carbonic anhydrase (Patel, Miao, Yetiskul, et al., 2020). The minor quantities of CO_2 additionally find its existence in the dissolved form in plasma and carbamino compounds in blood. The gradient flow of carbon dioxide between the blood and the air in the alveoli guides the reverse formation of bicarbonate (Fig. 7.1). The rate and amount of release is, however, dependent on the pH, which is regulated by various buffers inclusive of bicarbonate. The metabolism of carbon dioxide is fundamentally surmised by lungs, and in some quantity by kidneys.

7.6 Uses

Carbon dioxide is extensively generated and employed in the chemical and petrochemical industries. The major part is retrieved from the former source, where it is emanated as a by-product of enumerate chemical processes such as synthesis of ammonia, methanol, hydrogen, substitute natural gas, limestone calcification, and likewise. It also finds its applicability in hydrocracking and hydrotreating petroleum products and as a coolant. While in the food processing industry, it is handled in refrigeration, carbonation of beverages, and as a preservative. Additionally, it is utilized in fire extinguishers due to its oxygen displacing property. Hence, it lowers or anticipates burning caused due to the dearth of oxygen. It also find its application in welding to impede molten metal reacting with oxygen.

There is also therapeutic usefulness of carbon dioxide in moderate concentrations, as it proves to be beneficial for neutralizing the effects of oxygen inadequacy. In patients after a head injury or succeeding neurosurgery, the carbon dioxide concentration may be manipulated to modify cerebral blood vessel thickness and blood flow.

Carbon dioxide has also been used in insufflation of the peritoneal cavity to permit visualization of the abdominal viscera during gynecological and surgical laparoscopic procedures. In the preceding years, it was also used in the management of hiccups, anxiety, and other psychiatric disorders, however, the appropriateness of usage is unknown. Although, some practitioners still occasionally use this form of treatment and deaths have been reported.

7.7 Sources of contaminants

The carbon dioxide contamination could be due to either natural or human sources. Where the former includes decomposition, releases from ocean and organism respiration as well as decomposition, and later includes activities like cement production, deforestation, volcano eruption, forest fires, and burning of fossil fuels.

With the advent of the Industrial Revolution, the amount of carbon dioxide has reached devastating levels in the atmosphere with time. On the other side, natural sources had also contributed to the incremental atmospheric quantity of carbon dioxide. One of the prominent reasons behind this could be the removal of natural sinks, responsible for the atmospheric balance of this gas, from the environment. The man-made sources are not only limited to above mentioned activities, rather exposure of limestone to the atmosphere and weathering of rain also contribute toward the release of carbon dioxide.

7.8 Risk assessment

The risk assessment while handling carbon dioxide primarily is a challenge in itself. The subsistence of carbon dioxide in humans and the environment mitigates the need for risk analysis. Generally, carbon dioxide subsists as a gas or solid, but due to technical and economical reasons, it is principally condensed into liquid or even supercritical form before transit. Moreover, in case of any release of CO_2, upon expansion, this changes into either a vapor or a vapor/solid, depending upon the inventory pressure and temperature. The disengagement of gas in the air can result in Joule-Thomson (J-T) cooling culminating up to 90°C drop in temperature. Furthermore, release below the dew point temperature of the air can mainspring the moisture present in the air to freeze, hence, inducing the accumulation of thick and icy fog (Phillips, Shirvill, & Holt, 2011). Additionally, the gas released within the vicinity of the confined areas tend to stay there for a comparatively longer time as well as fabricate colder conditions, than open spaces (Engebo, Ahmed, Garstad, & Holt, 2013). This substantiates the need to assess the risk while transporting, capturing, and storing CO_2. The estimation of risk assessment constitutes two quantitative methods that is, probabilistic and deterministic. The former approach deals with the degree of uncertainty while the latter follows a well-known predictable rule (Glessner & Young, 2008).

Some of the major risk assessments in correlation with pipeline, capture, and storage of carbon dioxide are briefly discussed in the following section.

Pipeline Risk Assessment: One major element of risk assessment is pipeline, as a substantiate portion of CO_2 transmission from the capture site to the storage site takes place via them. The approximation of release from the pipeline is maximum, which could effectively blow away in case of high pressure. On the other side, corrosion, poor quality of weld seams, risk of hydrate formation, and risk of water freezing are some other major problems associated with these pipelines. There are some reports by Neele et al. concerning the recommendation of Quantitative Risk Assessment (QRA) methods to study the pipeline risks.

Capture and Storage Risk Assessment: Extensively, the carbon dioxide capture and storage areas are constituted within confined spaces (Koornneef, Ramirez, Turkenburg, & Faaij, 2012). The principal methods for CO_2 storage comprise geological storage, ocean storage, and mineral carbonation or mineral sequestration. The geological storage includes drained oil and gas reservoirs, unmineable coal deposits and likewise. Where the ocean storage is confided to the compressed isolation of gas and mineral carbonation is a chemical process used to form insoluble carbonates (GCCSI, 2011).

Some of the principal risk factors correlated with CO_2 capture processes comprehend with the produced vent gas in the facility and the presence of impurities such as SO_2, NO, H_2S, H_2, CO, CH_4, N_2, Ar, and O_2 (BRGM, 2005). The contemplation pondering categorical distribution of the risk resulting from carbon dioxide storage by Figueredo et al. include toxicological effects, environmental effects, subsurface trespass, induced seismic actions, and climate effects (Figueredo et al., 2012). The Det Norske Veritas (DNV) assessed the capture, transport, and storage of CO_2 by an expert panel/workshop approach that is, SWIFT (Structured What IF Technique) analysis, in identifying potential hazards and uncertainties.

There are several methods that can be utilized in risk assessment (Oraee-Mirzamani, Cockerill, & Makuch, 2013), some of them are briefly detailed in the ahead section (Table 7.2).

TABLE 7.2 Risk assessment methodologies.

Risk assessment method	Description	Aim
Fault Tree Analysis (FTA)	A visual model inculcated with cause and effect relationship	Apprehend CO_2 storage-related risks
Analytical Hierarchy Process (AHP)	A mathematical illustration that is employed to acquire ratio scales from paired comparisons	The output harvested from this method is ratio scales and a consistency index
Features, Events, Processes (FEP)	The attributes of this model include Feature, Events, and Processes	This method is extensively employed for illustrating the scenarios relevant to the safety assessment of the geological disposal of radioactive wastes
Drivers, Pressures, States, Impacts, and Responses (DPSIR)	This framework is a conceptual model to describe the relationships between the environment and society in a simplified manner	The method is aimed to help in selecting and structuring environmental indicators and provides insight into cause-effect relationships between them
Evidence Supported Logic (ESL)	The model works on TESLA software	The method gives primary judgment and technical overview correlating the evidence-based risk interpretation
Bow-tie Risk Assessment	The protocol gives insight into visual demonstration of risk associated with engineering facilities	This protocol can find its provisional demonstration of risks reduced to levels that are As Low As Reasonably Practicable (ALARP)

Fault Tree Analysis (FTA): It is a well-acknowledged visual model inculcated with easy-to-understand cause and effect relationship. The method is prepared in a way to guide across to apprehend the risks in correlation with the storage of carbon dioxide in an advanced way.

Analytical Hierarchy Process (AHP): This method is a mathematical illustration employed to acquire ratio scales from paired comparisons. The computation study is designed in a way to critically assess the associated risks. The output harvested from this method is ratio scales and a consistency index. The approach is dependent on the solution of an Eigenvalue problem.

Features, Events, Processes (FEP): This design illustrates the relevant safety measures concerning geological disposal of radioactive wastes (Hnottavange-Telleen, 2013). Here "Feature" stands for geological formation of CO_2 and its characteristics; "Events" for what may or will happen in the future, and "Processes" for the ongoing matters that influence the evolution of the system, like the erosion of the land surface. Varied CO_2 storage projects have successfully employed this method, such as Weyburn (PTRC, 2004) or Illinois Basin-Decateur project (Hnottavange-Telleen, Krapac, & Vivalda, 2009).

Drivers, Pressures, States, Impacts, and Responses (DPSIR): It is a conceptual model describing the correlation between the society and environment, developed by EEA (European Environmental Agency) (Smeets & Weterings, 1999). This method guides the selection of particular indicators responsible for affecting the environment and is in use by the Netherlands (Koornneef, Ramírez, Turkenburg, & Faaij, 2011).

Evidence Supported Logic (ESL): This model is supported by the TESLA that is, The Evidence Support Logic Application software. The computational approach is built on the logical hypothesis model whose attributes give primary judgment and a technical point of view correlating the evidence-based risk interpretation. The algorithmic approach allows time to time monitoring of risk assessment as well as all-related uncertainty (ESL guide, v2.1, n.d.).

Bow-tie Risk Assessment: The Bow-tie approach is laid on the cohesion between the causes of unwanted events, the escalation of such events to a range of possible outcomes, the controls preventing the event from occurring, and the mitigation measures in place to limit the consequences. The protocol provides a visual interpretation of the related risks, hence, applied in the engineering facilities. Therefore, this protocol can find its provisional demonstration of risks reduced to levels that are As Low As Reasonably Practicable (ALARP). This protocol has been in use by several CO_2 storage projects, such as the Goldeneye CCS project (Tucker, Holley, Metcalfe, & Hurst, 2013).

CO_2-PENS: It is a commercially mathematical model that runs on GoldSim system programing software. The GoldSim allows detailed simulations of transport of CO_2 and brine within an injection zone via a dynamically linked library.

There are few other available risk assessment tools namely NRAP Integrated Assessment Model—Carbon Storage (NRAP-IAM-CS), Reservoir Evaluation and Visualization (REV), Wellbore Leakage Analysis Tool (WLAT), Natural Seal ROM (NSealR), Aquifer Impact Model (AIM), Design for Risk Evaluation and Monitoring (DREAM) and Short-Term Seismic Forecasting (STSF).

7.9 Environmental effects

One of the reasons beneath the mitigation of devastating effects on global climatic alterations lies around the carbon capture and storage (CCS). The atmospheric level of carbon dioxide, the chief greenhouse gas, has escalated around 40% since the advent of the Industrial Revolution. Out of which the oceans take up nearly one-third of that CO_2, propagating a rise in acidification by forming carbonic acid. Whereas, under other conditions, the runoff from the soil, human and animal waste, and similar origins in the ocean and rivers also play a crucial role in polluting ocean water. Additionally, initiating the depletion of oxygen level in water by the breakdown of organic matter formed during algal blooms. The potential risks may also contribute to the propagation of degraded marine ecosystems (Strong, Kroeker, Teneva, Mease, & Kelly, 2014).

Though, carbon dioxide concentrations in ambient air do not have any lethal effect on plants, but the elevated concentrations can disturb their growth, causing yellowing and drying out. Indirectly it may also permute pH of the soil as well restrict root development (Rusin & Stolecka, 2014).

7.10 Hazardousness

The toxic behavior of carbon dioxide generally arises in presence of concentration greater than 5%, which consequently results in hypercapnia and respiratory acidosis. The interference by hydrolysis of acetylcholine by acetylcholinesterase aggravate the effects of parasympathetic nervous activity. Concentrations greater than 10% of carbon dioxide may cause convulsions, coma, and death (Langford, 2005; Zaba et al., 2011) and more than 30% can lead to rapid loss of consciousness within a fraction of time (Fig. 7.2). This clearly explains why victims of incidental intoxications often do not personate to resolve the situation (open a door, etc.) (Gill et al., 2014; Kettner, Ramsthaler, Juhnke, Bux, & Schmidt, 2013; Srisont, Chirachariyavej, & Peonim, 2009).

Studies have shown a vast variableness of CO_2 tolerance in blood ranging from 0.055 to 0.085 atm (41.8–64.6 mmHg) alluding to safe exposure level (Kettner et al., 2013), which reduces with age and in cigarette smokers. Concentrations of inevitable cases of carbon dioxide transmute between 14.1% and 26% CO_2 level and an accompanying O_2 level between 4.2% and 25% (Norimine, Ishizawa, Honda, & Uemura, 2009; Zaba et al., 2011). In literature, the effects of oxygen management, both normal and high concentration, have been documented (Gill et al., 2014; Spitz & Spitz, 2006).

The enhanced concentrations of oxygen elevate venous partial pressure minimizing the dissolvability of carbon dioxide in the blood. The process, in turn, induces venous partial pressure of CO_2 to rise in blood with no alteration in the metabolic requisites. A commencing augment of pCO_2 in the bloodstream is expected while giving oxygen to a hypoxic carbon dioxide intoxicated person. This effect is also so-called as Haldane effect. The high concentrations of oxygen may also result in the enhancement of dead space (Gill et al., 2014; Spitz & Spitz, 2006).

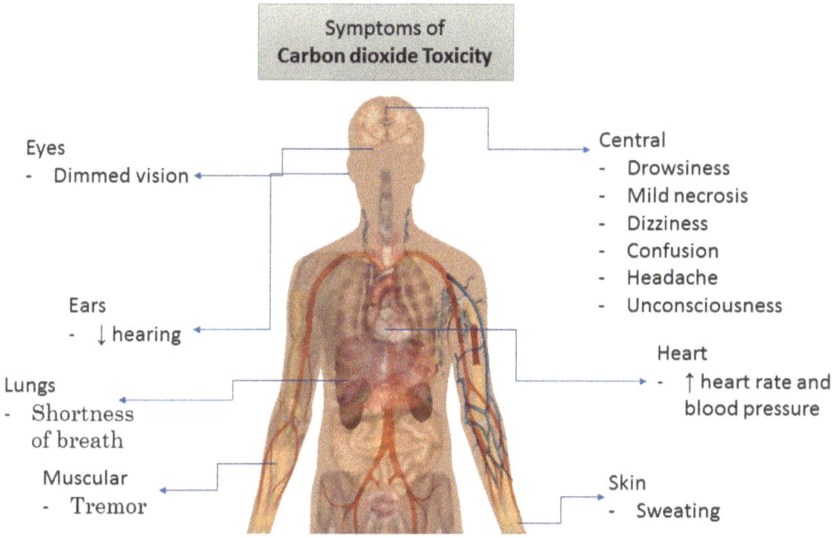

FIG. 7.2 Figure showing the main symptoms of carbon dioxide toxicity. *Credit: Malvika Chawla.*

7.11 Treatment

As such, there is no specific accessible antidote or treatment available for carbon toxicity management. Therefore, preferably symptoms-based treatment should be looked into as prime treatment. The affected person should be assisted for hypotension, dysrhythmias, electrolyte, and acid-base disturbances, and hypoxia (Langford, 2005).

However, in the case of dry ice burns, initially, further skin contact with ice should be minimized. Furthermore, the removal of nonadherent clothing; rapid-warming and splinting and elevation of the affected area; use of pain killers medicines; are some of the measures that can be taken care of (Biem, Koehncke, Classen, et al., 2003). Studies suggest both normal and high concentration oxygen can be given as primary management (Gill et al., 2014; Spitz & Spitz, 2006).

7.12 Risk management

The complete risk analysis substantiates the need to mitigate the risk control approach addressing the potential consequences. Extensively, the protocol is laid on risk elimination, application of substitution where possible, engineered controls, and administrative or managerial controls. It is an ongoing and iterative process aiming at continual improvement of

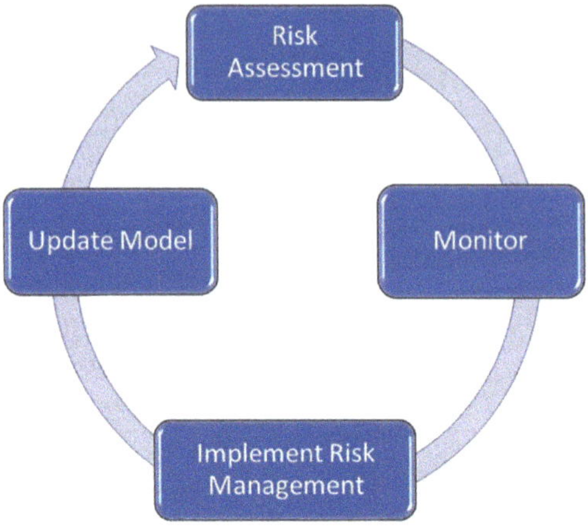

FIG. 7.3 Illustration depicting the Risk Management Process. *Credit: Malvika Chawla.*

risk analysis, throughout the CO_2 storage life (Fig. 7.3). Risk management bridges the gap between the potential risk uncertainties. Some of the risk management correlating carbon dioxide are discussed briefly in the following section.

Pipeline Risk Management: Measures to minimize the corrosion of CO_2 pipelines should be undertaken such as removal of impurities such as hydrogen sulfide or water; selection of corrosion-resistant materials like avoidance of elastomeric seals (Eldevik, Graver, & Torbergsen, 2009; Koornneef, Spruijt, Ramirez, Turkenburg, & Faaij, 2010; Rybicki & Łaciak, 2008); the restricted amount of free water that is, below 50ppm (Serpa, Morbee, & Tzimas, 2011). The two-phase flow could lead to the damage of compressors and other equipment, and hence should be avoided. Another crucial element of pipeline safety is the correct placement of safety valves together with automatic gas leak detection systems.

There are studies recommended to be followed such as DNV practice (DNV, 2010) and ISO 31000 on Risk Management (ISO 31000, 2010) of carbon dioxide. Koornneef et al. from Utrecht University reviewed the uncertainties regarding the quantitative risk assessment of CO_2 transport by pipelines. Therefore "*mitigation of risks should be focused on reducing the probability and consequences of large releases and less on reducing the probability and consequences of small scale leaks*" (Koornneef et al., 2010).

Capture and Storage Risk Management: In order to prevent corrosion and formation of hydrate during transportation, the removal of moisture should be primarily done (IPCC, 2005). In the case of carbon dioxide capture and storage-related risk management, the engineering facilities and other relevant techniques used by petrochemical facilities can be used such as HAZard Operability (HAZOP) studies (IPCC, 2005; Krzemień, Więckol-Ryk, Duda, & Koteras, 2013).

It is recommended to use the validated modeling methods as part of a risk assessment in order to prevent CO_2 releases such as PHAST, Computational Fluid Dynamics (CFD), ventilation modeling, PHAST In-Building Release model. In France, BRGM is one of the predominant institutes working on risk management of CO_2 storage (Bouc et al., 2009; Fabriol, 2009). Oxand, a company specialized in industrial risk management reportedly uses a classic risk management approach for storage (Identification, Estimation, Evaluation, Treatment). The (semi)integrated studies such as Australian and New Zealand Standard for Risk Management (AS/NZS 4360: 2004) defining a set of extreme risks associated with integrated have also been considered. A Joint Industry Project, CO2RISKMAN (DNV, 2012), has also investigated the applicability of current knowledge for managing risk to carbon capture and storage.

7.13 Current research

In a world-first breakthrough, researchers from the RMIT University, Melbourne developed a modern technique that can effectively transmute CO_2 from a gas into solid particles of carbon (Esrafilzadeh, Zavabeti, Jalili, et al., 2019). RMIT researcher Dr. Torben Daeneke aforesaid converting CO_2 into a solid could be a more endurable approach. It is an electrochemical technique developed to arrest and change atmospheric CO_2 to storable solid carbon. This could change the approach of carbon capturing and storing.

The liberated CO_2 is creating a cascade of chemical consequences that gets inconsiderable public care, which is advancing the unwanted deviation in ocean acidity. Sunda and Cai developed a computer model for the estimation of probable consequences of ocean acidification from this process both currently and with future advances in atmospheric CO_2 (Sunda & Cai, 2012). Recently, Digdaya and the research team have also reported an electrochemical system that can capture and consequently convert CO_2 from ocean water (Digdaya, Sullivan, Lin, et al., 2020).

The communications proposing the assessment of the relative effectiveness of several CO_2 mitigation routes by calculating the energy cost of carbon abatement (kilowatt-hour spent per kilogram CO_2-equivalent, or $kWh\,kgCO_2e^{-1}$) mitigation has also been under consideration (Babacan, De Causmaecker, Gambhir, et al., 2020).

With the arrival of the tech era, studies reporting processes dealing with the removal of alleviated CO_2 atmospheric levels are under consideration. One such study exploring the utilization of magnesite ($MgCO_3$) feedstock to repeatedly calcined and carbonated to remove CO_2 from the air has also turned the table toward CO_2-free atmosphere (McQueen, Kelemen, Dipple, et al., 2020).

7.14 Carbon dioxide toxicity: Case study

May 25, 2019: According to the reports, leakage of carbon dioxide on the Chinese cargo ship Jin Hai Xiang that was being repaired at a dock in Shandong province. Consequently, the incident killed 10 people, whereas 19 were reportedly found injured.

August 16, 2008: An incident took place post leakage of carbon dioxide in a gloss paint manufacturing plant in Mönchengladbach, Germany. The outbreak triggered the fire, propagating the formation of cloud, that intoxicated a total of 107 individuals.

7.15 Safety measures

- Wear chemical safety goggles.
- Always wear insulated protective clothing, if in contact with refrigerated gas.
- Wear cryogenic work gloves when handling carbon dioxide.
- Keep a check on any tempering with storage tank pressure relief devices.
- Only trained and authorized personnel should operate and maintain storage facilities and system components for carbon dioxide.
- Report excessive venting, deficiencies, or other safety concerns as soon as possible.
- Use level-sensing telemetry, if continued supply of carbon dioxide is critical to the facility's operation.

7.16 Storage and handling

7.16.1 Handling of dry ice

Few safety particulars must be taken care of while handling dry ice due to the tendency of forming gaseous CO_2 and its low temperature, such as:

- As it is not a consumable so its presence is strictly prohibited in beverages. Dry ice should be kept away from children's reason might be its offensive effects on the human body.
- Low temperature makes dry ice handling impossible therefore use of gloves or tongs is suggested.
- At the time of manual chopping, the eyes should be protected from flying particles.
- The pressurized containers must be carefully stored and transported due to high chances of bursting.
- Well ventilated rooms should be secured and only trained personnel should be allowed to enter the premises.

7.16.2 Handling

- The cylinder should be kept in an upright position
- Avoidance of contact with incompatible chemicals.
- Hand truck should be preferred while carrying cylinders.
- Usage of pressure regulator for cylinder should be done.

7.16.3 Storage

Store carbon dioxide containing containers in cool, dry, well-ventilated, out of direct sunlight and heat source, preferably in an isolated area. It can be stored in numerous ways such as compressed gas cylinders, on-site liquid storage tanks, portable liquid containers, and wrapped dry-ice blocks.

It is generally stored below the critical temperature, 87.8°F (31°C) when stored in compressed gas cylinders. These cylinders are constructed as per DOT specifications 3A, 3AA, 3AL, or 3E and valves with CGA-320 connections. In case of on-site liquid containers, ensure that the system can accommodate the enormous increase in volume as liquid carbon dioxide vaporizes. In addition, specially designed carbon dioxide storage tanks are, generally, made up of carbon steel. The tanks are insulated to reduce heat transfer and to prevent frost build-up on the outer surface. Moreover, storage tanks consist of in-build mechanical refrigeration units, pressure relief devices, and pressure-building vaporizers to maintain internal temperature and pressure. Storage temperatures for liquid carbon dioxide range from –20°F (–28.9°C) to 4°F (–15.6°C) with corresponding vapor pressures from 200 psig (1378 kPa) to 312 psig (2151 kPa).

7.17 Precautions

- Proper ventilation must be maintained in rooms containing CO_2.
- The provision of a self-contained breathing apparatus should be there in the area with CO_2.
- Prefer primary evacuation of personnel at low lying areas, in case of release.
- Operational CO_2 extinguishing systems must be installed.
- Do not taste carbon dioxide.
- Avoid touching frosted pipes or valves with bare skin.

References

Babacan, O., De Causmaecker, S., Gambhir, A. , et al. (2020). Assessing the feasibility of carbon dioxide mitigation options in terms of energy usage. *Nature Energy, 5*, 720–728. https://doi.org/10.1038/s41560-020-0646-1.

Biem, J., Koehncke, N., Classen, D. , et al. (2003). Out of the cold: Management of hypothermia and frostbite. *CMAJ, 168*, 305–311.

Bouc, O., Audigane, P., Bellenfant, G., Fabriol, H., Gastine, M., Rohmer, J. , et al. (2009). Determining safety criteria for CO_2 geological storage. *Energy Procedia, 1*, 2439–2446.

BRGM (2005). *Stockage Géologique du CO_2: Analyses des risqué, surveillance et mesures.* Rapport final.

Digdaya, I. A., Sullivan, I., Lin, M. , et al. (2020). A direct coupled electrochemical system for capture and conversion of CO_2 from oceanwater. *Nature Communications, 11*, 4412. https://doi.org/10.1038/s41467-020-18232-y.

DNV (2010). *Design and operation of CO_2 pipelines recommended practice DNV-RP-J202.* Norway: Det Norske Veritas.

DNV (2012). *CO2RISKMAN guidance on effective risk management of safety and environmental major accident hazards from CCS CO_2 handling systems.* Retrieved from: www.dnv.com.

Eldevik, F., Graver, B., & Torbergsen, L. (2009). Development of a guideline for safe, reliable and cost efficient transmission of CO_2 in pipelines. *Energy Procedia, 1*, 1579–1585.

Engebo, A., Ahmed, N., Garstad, J. J., & Holt, H. (2013). Risk assessment and management for CO_2 capture and transport facilities. *Energy Procedia, 37*, 2783–2793.

ESL guide, v2.1. (n.d.). Retrieved from Quintessa: http://www.quintessa-online.com/TESLA/ESLGuide.pdf.

Esrafilzadeh, D., Zavabeti, A., Jalili, R. , et al. (2019). Room temperature CO_2 reduction to solid carbon species on liquid metals featuring atomically thin ceria interfaces. *Nature Communications, 10*, 865. https://doi.org/10.1038/s41467-019-08824-8.

Fabriol, H. (2009). Risk management and uncertainties: Ensuring CO_2 storage safety. In *3rd international symposium: Capture and geological storage of CO_2, 5–6 November 2009, Paris.*

Figueredo, M. , et al. (2012). *The liability of carbon storage.* Retrieved from: http://sequestration.mit.edu/pdf/GHGT8_deFigueiredo.pdf.

GCCSI (2011). *Accelerating the uptake of CCS: Industrial use of captured carbon dioxide.* Global CCS Institute.

Germann, F. E. E. (1938). The occurrence of carbon dioxide with notes on the origin and relative importance of subterranean carbon dioxide. *Science, 87*(2267), 513–521.

Gill, M., Natoli, M. J., Vacchiano, C., MacLeod, D. B., Ikeda, K., Qin, M. , et al. (2014). Effects of elevated oxygen and carbon dioxide partial pressures on respiratory function and cognitive performance. *Journal of Applied Physiology, 117*(4), 406–412.

Glessner, M. M., & Young, J. E. (2008). Carbon capture and storage. *Chemical Engineering, 115*(5). 28–34, 36, 38.

Hnottavange-Telleen, K. (2013). Common themes in risk evaluation among eight geosequestration projects. *Energy Procedia, 37*, 2794–2801. https://doi.org/10.1016/j.egypro.2013.06.164.

Hnottavange-Telleen, K., Krapac, I., & Vivalda, C. (2009). Illinois Basin-Decateur Project: Initial risk assessment results and framework for evaluating site performance. *Energy Procedia, 1*, 2431–2438.

IPCC (2005). *Special report on carbon dioxide capture and storage, Intergovernmental Panel on Climate Change.* London: Cambridge University Press.

ISO 31000 (2010). *Risk management, principles and guidelines.* January 2010, NF ISO 31000.

Kettner, M., Ramsthaler, F., Juhnke, C., Bux, R., & Schmidt, P. (2013). A fatal case of CO_2 intoxication in a fermentation tank. *Journal of Forensic Sciences, 58*(2), 556–558.

Koornneef, K., Spruijt, M., Ramirez, A., Turkenburg, W., & Faaij, A. (2010). Quantitative risk assessment of CO_2 transport by pipelines—A review of uncertainties and their impacts. *Journal of Hazardous Materials, 177*, 12–27.

Koornneef, J., Ramírez, A., Turkenburg, W., & Faaij, A. (2011). The environmental impact and risk assessment of CO_2 capture, transport and storage—An evaluation of the knowledge base using the DPSIR framework. *Energy Procedia, 4*, 2293–2300.

Koornneef, J., Ramirez, A., Turkenburg, W., & Faaij, A. (2012). The environmental impact and risk assessment of CO_2 capture, transport and storage—An evaluation of the knowledge base. *Progress in Energy and Combustion Science, 38*, 62–86. https://doi.org/10.1016/j.pecs.2011.05.002.

Krzemień, A., Więckol-Ryk, A., Duda, A., & Koteras, A. (2013). Risk assessment of a post-combustion and amine-based CO_2 capture ready process. *Journal of Sustainable Mining, 12*(4), 18–23.

Langford, N. J. (2005). Carbon dioxide poisoning. *Toxicological Reviews, 24*(4), 229–235.

McQueen, N., Kelemen, P., Dipple, G., et al. (2020). Ambient weathering of magnesium oxide for CO_2 removal from air. *Nature Communications, 11*, 3299. https://doi.org/10.1038/s41467-020-16510-3.

National Center for Biotechnology Information (2020). *PubChem compound summary for CID 280, carbon dioxide.* https://pubchem.ncbi.nlm.nih.gov/compound/Carbon-dioxide. (Accessed 13 October 2020).

Norimine, E., Ishizawa, F., Honda, K., & Uemura, S. (2009). Suicide case of carbon dioxide poisoning using dry ice. *Chudoku Kenkyu, 22*(2), 121–124.

Oraee-Mirzamani, B., Cockerill, T., & Makuch, Z. (2013). Risk assessment and management associated with CCS. *Energy Procedia, 37*, 4757–4764.

Patel, S., Miao, J. H., Yetiskul, E., et al. (2020). Physiology, carbon dioxide retention. In *StatPearls*. Treasure Island, FL: StatPearls Publishing.

Phillips, L., Shirvill, L., & Holt, H. (2011). Shells CO_2 release program and its input into DNV's CO2PIPETRANS JIP. In: *IChemE hazards XXII conference.*

PTRC (2004). IEA GHG Weyburn CO_2 monitoring & storage project: Summary report 2000–2004. In *Proceedings of the 7th international conference on greenhouse gas control technologies, September 5–9, 2004, Vancouver, Canada.*

Rusin, A., & Stolecka, K. (2014). Assessment of risk related to transport of carbon dioxide. *Journal of Power Technologies, 94*(4), 323–328.

Rybicki, C., & Łaciak, M. (2008). Transport rurociągowy CO_2 [pipeline transport of carbon dioxide], Rurociągi—Polish. *Pipeline Journal, 4*, 2–5.

Serpa, J., Morbee, J., & Tzimas, E. (2011). *Technical and economic characteristics of a CO_2 transmission pipeline infrastructure.* JRC report European Union. Retrieved from: http://publications.jrc.ec.europa.eu/repository/bitstream/111111111/16038/1/reqno_jrc62502_aspublished.pdf.

Smeets, E., & Weterings, R. (1999). *Environmental indicators: Typology and overview.* Copenhagen: European Environment Agency.

Spitz, W. U., & Spitz, D. J. (2006). Asphyxia. In W. U. Spitz (Ed.), *Spitz and Fischer's medicolegal investigation of death* (pp. 783–845). Springfield: Charles C Thomas.

Srisont, S., Chirachariyavej, T., & Peonim, A. V. (2009). A carbon dioxide fatality from dry ice. *Journal of Forensic Sciences, 54*(4), 961–962.

Strong, A. L., Kroeker, K. J., Teneva, L. T., Mease, L. A., & Kelly, R. P. (2014). Ocean acidification 2.0: Managing our changing coastal ocean chemistry. *Bioscience, 64*, 581–592. https://doi.org/10.1093/biosci/biu072.

Sunda, W. G., & Cai, W.-J. (2012). Eutrophication induced CO_2-acidification of subsurface coastal waters: Interactive effects of temperature, salinity, and atmospheric pCO_2. *Environmental Science & Technology, 46*(19), 10651–10659.

Tucker, O., Holley, M., Metcalfe, R., & Hurst, S. (2013). Containment risk management for CO_2 storage in a depleted gas field, UK North Sea. *Energy Procedia, 37*, 4804–4817. https://doi.org/10.1016/j.egypro.2013.06.390.

Zaba, C., Marcinkowski, J. T., Wojtyla, A., Tezyk, A., Tobolski, J., & Zaba, Z. (2011). Acute collective gas poisoning at work in a manure storage tank. *Annals of Agricultural and Environmental Medicine, 18*(2).

CHAPTER

8

Carbon monoxide: Risk assessment, environmental, and health hazard

Raj Nath Yadava[a] *and Vipul Bhatt*[b]

[a]Purnea University, Purnia, Bihar, India

[b]Gurukula Kangri (Deemed to be University), Haridwar, Uttarakhand, India

Abbreviations

ALARP as low as reasonably practicable
ATA atmosphere absolute
CAA Clean Air Act
CNS central nervous system
DNS delayed neurologic syndrome
ECG electrocardiograph
FDA Food and Drug Administration
HbCO carboxyhaemoglobin
HBO hyperbaric oxygen therapy
HbO_2 oxyhaemoglobin
NRC National Research Council
OSHA Occupational Safety and Health Administration
TSSA Technical Standards and Safety Authority
EPA Environmental Protection Agency
VOC volatile organic compounds

8.1 Introduction

Carbon monoxide (CO) is one of the frequently encountered lethal gas in the atmosphere and a foremost cause of worldwide hazardousness (Cobb & Etzel, 1991; National Center for Health Statistics, 1988; Raub, Mathieu-Nolf, Hampson, & Thom, 2000) that can be acute, subacute, or

https://doi.org/10.1016/B978-0-323-89857-7.00030-X

chronic (Grace & Platt, 1981). It is a nonirritating, colorless, odorless, and tasteless gas generated by inexhaustive burning of carbon-containing fossil fuels, such as gasoline, natural gas, oil, coal, and wood. Alike other pollutants, the primary origin of carbon monoxide comprehends volcano and forest fires, vehicle emissions, biomass burning, photochemical oxidation of methane in surface waters (lakes, streams, rivers, oceans), surface soils and volatile organic compounds (VOCs), a metabolic by-product from vegetation or some industrial activities.

Consequently, the inhaled CO tends to bind with hemoglobin replacing the oxygen, hence, causing severe health implications such as headache, cardiovascular issues, or impaired nervous system. Each year it causes thousands of uncalled deaths and those who outlive the initial poisoning generally unfolds retard neurologic dysfunction. In acknowledgment of the above stated health issues, US EPA in 1971 worked upon the health-based air quality standards for CO, as guided by the Title I of Clean Air Act (CAA) (EPA, 2000). Additionally, EPA has also established an environmental limit of $10\,mg/m^3$ whereas, OSHA has set a legal limit of $55\,mg/m^3$ in air averaged over 8h.

8.2 Occurrence

Numerous postulations correlating the occurrence of CO in the atmosphere directing oxidation of methane as a prime source (Levy, 1972; McConnell, MCelroy, & Wofsy, 1971; Stevens et al., 1972; Weinstock & Niki, 1972) has been made. While, in 1974 Duley anticipated the presence of CO in interstellar grain mantles, which was confirmed observationally in 1984 by Lact et al (Whittet & Walker, 1991).

Also, the reports discussing the occurrence of CO from surface layers of the ocean have also proved to be some of the ground-breaking research of all times (Linnenbom, Swinnerton, & Lamontagne, 1973; Liss & Slater, 1974). On the other hand, some of the natural sources of CO comprehend forest and grass fires, volcanoes, marsh gases, electric storms, photodissociation of carbon dioxide (Altshuller & Bufalini, 1965; Bates & Witherspoon, 1952), rainwater (Swinnerton, Lamontagne, & Linnenbom, 1971), degradation of rutin by *Aspergillus flavus* (Westlake, Roxburgh, & Talbot, 1961), etc.

Carbon monoxide is also formed by microbial action on vegetation (Delwiche, 1970; Siegel, Renwick, & Rosen, 1962; Wilks, 1959) as well as a metabolic byproduct. There are comprehensive details showing the presence of around 800ppm CO on kelp (Chapman & Tocher, 1966).

8.3 Historical background

The 18th century witnessed the recognition of carbon monoxide; the achievement had been credited to an English chemist Joseph Priestley. Considering the fact, the deadliness of gas was reported by Claude Bernard in 1870, who was accounted for an insightful detail on the relationship between CO and hemoglobin, directing towards the replacement of oxygen and the formation of carboxyhemoglobin. Moreover, in 1895, J.S. Haldane clearly illustrated the existence of carboxyhemoglobin, which exemplified the relative partial pressures of carbon monoxide and oxygen in inhaled gas (Dawson & Snyder, 1994; World Health Organization, 1979).

8.4 Physicochemical properties

Carbon monoxide is moderately lighter than air, highly toxicant, colorless, odorless, tasteless, and flammable gas (George, 2001) which burns with a bright blue flame (O'Neil, Heckelman, Koch, et al., 2006). It turns liquid at 81.62 K (−191.53°C) and is insoluble in water above 70°C (George, 2001). Carbon monoxide absorbs electromagnetic radiation in the IR region at 4.67 μm, which is characteristically utilized for its measurement in air. While it is chemically inert under STP but becomes reactive at higher temperatures and can act as a strong reducing agent. Substantially, the considerable amount of CO is peculiarly oxidized to form carbon dioxide (Ref. Chapter 7).

Additionally, the inspired quantity of CO reacts with the iron in protohaem (a constituent of hemoglobin) forming strong coordination bond of carboxyhemoglobin (HbCO). Thus, HbCO is comparatively more stable than oxyhemoglobin (HbO_2) entailing the CO toxicity. Some of the physical and chemical properties of carbon monoxide are discussed in Table 8.1 (National Center for Biotechnology Information, 2020).

8.5 Physiological role

There are primarily two pathophysiologic mechanisms related to CO that is, hypoxic and nonhypoxic. After inhalation, carbon monoxide remains unchanged in the lungs where it crosses the alveolar epithelium to reach the blood, from which 90% binds to hemoglobin, 10% to myoglobin and cytochrome C-oxidase, around 1% gets dissolved in plasma, and less than 1% is oxidized to carbon dioxide (Fig. 8.1). The 200 times greater affinity of CO for hemoglobin results in the incremental release of O_2 from the hemoglobin to the body tissues, and also decreases the oxygen-carrying capacity of the blood, consequently causing oxygen

TABLE 8.1 Physical and chemical properties of carbon monoxide.

Molecular formula	CO
Synonyms	Carbon monoxide Carbon monoxide Carbonic oxide Carbon(II) oxide
Molecular weight	28.01 g/mol
Boiling point	−312.7°F at 760 mmHg
Melting point	−326°F
Density	0.791 at −312.7°F
Vapor density	0.968 (Air = 1)
Vapor pressure	>35 atm
Viscosity (at 273 K)	16.62 uN s/m^2
Heat of combustion	−2412 cal/g
Heat of vaporization	51.6 cal/g
Surface tension	9.8 mN/m (of the liquid at 80 K)
Ionization potential	14.01 eV

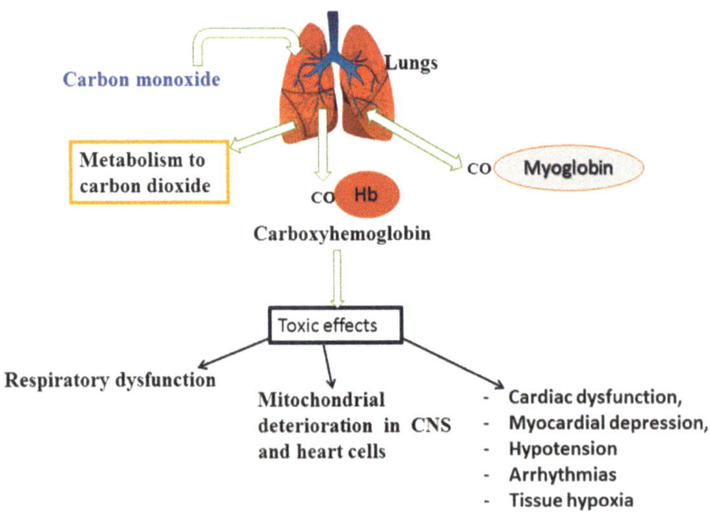

FIG. 8.1 Physiology of carbon monoxide depicting the health effects. *Credit: Raj Nath Yadava.*

toxicity (Chakraborty, Balakotalah, & Bidani, 2004). On the other hand, the nonhypoxic mechanisms have been attributed to the binding of CO to heme in proteins. Notably, CO also triggers some other essential physiological mechanistic pathways such as brain and muscle oxygen storage and utilization; nitric oxide and prostaglandin cell signaling pathway, and likewise. Additionally, the carbon monoxide is also endogenously produced from various endogenous as well as exogenous precursors such as heme degradation, auto-oxidation of phenols, etc. The elimination of carbon monoxide follows exhalation and oxidative metabolism, where the former is the dominant route of elimination from the body.

CO has a tendency of inducing cellular injury that is discussed in the below section:

A. It deteriorates both the oxygen-carrying capacity of blood and oxygen diffusion.
  1. CO causes the Haldane effect as it has 200 times more affinity than oxygen towards hemoglobin.
  2. With a high binding affinity towards hemoglobin, CO shifts the oxygen-hemoglobin dissociation curve. This shift hinders the capacity of hemoglobin to deliver oxygen to tissues. The resultant effect may lead to the development of tissue hypoxia.

B. CO has a direct effect on causing tissue injury.
As per communications, it has already been detailed that CO has a more direct effect than the lowering of oxygen-carrying capacity of hemoglobin (Goldbaum, Orellano, & Dergal, 1976).
  1. CO weakens the respiratory performance of cells by irreversibly binding to hemeproteins (cytochrome a-3 and myoglobin) that carry oxygen within the cell. This concludes the mitochondrial deterioration in CNS and heart cells, cellular damage, and ultimately causes tissue damage (Turner, Hamilton-Farrell, & Clark, 1999).
  2. CO has 60 times more binding affining with myoglobin causing myocardial injury. Binding induces cardiac dysfunction, myocardial depression, hypotension and arrhythmias, and tissue hypoxia (Zhang & Piantadosi, 1992). It also persuades re-oxygenation injury in CNS.
  3. CO promotes the formation of oxygen free radicals resulting in reversible demyelination in the brain (Chang, Han, Kim, Wie, & Han, 1992; Thom, 1990).

8.6 Uses

Carbon monoxide is used in a variety of industries for a wide range of applications. For industrial and domestic heating, it is generally employed in fuel gas mixtures with hydrogen and other gases. It is also utilized in the synthesis of a variety of chemicals such as acids, esters, alcohols, nickel carbonyl. Carbon monoxide is a major component for the synthesis of aldehydes, synthesis of acetic acid by Monsanto process, purification of nickel in mond process, and water-gas shift reaction producing hydrogen. Being a robust reducing agent, it has been used in pyrometallurgy to reduce metals from ores. This ideal blast furnace procedure involves the formation of carbon monoxide in the presence of oxygen-carrying ore, a carboniferous agent such as coke, at high temperature. It has also been employed in several pharmaceutical manufacturing procedures, electronics, and semiconductor applications as well as in the synthesis of high-powered infrared lasers. The 0.4%–0.5% of carbon monoxide is used in modified atmosphere packaging systems, chiefly with fresh meat products such as beef, pork, and fish to keep them appear fresh. The formation of carboxymyoglobin on the reaction of CO with myoglobin, a bright-cherry-red pigment, keep meat products stable for a longer period. Although this technology is banned in many countries, including Japan, Singapore, and the European Union after confirmation by FDA regarding its disguise spoilage odor in food items.

Niche uses: It has been intended for use as a fuel on Mars. Carbon monoxide/oxygen engines have been suggested for early surface transportation use as both can be straightforwardly generated from the atmosphere of Mars by zirconia electrolysis. Likewise, blast furnace gas collected at the top of the blast furnace still holds some 10%–30% of carbon monoxide and is applied as fuel on Cowper stoves and Siemens-Martin furnaces on open-hearth steelmaking.

8.7 Sources of contaminants

The major recurring cause behind the propagation of carbon monoxide is inexhaustive incineration of carbon in liquid, solid, and gaseous fuels (George, 2001). The commercial industrial-scale production involves either gasification of coal and coke or partial oxidation of hydrocarbon gases from natural gas. While heating of calcium carbonate with zinc dust or dehydration of formic acid leads to lab-scale preparation of carbon monoxide (O'Neil et al., 2006; Lewis, 2007), a considerable amount is procured from the chemical synthesis site and steel manufacturing plant (George, 2001). The human body also yields carbon monoxide as a byproduct of hemoglobin degradation, resulting in baseline carboxyhemoglobin (COHb) saturation of 1%–3% in nonsmokers, which advances to 10%–15% among heavy smokers. It has also been seen that high-altitude areas where army officials reside are more prone to this poisoning as they often sleep in closed tents with burning bhukharis (charcoal/kerosene) kept inside (Mehta et al., 2001).

Some of the other origins of contamination are as follow:

(1) **Endogenous production:** Human metabolism of hemoglobin propagates CO endogenously. It is produced 0%–5% in every individual, 3%–7% in infants, 5%–10% in smokers and patients with hemolytic anemia. Under physiological circumstances, CO saturates nearly 0.5% of hemoglobin, known as COHb (Blumenthal, 2001).
(2) **Hydrocarbons:** The enhanced concentration of CO in the urban atmosphere is due to the partial burning of compounds with carbon components, such as coal, wood, petroleum, fertilizers, dried dung, and natural gas.
(3) **Exhaust gases:** One of the mortal sources of CO is ejected exhaust gases from motor vehicles. Potentially fatal exposure may also occur in semiclosed garage or areas adjacent to the garage (Meredith & Vale, 1988).
(4) **Fire:** Combustion of any material leads to the liberation of CO.
(5) **Propane and methane:** Incomplete combustion of gases propagates CO. Leaks from poorly maintained or improperly used heating systems can be an origin as well as the petrochemical industry using natural gas and crude oil.
(6) **Methylene chloride:** Vapor of methylene chloride, component of thinners and other solvents, enters the skin, is inhaled through the lungs and transported to the liver via blood circulation, where it is metabolized, resulting in the release of CO.
(7) **Cigarette smoke:** Baseline COHb level has been found to be 10% in smokers and 2% in nonsmokers, respectively (Meredith & Vale, 1988).

8.8 Risk assessment

On account of higher indoor intoxication of CO across the globe, specifically during winters, a large count of individual die accidentally within their premises. Considering the toxicity of carbon monoxide, the need to determine the exposure risk level and identify the process accurately befit an indispensable part. Hence, the risk assessment dealing with CO chiefly revolves around the mitigation of concurrent studies mostly conducted concerning the indoor facilities. In accordance with the encountered challenges, National Research Council (NRC) has established NAAQS (National Ambient Air Quality Standards) in order to scrutinize the CO associated issues. The committee analyzes potential approaches to predict, assess, and manage episodes of CO poisoning; contemplate interrelationships among the major sources and control measures; reviews level and impact of human exposure as well as explores more advanced measures. Some of the risk assessment protocols employed in diverse facilities intended to deal with CO-related risks are detailed in the sections ahead.

A comprehensive study done at welding workshops suggests the usage of a semi-quantitative chemical risk assessment (SQRCA) approach. The outlay of the method suggests the determination of pollutants (such as CO in this case), level of exposure risk, exposure risk (ER), and hazardous risk (HR). Furthermore, the hazard coefficients and exposure of gas were evaluated and computed using the following formula.

$$\text{Risk} = \sqrt{\text{HR} \times \text{ER}}$$

The final calculated value can be used for the risk analysis considering the five levels, that is negligible (N), low (L), moderate (M), high (H), and very high (VH) (Mehrifar, Zeverdegani, Faraji, & Rismanchian, 2018).

Whereas, in steel industries, where propagation of CO in the air is comparatively higher, primary hazard analysis and further quantification, interdependent on the storage area and processes involved in the facility are carried out. The risk for the most part is meticulously examined and deduced based on vulnerability. The fire, explosion, and toxicity indexing (FE & TI) is one such method utilized to carry out a quick risk assessment. Consequently, obtained values from the method guide about the degree of hazard that is, 0–60 as Light; 61–96 as Moderate; 97–127 as Intermediate;128–158 as Heavy; and 159-up as Severe hazard.

A comprehensive approach has also been worked out to mitigate the risk associated with the quantification of carbon monoxide dwelling at residential places. The simulated framework is built upon the attributes such as injury burden and the risk of injury or fatality (RIF), measured in fatality-equivalents/million people/year (FE/mp/y) (Veeramany & Srikanth, 2014). The former factor reflects the impact on human health computed on yearly basis and expressed as disability-adjusted life years (DALYs) (Veeramany & Srikanth, 2014). While WHO introduced DALY as a single measure to quantify the burden of diseases and injuries. On the other hand, the latter attribute predicts the expected injury burden in the future. For evaluation of such risks with respect to the general population, in Ontario, a specific TSSA (Technical Standards and Safety Authority) uses the criterion of 1.0 fatality/million people/year. Additionally, ALARP (as low as reasonably practicable) principles have also been adapted to map associated risks (Sridharan & Mangalam, 2017).

8.9 Environmental effects

Carbon monoxide can last in the surrounding approximately for a month and can be transported to long distances. Several greenhouse gases including methane, tropospheric ozone, and carbon dioxide can be affected by the concentration of CO. It readily reacts with the hydroxyl radical forming carbon dioxide, in turn increasing the concentrations of methane. As documented by NASA, CO is accountable for a 13% decrease in hydroxyl concentrations and a 9% drop in sulfate concentrations.

8.10 Hazardousness

Although all tissues are vulnerable to CO, due to high oxygen requirements in the brain and heart, they are severely affected by the induced injury. Some of the major hazardous effects are detailed in the aforementioned section.

Central nervous system effects: Headache is the foremost symptom of CO poisoning. Other general neurological symptoms include dizziness, lightheadedness, weakness, and confusion. Severely poisoned patients can have reduced levels of perception advancing to coma, seizures, and acute stroke. Brain computed tomography and magnetic resonance imaging of severely poisoned patients may disclose low density of the white matter and necrosis of the basal ganglia and cerebral gyri.

Cardiovascular effects: Patients with CO poisoning may confer with symptoms and electrocardiograph (ECG) changes with myocardial ischemia or infarction, particularly if they have basic coronary artery disease. Cardiac troponin isoenzymes are often uplifted even in the absence of ECG changes. Supraventricular dysrhythmias and intraventricular conduction delay, mortality due to hypoxia-induced ventricular dysrhythmias have also been observed.

Respiratory effects: CO-induced hypoxia can arise in a compensative respiratory alkalosis that in severe cases is further fueled by the ventilator response to lactic acidosis. About 20%–30% of intubated CO-poisoned patients develop pulmonary edema presumably due to left ventricular failure.

Renal effects: Severely poisoned patients seldom develop acute renal failure secondary to nontraumatic rhabdomyolysis. Although CO may have some direct toxicity to the renal tubules, nearly all reported cases of acute renal failure in CO poisoning have been assisted by muscle necrosis.

Dermatologic effects: Cutaneous blisters are common in severe CO poisoning, likely secondary to pressure necrosis. The "classic" finding of "cherry red skin" is a rare and usually postmortem observation likely due to the combination of CO-induced vasodilatation and tissue ischemia.

Delayed neurologic syndrome: Totally, 1%–10% of victims of CO poisoning develop delayed neurologic syndrome (DNS). These symptoms classically comprise mental deterioration (dementia, personality change, and primitive reflexes), urinary incontinence, gait disturbance, a myriad of neurologic symptoms, such as Parkinsonism, amnesia, autism, psychosis, paralysis, tremor, peripheral neuropathy, and flaccid paralysis. The probability of DNS increases with the intensity of poisoning and the patient's age. Although up to 75% of patients with DNS completely or fractionally recover, many victims of CO poisoning suffer permanent changes in personality, affect, and cognition.

8.11 Treatment

Hyperbaric oxygen therapy: Hyperbaric oxygen therapy (HBO) involves enclosing the patient in a pressure chamber and administering 100% oxygen at a pressure greater than 1 atm absolute (ATA). HBO at 2.5 ATA lowers the half-life of COHb to nearly 20 min. HBO has been used in CO poisoning administration since 1962 (Harper & Croft Baker, 2004). In adjunct to the accelerated rate of CO elimination from hemoglobin, removal from intracellular binding sites is also enhanced by HBO treatment. The early administration of HBO anticipates neuronal injury, prevents delayed neuropsychological sequelae, and ceases the biochemical deterioration (Ernst & Zibrak, 1998; Hark & Kennedy, 1998; Harper & Croft Baker, 2004).

Other therapy: Adjunctive therapy such as administration of corticosteroids, mannitol, hypothermia, and hyperventilation has been commended in serious cases of CO poisoning but has not much verification.

8.12 Risk management

The risk management considering the adverse effects of carbon monoxide plays a pivotal role. For governing risks in a residential area, various aspects should be addressed such as adequate ventilation; the usage of approved appliances, which are periodically inspected; carbon monoxide monitors; and focus should be centralized on the regular conductance of public awareness campaigns.

There are contemplations done in ice arenas for the usage of an electric resurfacer, installation of a CO monitoring and alarm system, mechanical ventilation, and well-trained personnel in order to mitigate the risks (Salonen et al., 2008).

8.13 Current research

The U.S. Food and Drug Administration on 14 March 2019 permitted the marketing of a modern device, ClearMate, designed to be employed in an emergency room settled to help and treat patients suffering from carbon monoxide poisoning. The device uses a novel process for rapidly removing carbon monoxide from the body by increasing the patient's rate of respiration.

ClearMate consists of a gas mixer, valves, meters, breathing circuits, an oxygen reservoir, a mask, and hoses; works by accelerating up the elimination of carbon monoxide from the body. It gives forth both 100% oxygen to the patient, as well as a mixture of oxygen and carbon dioxide, inducing the patient to breathe faster. The increased breathing speed up the rate at which the carbon monoxide leaves the patient's body, permitting a normal amount of oxygen to attach to hemoglobin and be carried where it is needed throughout the body.

June 20, 2019: For the critical detection of carbon monoxide (CO) and carbon dioxide (CO_2) Restek's has come up with a better improved version of methanizer (CH_4izer), a cost-effective instrument for Thermo TRACE 1300/1310 GCs. For efficient and complete conversion of CO and CO_2 to CH_4, the methanizer is set at 380°C, which can be easily adjusted accordingly. It has easy installation, fast catalyst tube replacement, and less cumbersome features making it easy to handle.

8.14 Carbon monoxide toxicity: Case study

The body of an elderly Afro-Caribbean man revealed noticeable proof of postmortem decay and gas bloating. The body was found lying supine on the bed. Due to his being Afro-Caribbean and the state of decay of the body, there was no signal of the classic cherry-red coloring of carbon monoxide poisoning, due to the formation of carboxyhemoglobin that might be seen in a Caucasian. Libido or any cherry pink hypostasis was impracticable to determine with no sign of the body having been moved (Henssge & Madea, 2007).

During an evening memorial service at a church hall, having 60 adults in the arena. A major accident was declared, as most of the person present during the service experienced symptoms, including lightheadedness, headache, and nausea. After examination of the admitted people in the hospital, carbon monoxide poisoning was confirmed with 35% carboxyhemoglobin concentration.

8.15 Safety measures

- Fuel-burning appliances should be periodically inspected by trained personnel.
- The in-house appliances should be installed keeping ventilation as a priority.

- Use process enclosures, local exhaust ventilation, or other engineering controls to keep worker exposure to airborne contaminants below any recommended or statutory limits.
- The emissions from ventilation should be timely checked to ensure that they comply with the environmental protection limits set by competent bodies.
- Ensure usage of appropriate chemical resistant gloves.
- Based on hazard and potential for exposure, the certified respirator should be available in functional condition at vulnerable zones.

8.16 Handling and storage

- Carbon monoxide is shipped and stored in high-pressure cylinders, tube trailers, or ISO modules, depending on the quantity required by the consumer. These containers are manufactured according to applicable codes and specifications for the temperatures and pressures involved.
- Cylinders should be secured in an upright position, stored in a well-ventilated area, and protected from the weather changes.
- All precautions necessary for the safe handling of flammable gases must be observed with carbon monoxide.
- Avoid storing with oxidizers or other flammable materials.
- Carbon monoxide detectors with audible and visual alarms are highly recommended for the protection of personnel.
- The storage area should be secured with limited access. Storage area temperatures should not exceed 52°C (125°F), and should be free from combustible materials and ignition sources.
- Storage should be away from heavily traveled areas and emergency exits.
- Avoid areas where salt or other corrosive materials are present.
- Valve protection caps and valve outlet seals must remain on any unconnected cylinder.
- When returning a cylinder to the storage, the valve outlet seal must be installed leak-tight and the cylinder cap secured.
- Visually inspect stored cylinders on a routine basis, at least weekly, for any indication of leakage or other problems.
- The use of "Full", "In Use", and "Empty" tags is highly recommended.
- Storage areas must be posted with the proper signage, such as "No Smoking", "No Open Flames", or NFPA 704 ratings.
- Ensure that the cylinder valve is properly closed, the valve outlet seal has been reinstalled leak-tight, and the valve protection cap is installed before returning to the storage, moving, or shipping the cylinder.

8.17 Precautions

- Ensure regular inspection of the vehicle's muffler and tailpipe.
- Avoid sleeping in a car with the engine running.
- Ensure routine services of installed heating appliances.
- Install a CO detector in the home and garage.
- Check for proper ventilation in case of any indoor fuel-burning.
- Avoid usage of heaters while sleeping.

References

Altshuller, A. P., & Bufalini, J. J. (1965). Photochemical aspects of air pollution: A review. *Photochemistry and Photobiology, 4*, 97–146.

Bates, D. R., & Witherspoon, A. E. (1952). The photo-chemistry of some minor constituents of the earth's atmosphere (CO_2, CO, CH_4, N_2O). *Monthly Notices of the Royal Astronomical Society (London), 112*, 101–124.

Blumenthal, I. (2001). Carbon monoxide poisoning. *Journal of the Royal Society of Medicine, 94*(6), 270–272.

Chakraborty, S., Balakotalah, V., & Bidani, A. (2004). Diffusing capacity reexamined: Relative roles of diffusion and chemical reaction in red cell uptake of O_2, CO, CO_2, and NO. *Journal of Applied Physiology, 97*, 2284–2302.

Chang, K. H., Han, M. H., Kim, H. S., Wie, B. A., & Han, M. C. (1992). Delayed encephalopathy after acute carbon monoxide intoxication: MR imaging features and distribution of cerebral white matter lesions. *Radiology, 184*(1), 117–122.

Chapman, D. J., & Tocher, R. D. (1966). Occurrence and production of carbon monoxide in some brown algae. *Canadian Journal of Botany, 44*, 1438–1442.

Cobb, N., & Etzel, R. A. (1991). Unintentional carbon monoxide—Related deaths in the United States, 1979 through 1988. *JAMA, 266*(5), 659–663.

Dawson, T. M., & Snyder, S. H. (1994). Gases as biological messengers: Nitric oxide and carbon monoxide in the brain. *Journal of Neuroscience, 14*(9), 5147–5159.

Delwiche, C. C. (1970). Carbon monoxide production and utilization by higher plants. *Annals of the New York Academy of Sciences, 174*, 116–121.

EPA. (2000). *Air quality criteria for carbon monoxide*. Washington, DC: U.S. Environmental Protection Agency. EPA600P66001F. Available from http://www.epa.gov/NCEA/pdfs/coaqcd.pdf.

Ernst, A., & Zibrak, J. D. (1998). Carbon monoxide poisoning. *The New England Journal of Medicine, 339*, 1603–1608.

George, C. (2001). Carbon monoxide. In *Kirk-Othmer encyclopedia of chemical technology*. John Wiley & Sons, Inc.

Goldbaum, L. R., Orellano, T., & Dergal, E. (1976). Mechanism of the toxic action of carbon monoxide. *Annals of Clinical & Laboratory Science, 6*(4), 372–376.

Grace, T. W., & Platt, F. W. (1981). Subacute carbon monoxide poisoning: Another great imitator. *JAMA, 246*(15), 1698–1700.

Hark, I. K., & Kennedy, P. G. E. (1998). Neurological manifestation of carbon monoxide poisoning. *Postgraduate Medical Journal, 64*, 213–216.

Harper, A., & Croft Baker, J. (2004). Carbon monoxide poisoning: Undetected by both patients and their doctors. *Age and Ageing, 33*, 105–109.

Henssge, C., & Madea, B. (2007). Estimation of the time since death. *Forensic Science International, 165*(2–3), 182–184.

Levy, H. H. (1972). Photochemistry of the lower troposphere planet. *Space Science, 20*, 919–935.

Lewis, R. J. (2007). *Hawley's condensed chemical dictionary* (15th ed., p. 235). Hoboken, NJ: John Wiley & Sons, Inc.
Linnenbom, V. J., Swinnerton, J. W., & Lamontagne, R. A. (1973). The ocean as a source for atmospheric carbon monoxide. *Journal of Geophysical Research, 78*(24), 5333–5340.
Liss, P. S., & Slater, P. G. (1974). Flux of gases across the air-sea interface. *Nature (London), 247*, 181–184.
McConnell, J. C., MCelroy, M. B., & Wofsy, S. C. (1971). Natural sources of atmospheric CO. *Nature (London), 233*, 187–188.
Mehrifar, Y., Zeverdegani, S. K., Faraji, M., & Rismanchian, M. (2018). Risk assessment of welders exposure to the released contaminated gases in different types of welding processes in a steel industry. *Health Scope, 8*(1).
Mehta, S. R., Niyogi, M., Kasthuri, A. S., Dubal, U., Bindra, S., Prasad, D., et al. (2001). Carbon monoxide poisoning. *The Journal of the Association of Physicians of India, 49*, 622–625.
Meredith, T., & Vale, A. (1988). Carbon monoxide poisoning. *British Medical Journal (Clinical Research Ed), 296*(6615), 77.
National Center for Biotechnology Information. (2020). *PubChem compound summary for CID 281, carbon monoxide.* https://pubchem.ncbi.nlm.nih.gov/compound/Carbon-monoxide. (Accessed 23 October 2020).
National Center for Health Statistics (1988). *Vital statistics of the United States*: (pp. 89–1102). Washington, DC: Government Printing Office. 1991.
O'Neil, M. J., Heckelman, P. E., Koch, C. B., et al. (2006). *The Merck index. An encyclopedia of chemicals, drugs, and biologicals* (14th ed., p. 294). Whitehouse Station, NJ: Merck & Co., Inc.
Raub, J. A., Mathieu-Nolf, M., Hampson, N. B., & Thom, S. R. (2000). Carbon monoxide poisoning—A public health perspective. *Toxicology, 145*(1), 1–14.
Salonen, R. O., Pennanen, A. S., Vahteristo, M., Korkeila, P., Alm, S., & Randell, J. T. (2008). Health risk assessment of indoor air pollution in Finnish ice arenas. *Environment International, 34*(1), 51–57.
Siegel, S. M., Renwick, G., & Rosen, L. A. (1962). Formation of carbon monoxide during seed germination and seedling growth. *Science, 137*, 683–684.
Sridharan, S., & Mangalam, S. (2017). Carbon monoxide risks and implications on maintenance-intensive fuel-burning appliances—A regulatory perspective. In *Annual reliability and maintainability symposium (RAMS)*.
Stevens, C. M., Krout, L., Walling, D., Venters, A., Engelkemeir, A., & Ross, L. E. (1972). The isotopic composition of atmospheric carbon monoxide. *Earth and Planetary Science Letters, 16*, 147–165.
Swinnerton, J. W., Lamontagne, R. A., & Linnenbom, V. J. (1971). Carbon monoxide in rain water. *Science, 172*, 943–945.
Thom, S. R. (1990). Carbon monoxide-mediated brain lipid peroxidation in the rat. *Journal of Applied Physiology, 68*(3), 997–1003.
Turner, M., Hamilton-Farrell, M. R., & Clark, R. J. (1999). Carbon monoxide poisoning: An update. *Journal of Accident & Emergency Medicine, 16*(2), 92.
Veeramany, A., & Srikanth, M. (2014). Application of disability-adjusted life years to predict the burden of injuries and fatalities due to public exposure to engineering technologies. *Population Health Metrics, 12*(1).
Weinstock, B., & Niki, H. (1972). Carbon monoxide balance in nature. *Science, 176*, 290–292.
Westlake, D., Roxburgh, J., & Talbot, G. (1961). Microbial production of carbon monoxide from flavonoids. *Nature, 189*, 510–511. https://doi.org/10.1038/189510a0.
Whittet, D. C. B., & Walker, H. J. (1991). On the occurrence of carbon dioxide in interstellar grain mantles. *Monthy Notices of the Royal Astronomical Society, 252*, 63–67.
Wilks, S. S. (1959). Carbon monoxide in green plants. *Science, 129*, 964–966.
World Health Organization. (1979). *Environmental health criteria 13: Carbon monoxide.* Geneva: WHO.
Zhang, J., & Piantadosi, C. A. (1992). Mitochondrial oxidative stress after carbon monoxide hypoxia in the rat brain. *The Journal of Clinical Investigation, 90*(4), 1193–1199.

CHAPTER

9

Chlorine: Risk assessment, environmental, and health hazard

Malvika Chawla

Department of Chemistry, Gurukula Kangri (Deemed to be University), Haridwar, Uttarakhand, India

Abbreviations

ACGIH	American Conference of Governmental Industrial Hygienists
ALARP	as low as reasonably practicable
ALOHA	areal locations of hazardous atmospheres
ATD	atmospheric transport and dispersion
AWWA	American Water Works Association
CPQRA	chemical process quantitative risk analysis
DEGADIS	dense gas dispersion model
EPA	Environment Protection Agency
HAZOP	hazard and operability
HEM	homogeneous equilibrium model
MCL	maximum contaminant level
MLM	metastable liquid model
MRDL	maximum residual disinfectant level
NIOSH	National Institute for Occupational Safety and Health
OSHA	Occupational Safety and Health Administration
PEL	permissible exposure limit
PSM	process safety management
PVC	polyvinyl chloride
QRA	quantitative risk analysis
REL	recommended airborne exposure limit
RMP	risk management plan
STEL	short-term exposure limit
TLV	threshold limit value
TOR	tolerability of risk
UDM	unified dispersion model

Hazardous Gases
https://doi.org/10.1016/B978-0-323-89857-7.00004-9

9.1 Introduction

Chlorine is an unstable and highly reactive gas, therefore does not exists as a monoatomic molecule. The name chlorine is derived from the Greek word "khloros" on account of greenish-yellow color. It is the 17th element belonging to Group VII on the periodic table and classified as halogen. It mainly exists either in form of chlorine (Cl_2) or chloride ion (Cl^-) and has an irritating odor. Both the forms subsist in different ways with varying effects when exposed to humans, where the former exists as a diatomic molecule. In total it is 0.14% of the concentration of elements present in the human body (Ila, 2006). It is also a major product of electrolysis, as given in Eq. (9.1).

$$2NaCl + 2H_2O \rightarrow 2NaOH + Cl_2 + H_2 \quad (9.1)$$

Conversely, chloride prevails as 70% of the body's total negative ion content, with a daily intake of roundabout 750–900 mg/day (Ila, 2006). It exists in the form of salts such as NaCl, KCl, etc. Chloride is broadly responsible for the maintenance of electrical balance in the CNS as well as plays a vital role in intracellular and extracellular transport. It manages several biological functions like osmotic pressure, membrane potentials, condensation of polyelectrolytes, and ionic strength for activity (Selinus et al., 2005). This ion also combines with hydrogen present in the body to form HCl that breaks down proteins for the absorption of other metallic minerals. It also actively maintains electrical neutrality across the stomach membrane.

Chlorine is extensively found in numerous chemicals that are used on daily basis such as table salt (sodium chloride), household bleach (sodium hypochlorite), and hydrochloric acid (hydrogen chloride). Some of the other applications apart from cleaning and bleaching include the preparation of paper, solvents, insecticides, paints, medicines, plastics, and textiles. It is widely utilized in the purification of water supplies and pools. Although there's a familiar misbelieve stating that molecular chlorine (Cl_2) is present in chlorinated water. Rather it is the presence of hypochlorous acid and hypochlorite anion generally coined as "free chlorine" and "aqueous chlorine" that are responsible for disinfecting water.

At salient concentrations, the exposure to chlorine may adversely affect human as well as marine life. The postexposure health effects vary greatly on reaction with water present in tissues that is responsible for the formation of hydrochloric acid. Likewise, effects include irritation to the airways, lungs, and eyes.

9.2 Occurrence

Chlorine is ranked 20th among the elements with an abundance of about 2% in seawater and 100–300 ppm in the Earth's crust. The study suggests the existence of a figure of about 4.8 ppm chlorine in the solar photosphere (Lodders, 2003). The continental crust contains 130–470 ppm Cl compared to ~1–5 ppm Cl in the depleted mantle (Rudnick & Gao, 2003; Shimizu et al., 2016). Primitive mid-ocean ridge and ocean island basalt glasses contain ~30–2000 ppm Cl (Kendrick, 2018; Michael & Cornell, 1998; Schilling, Unni, & Bender, 1978).

The weathering process is of paramount importance that directs in the formation of chloride from various rocks into soil and water for years (Ila, 2006). Furthermore, chloride is transmuted to oceans or closed basins from where it is extracted. Over millions of years, the required chloride has been extracted from the oceans. A large amount of chlorine has been mined from underground salt beds after drying oceans and further packed to form huge salt domes. It is also found in the stomach in form of HCl, where it is responsible for maintaining the hydrogen ion concentration.

9.3 Historical background

For ages, chlorine compounds have been an essential part of human existence. This got its recognition as an element in 1774 when Scheele found the released gas as chlorine gas while studying the mineral pyrolusite. During experiments, Scheele mixed pyrolusite, primarily manganese dioxide (MnO_2) with hydrochloric acid (HCl), and called it spiritussalis. There he found the release of an offensive suffocating greenish-yellow gas.

He gave it a complex name of dephlogisticated marine acid based of its properties such as reaction with metals, slight dissolvability in water, and affinity towards bleaching flowers and leaves. For several years the nature of his discovery remained misunderstood as few scientists argued that his dephlogisticated marine acid was a compound of a new element and oxygen. In 1807, an English scientist Sir Humphry Davy cleared this confusion by proving that Scheele's substance was a pure element and finally suggested the name chlorine for the element (Davy, 1811).

Furthermore, around the 1830s Davy's lab assistant Michael Faraday bestowed the world with his finest work on the electrolytic synthesis of chlorine and its liquefaction procedure. However, the first patent analogous to the preparation of chlorine in an electrolytic cell was acquired by Charles Watt in 1851. And, with the advent of the industrial revolution came the widespread applicability of chlorine around the globe (Occidental Chemical Corporation (OxyChem), 2012).

9.4 Physicochemical properties

Chlorine is a greenish-yellow color gas with a characteristic odor that is detectable at a concentration as low as 0.2–0.4 ppm. It persists with an atomic weight of 35.453, a molecular weight of 70.906, and an atomic number of 17, respectively. It is comparatively denser than air with a density of 3.21 g/L. Chlorine converts from a gas into a liquid at a temperature of −34.05°C (−29.29°F) and from a liquid to a solid at −101.00°C (−149.80°F). The gas is dissolvable in water and reacts considerably to form hydrochloric acid (HCl) and hypochlorous acid (HOCl), Eq. (9.2).

$$Cl_2 + H_2O \leftrightarrow HOCl + HCl \tag{9.2}$$

Furthermore, hypochlorous acid forms hydrogen and hypochlorite ions after dissociating in water (Eq. 9.3). While the degree of dissociation is dependent on the pH and temperature of the water.

$$HOCI \leftrightarrow H^+ + OCl^- \tag{9.3}$$

It is a strong oxidizing agent and does not burn but help other substances burn. In general, it reacts with all elements except noble gases. It may also react violently with many substances. The dry chlorinating chemicals like calcium hypochlorite refer to dry chlorine. While wet chlorine is more devastating than dry chlorine.

9.5 Physiological role

Regarding the metabolism of chlorine very little information is accessible. The chlorinated compounds are the product of a reaction between hypochlorous acid (Fukayama, Tan, Wheeler, & Wei, 1986) and proteins, amino acids, and unsaturated lipids, whereas products with carbohydrates yield oxidation products. There are animal study reports for the resultant production of *N*-chloroalanine, *N*-chloroglycine, and *N*-chlorophenyl alanine by chlorination in the stomach. Studies have also investigated the in vivo formation of chemicals such as chloroform, dichloroacetonitrile, and di-and trichloroacetic acids in the stomach (Mink, Coleman, Munch, Kaylor, & Ringhand, 1983).

By the employment of a noninvasive process, the fragment of respired chlorine by nonsmoker males and females has been measured (Nodelman & Ultman, 1999a,1999b). In accordance with the data collected it was cleared that irrespective of the breathing done by nose or mouth the inhaled chlorine gets absorbed in the upper airways (above the cords) only. This is illustrated in Fig. 9.1.

9.6 Uses

Chlorine is widely acknowledged for purification of water, in the paper and pulp industry as bleach, and for the synthesis of several chemicals like ethylene dichloride. About 90% of ethylene dichloride is chiefly employed

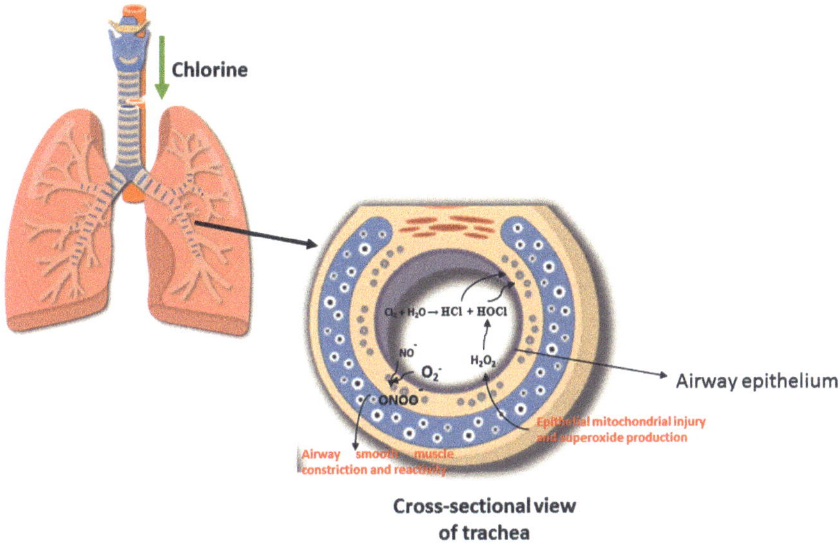

FIG. 9.1 Postulated mechanisms for airway injury due to chlorine inhalation. *Credit: Malvika Chawla.*

in the manufacturing of polyvinyl chloride (PVC), which is further utilized for preparing piping, tubing, flooring, siding, film, coatings, and many other products.

The water disinfection process can be classified according to their use, as discussed in the below section:

Drinking water: The chlorine is principally laid down in the large tanks or small to medium sized water channels that are meant to supply water for household work. The process is accountable to produce bacteria-free water and eliminate algae and slime. Also, chlorination works on the effective abolition of hydrogen sulfide and iron bacteria (crenothrix).

Chilling water: One of the vital aspects of processed food is its life and in order to achieve this purpose chlorination substantiates as a mandatory step. Basically, before vacuum packaging, any eatables like meat, fish, and poultry need to undergo chlorination of chilling water. This prevents food items from any bacterial contamination.

Swimming pools: The process of gas chlorination is quite popular in disinfecting swimming pools at societal places. It is among the cheapest and effective methods. To safeguard against any harm caused to human or marine life, mainly hydro gas chlorinators are used at such places.

Sanitary sewage: For the purpose of killing harmful bacteria, the sewage is subjected to undergo chlorination prior entering into the treatment process. Not only this, but the process of addition of chlorine also shows a synergistic increment in solid concentration and sedimentation as well as manages the disagreeable odor. The utilization of hydro gas chlorinator proves to be an epitome for packaged sewage treatment plants.

Process wastewater: Most wastewater treatment facilities use chlorination to reduce a load of impurities before dumping in common sewage. Some premises also reuse the treated water for the purpose of washing or rinsing.

9.7 Sources of contaminants

Workers along with the civilians staying around the industries particularly food and paper industries, where chlorine is produced or used are at increased risk of exposure to chlorine (US EPA, 1992). The vulnerability to chlorine may often happen from places disinfected with chlorine such as a swimming pool (US EPA, 1999) or incidental releases such as a spill or tank rupture (Horton, Berkowitz, & Kaye, 2002; US EPA, 1992).

9.8 Risk assessment

Chlorine is a harmful element that can be a potential threat to both facility employees and the surrounding community. Due to its wide applicability in water disinfection in urban and rural areas, the execution of periodic risk assessment is recommended. The risk assessment refers to the methodology implemented for the estimation of risks and their qualitative and quantitative evaluation. Process hazard analysis and quantitative risk assessment provide valuable input both for the initial design and for the continuing operation of chemical processing facilities.

To prevent and mitigate the risks, numerous competent agencies have guided facilities dealing with chlorine to implement emergency plans. The Process Safety Management (PSM) regulations provided by OSHA, Risk Management Plan (RMP) by EPA, The Risk Assessment Methodology for Water Utilities by the union of EPA and Research Foundation of the American Water Works Association (AWWA) are some of the documented risk assessments that are globally followed.

Qualitative hazard analysis tools including Hazard and Operability (HAZOP) studies and quantitative methods such as fault tree analysis (See Chapter 7, Section 7.8), accident consequence models, and risk estimation models, have been used to determine the most appropriate of many available options for improving the design of the existing system. Among the mentioned methods, FTA is one of the powerful tools, which is widely used for the assessment of major root causes. Also, some researchers have suggested that slight variations in the fault tree method can prove to be prolific. The variations could be the application of fuzzy set theory (Shu, Cheng, & Chang, 2006), computer-aided fuzzy logic (Antonio, 1999; Dash, Pradhan, & Singh, 2018; Ferdous, Khan, Veitch, & Amyotte, 2009; Khan & Abbasi, 2000), a combination of task analysis for accident and incident analysis (Doytchev & Szwillus, 2009), or the introduction of components of fuzzy failure probability (Chanda & Bhattacharjee, 1998). However, studies demonstrating the usage of FTA at titanium tetrachloride plant has also been conducted (Roy, Bhatt, & Rajagopal, 2003). Renjith, Madhu, Nayagam, and Bhasi (2010) evaluated the probability of chlorine release using Two-dimensional fuzzy fault tree analysis.

Some manufacturing units also bifurcate risks on the basis of the ALARP principle. This concept helps in the identification of risk based on the Tolerability of Risk (TOR) framework.

The facilities involved in the handling of chlorine frequently conduct a series of reviews and analysis including:

(1) **Preliminary design reviews** to define issues of concern and potential alternatives among the plant and design personnel.
(2) **Safety, health, and environmental (SHE) review** is based on the participation of design and operating personnel as well as safety, health, and environmental professionals from the local operating plant and corporate staff. One of the major functions of this review is to identify areas where more sophisticated hazard evaluation techniques are justified.
(3) **Hazard and operability (HAZOP)** review is intended to identify specific potential accident scenarios associated with the chlorine facility as designed, qualitatively evaluate them, identify potential corrective actions, and address the concerns identified. This broadly identifies a number of specific improvements in piping, control and instrumentation, and operating procedures.
(4) **Initial quantitative risk assessment** may evaluate whether or not the design objective of reducing risk had been met. This assessment uses the results of the SHE Review and HAZOP in order to identify release incidents; fault tree analysis to estimate their frequency and various releases, and vapor dispersion and toxicity models to determine the extent of the release.

Overall, risk analysis guides the implementation of modified engineering design and operating systems, mutual assistance, and technical expertise to mitigate any incur harm. And, also lead to an enhanced and improved understanding of the sources of risk which might cause additional risk reduction. For instance, on some premises, the quantitative risk analysis is followed by the chemical process quantitative risk analysis (CPQRA) methodology. One example of initial QRA followed at facility handling chlorine is illustrated in Fig. 9.2. Here, the societal risk has been

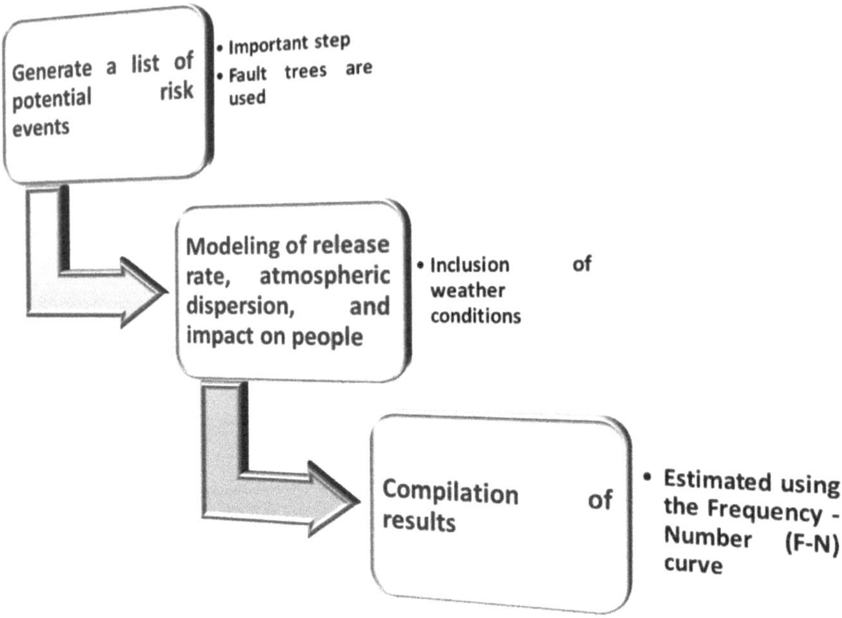

FIG. 9.2 Flow chart depicting the stages of initial quantitative risk analysis at chlorine handling facility. *Credit: Malvika Chawla.*

estimated using the F-N curve as the risk measure, the importance may vary with N, the number of people affected. For instance, a specific small pipe break may be the major contributor to the risk of incidents affecting one or two people, because of a relatively high frequency.

The studies investigating the quantification of risk postdisinfection by chlorine have proved to be a game-changer in the determination of vulnerable areas with respect to elevated THMs quantities. The contemplations have parameterized the risk based on the threshold concentration and extent of exposure. As a result, the attributes origin from the optimization of the chlorination process (Di Cristo, Leopardi, Quintiliani, & De Marinis, 2015). Similar studies have scrutinized potential hazards correlated with the water disinfection process (Falakh & Setiani, 2018).

Artimani et al. studied a dispersion model ALOHA for the efficient determination of the extent of chlorine emission. The model is based on the estimation of the emitted pollutants and their dispersion in the premise's atmosphere. The concept is laid on the foundation of evaluation of distance traveled by emitted pollutant, using its physical properties and toxicity rate; imaginary lining of the expected exposed areas; computing quantity of the pollutant; and speculating the observations at specific time intervals (Artimani, Arjmand, & Kalaei, 2012). Moreover, there are similar studies discussing the dispersion method, where researchers have chosen the Gauss model for the calculation of leakage dispersion concentration and distance and Matlab for the simulations (Zhou & Feng, 2016).

Some investigations have also revealed the use of comprehensive globally accepted computed models such as the Unified Dispersion Model (UDM) of Phast (Pandya, Gabas, & Marsden, 2012), Atmospheric Transport and Dispersion (ATD) model (Spicer & Tickle, 2020). Many other approaches have been considered to predict the chlorine release in past, for instance, the Homogeneous Equilibrium Model (HEM) and the Metastable Liquid Model (MLM) (Spicer & Miller, 2018). This stereotype guides the approximation of accidental releases. To prevent hazards from leakage caused during the transportation of chlorine, researchers have developed a new method. The developed protocol, DEGADIS, identifies handling uncertainty under the aegis of the algorithmic system (Scenna & Santa Cruz, 2005).

9.9 Environmental effects

The major origin of chlorine is seawater (California Environmental Protection Agency, 2002; Chang, Liu, & Jeng, 2004; Knipping & Dabdub, 2003). As per the EPA National Priorities List, the hazardous waste sites are rarely contaminated with chlorine as it is highly unstable and reactive (HazDat, 2007).

As per EPA, the impact of chlorine on the environment depends largely on how long the exposure period was and how a large amount of chlorine was present. Chlorine conventionally combines with chemicals such as dioxins that pollute water, contaminate fish, and transfer to humans and larger animals that eat the fish. This process is known as biomagnification, in which levels of accumulated substances concentrate on animals as you move up the food chain. Chlorine is an important ingredient in the organic compound DDT, the chemical that caused population levels in several bird species to plummet in the 1950s and led to Rachel Carson's seminal work "Silent Spring." Dioxins can change hormone levels by mimicking natural substances such as estrogen. They can also pose a risk of cancer, suppress immune systems, harm fetuses, and cause skin diseases. Organisms living in soil may also be affected by doses of chlorine.

9.10 Hazardousness

The level of toxicity caused by chlorine highly depends on its quantity, course of action, and duration of exposure. In humans, exposure to chlorine mainly impairs NaCl metabolism, which can result in excessive loss of sodium resulting in a life-threatening condition. The deficient condition may also cause loss of appetite, lethargy, and muscle weakness. Moreover, other health conditions such as hypochloremia may also originate from the conditions. Inhalation of chlorine may irritate the lungs and cause respiratory disorders, as well as burning eyes and skin.

There are also some reports documenting symptoms like tickling, itching, or dryness of the nose, and throat, shortness of breath, and headaches (Calabrese & Kenyon, 1991). Higher levels of chlorine may result in mild mucous membrane irritation, chest pain, vomiting, dyspnea, cough, skin irritation or burns, toxic pneumonitis, and pulmonary edema (Hazardous Substances Data Bank, 1993). The enhanced level of chlorine may exhibit respiratory effects, such as eye and throat irritation and airflow obstruction (California Environmental Protection Agency (CalEPA), 1999). Some researchers have confirmed decreased body weight gain, eye, and nose irritation and nonneoplastic lesions, and respiratory epithelial hyperplasia from chronic inhalation exposure to chlorine (Calabrese & Kenyon, 1991; California Environmental Protection Agency (CalEPA), 1999; US EPA, 1992). Refer to Fig. 9.3 for chlorine exposure thresholds and reported responses.

9.11 Treatment

The toxic effect of chlorine is mainly treated symptomatically, which is in synchronization with the reported signs and symptoms. However, overages chlorine can be treated principally with humidified oxygen.

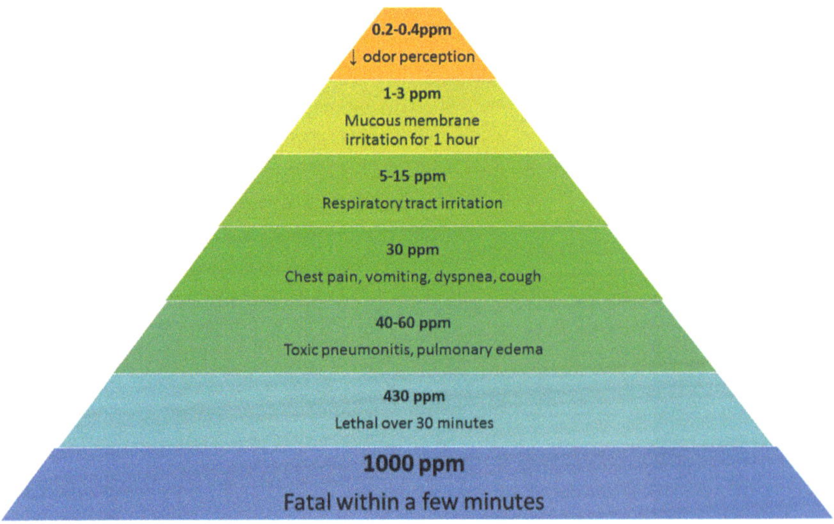

FIG. 9.3 Chlorine exposure thresholds and reported responses. *Credit: Malvika Chawla.*

This may soothe the irritation of the mucous membrane caused due to chlorine. Some of the main chlorine therapeutics comprehend inhaled bicarbonate (Aslan et al., 2006; Babu, Cardenas Jr, & Sharma, 2008; Bosse, 1994; Martinez & Long, 1995), glucocorticoids, systemic corticosteroid (Wang, Winskog, Edston, & Walther, 2005), supplemental humidified oxygen, besides, use of inhaled *b*-adrenergic agents (Wang, Zhang, & Walther, 2004) and inhaled agents such as budesonide (Wang, Zhang, & Walther, 2002). Although the effect of these therapies is still under research.

9.12 Risk management

The risk management framework is principally laid on the foundation of identification and estimation of risk and its prevention process. The main objective is to prevent any unwanted release or damage that may incur undesirable human and environmental hazards. The addressed risk management specifically involved in the facilities require proper design, technology, operational and maintenance activities, and routine audits, and training programs. The effective risk analysis process anticipates modification in the component of risk; such as modified pipe strength, improved installation, more protective transfer lines, inventory management, adequate ventilation, and likewise. The area specifically designed for chlorine storage must be purveyed with leak identification alarms and a chlorine institute-approved leak repair kit. They must strictly abide by the given competent authority guidelines and recommendations. For the ease of locating the possible leakage at fittings or pipes, a rag soaked in concentrated ammonia solution can be used. Moreover, the employee should look up for the presence of any white cloud or vapor.

On the other hand, in water and wastewater treatment operations, the movement of gaseous chlorine is given utmost importance. The process is generally controlled by using a vacuum-operated, gas feed chlorinator. The generated vacuum mixes chlorine with water, which is piped and diffused into the water or wastewater to provide the required chlorine dosage.

The chlorine premises must follow the regulations and limits set by federal or local authorities. Some of the regulations for chlorine by competent authorities are given in Table 9.1.

9.13 Current research

The subsistence of life is principally based on water. It is a known fact that the freshwater accounts for roundabout 1% of the total available water on the Earth, which is continuously depleting with every passing second.

TABLE 9.1 Recommended limits set by various federal agencies.

Authority	Limit
Levels in the air set by EPA	0.5 ppm
Levels in workplace air set by OSHA	Permissible exposure limit (PEL) is 1.0 ppm, not to be exceeded at any time
Levels in drinking water set by EPA	Maximum contaminant level (MCL) and maximum residual disinfectant level (MRDL) of 4 mg/L
NIOSH	Recommended airborne exposure limit (REL) is 0.5 ppm, which should not be exceeded during any 15-min work period
ACGIH	Threshold limit value (TLV) is 0.5 ppm averaged over an 8-h work shift and 1 ppm as a STEL (short-term exposure limit)

The consequences of this could be disastrous for existence of life on this planet. Hence, the present as well as future research programs are mainly directed to answer similar queries. In this direction, wastewater treatment is one of the crucial processes. In water-stressed areas, reused and treated wastewater has found increased interest.

Chlorine is one of the cheap and efficient universally used water disinfectants. A novel method has been developed by Purdue University researchers for water detoxification using chlorine and ultraviolet radiation, which may not only provide new hope for water-stressed areas but also help promote the reuse of wastewater. The team has developed a procedure for selectively degrading and detoxifying amines, organic compounds derived from ammonia that are conventionally available in water supplies. This method allows for the selective degradation of amines. The degradation of the compounds and the reduction of toxicity are activated by prechlorination with subsequent UV exposure.

Moreover, they are postulations illustrating the use of UV-activated mixed $HOCl/NH_2Cl$ in order to enhance the consumption of free chlorine and NH_2Cl in water (Liu et al., 2020). On the other hand, Kong et al. critically investigated the use of solar/chlorine process to detoxify the typical herbicides atrazine and mecoprop (methylchlorophenoxypropionic acid). For the evaluation of the synergistic effect of chlorine and solar radiation, the studies used a simulation approach (Kong et al., 2020). While Gao, Zhang, Li, Tian, and Gao (2020) looked upon the after-effects of UV/chlorine and UV/H_2O_2 pretreatment, that leads to the increased formation of disinfection by-product (DBP). Also, there are contemplations pondered for the sustainable recovery of gold from gold-bearing tailings via chlorination heat treatment (Li, Peng, Long, Li, & Zhang, 2020).

9.14 Chlorine toxicity: Case study

Before mentioned are some examples of incidents that happened post-release of chlorine gas.

Case 1: On the afternoon of May 27, 2008, the people of Jamshedpur, India were exposed to a dense, pale green, pungent, and poisonous gas chlorine, leaked from the unused cylinder lying in the Tata Motor's water treatment plant for the past 10 years. The exposure severely affected the 150–200 people including employees and their family members.

Case 2: On January 6, 2005, a clash between two trains among which one was carrying sodium hydroxide, cresol, and liquefied chlorine in Graniteville, South Carolina resulted in the release of approximately 120,000 pounds of chlorine gas to the surrounding atmosphere (NTSB, 2005).

Case 3: On June 28, 2004, a westbound Union Pacific Railroad freight train collided with an eastbound freight train in Macdona, Texas (NTSB, 2006). This incident resulted in the puncture of a tank car carrying 180,000 pounds of liquefied chlorine, which consequently released chlorine gas to the surrounding area (NTSB, 2006).

Case 4: In December 2002, there was a report on chlorine gas seepage at Indian Petrochemical Corporation Limited Gandahar complex near Dahej in Bharuch district, India that led to a major scare in the village encircling the plant, around 250 people were affected by this leakage.

Case 5: On October 22, 1998 during swimming classes, an emission of chlorine vapors occurred from the closed chlorinating maintenance procedures room of a recreational center in Rome. The chemicals from separated tanks used for water disinfection were mixed before circulating in the pool in controlled concentrations. The accident happened when an unknown amount of chlorine fumes was released post breakage of the tube carrying the hydrochloric acid to the mixing container. This leakage spread across the entire surrounding affecting swimmers and others. A total of 116 people were affected and referred to the emergency department of different hospitals in Rome, mostly complaining of eye and throat irritation, cough, shortness of breath, and anxiety (Agabiti et al., 2001).

Case 6: A freight train derailment, which took place on April 11, 1996, 2 miles west of Alberton, Montana resulted in a spill releasing 64.8 tons of chlorine into the environment (Agency for Toxic Substances and Disease Registry, 1998; NTSB, 1998).

9.15 Safety measures

- The gas chlorinator should be stipulated with pressure piping on the assumption that no liquid chlorine enters the chamber, as it can damage the chlorinator and could cause serious safety problems.
- When handling relatively concentrated chlorine bleach right out of the container, eye protection such as wrap-around safety glasses and/or

goggles, rubber household gloves or nitrile gloves, clothing covering skin, chemical protective aprons, the disposable protective suit must be worn by the handler.
- The storage area should be on the downward side of the building.
- The room where containers are kept should have a nonslip floor, ventilating-fans, emergency doors, vandal-proof switches, and motorized louvers.
- The person handling chlorine should wash hands thoroughly with soap and shampoo prior to and after usage of this gas as well as take shower at the end of the day.
- The eyewash stations and first-aid should be available in the proximity of the facility.
- The chlorine premises should be devoid of any other cleaning agents like ammonia as it may explode.
- All applicable safety precautions and instructions should be strictly followed.

9.16 Handling and storage

Storage: For ease of transportation, chlorine is mainly stored in a compressed form in containers. The size of containers varies according to the quantity of chlorine that needs to be transmuted. The valves of the cylinders ought to be housed securely when not in usage in order to prevent any damage. The accustomed cylinder should have foot-ring, bumped-bottom, and double-bottom construction and one single opening. However, valves should be designed with a fusible metal plug to melt between 158°F and 165°F to release any created pressure. This may prevent any possible injuries. Chlorine containing rooms should have easy access, lock, and equipment with adequate shelving (spill barriers and seismic considerations), spill controls and containment, as well as ventilation and respiratory protection for workers. Chlorine must be separated from incompatible materials by at least 20 ft or stored in separate secondary containment and kept dry.

Specification of containers: Broadly, three types of containers are utilized in the transportation of chlorine that is, ton containers, cargo tanks, and tank cars. First, for transfer of around 907 kg of Cl, steel concave/convex welded ton containers are used. They are easy to carry and have two valves, thus are capable of supplying both liquid and gas. Second, there are cargo tanks with an adequate capacity of 13,600–20,000 kg chlorine. They are specifically designed with four valves. And third, there are tank cars which are similar to the railroad. They have a variable capacity ranging between 55 and 90 tons of chlorine. All tank cars consist of four valves provided with one top-notch opening.

Transportation: At the time of transfers, the cylinders must be secured in an upright position with the utmost care. At the primary stage, all cylinders should be properly stamped and then moved forward for quality check. In accordance with the U.S. Department of Transportation (DOT) or Canadian Transportation of Dangerous Goods Regulations, all cylinder's mandates to be hydrostatically tested on a time interim of 5 years. The availability of placards on board is an imperative process. The person involved in the handling must be well trained.

Loading/unloading: For unloading or loading cylinders, a loading dock or a hydraulic tailgate should be preferably utilized. During this protocol, the cylinders must be either chained or secured before relocation. The loaded cylinders should be fastened with a storage rack, which should be lifted with the help of a forklift. The hoist of sufficient capacity may be used to offload ton containers. Also, tank cars can be emptied into a large storage tank or storage facilities.

9.17 Precautions

The following recommendations should be paid attention to prevent any manifold and additional concerns:

1. Use hazmat placards.
2. Tightly secure empty containers.
3. Do not drag a cylinder.
4. Keep cylinders away from heat.
5. Follow proper cylinder handling protocol.
6. Avoid knocking or dropping over the cylinder.
7. Never tamper with a fusible plug.

References

Agabiti, N., Ancona, C., Forastiere, F., Di Napoli, A., Presti, E. L., Corbo, G. M., et al. (2001). Short term respiratory effects of acute exposure to chlorine due to a swimming pool accident. *Occupational and Environmental Medicine, 58*(6), 399–404.

Agency for Toxic Substances and Disease Registry. (1998). *Alberton chlorine spill; Alberton, Montana. Phase 1 study report*. Atlanta, GA: Agency for Toxic Substances and Disease Registry.

Antonio, C. F. (1999). Fuzzy FTA: A fuzzy fault tree system for uncertainty analysis. *Annals of Nuclear Energy, 26*(6), 523–532.

Artimani, J. S., Arjmand, M., & Kalaei, M. Z. (2012). Modeling and assessing risk analysis of chlorine gas in water treatment plants. *European Journal of Experimental Biology, 2*(6), 2151–2157.

Aslan, S., Kandiş, H., Akgun, M., Cakır, Z., Inandı, T., & Görgüner, M. (2006). The effect of nebulized $NaHCO_3$ treatment on "RADS" due to chlorine gas inhalation. *Inhalation Toxicology, 18*(11), 895–900.

Babu, R. V., Cardenas, V., Jr., & Sharma, G. (2008). Acute respiratory distress syndrome from chlorine inhalation during a swimming pool accident: A case report and review of the literature. *Journal of Intensive Care Medicine, 23*(4), 275–280.

Bosse, G. M. (1994). Nebulized sodium bicarbonate in the treatment of chlorine gas inhalation. *Journal of Toxicology: Clinical Toxicology, 32*(3), 233–241.

Calabrese, E. J., & Kenyon, E. (1991). *Air toxics and risk assessment*. CRC Press.

California Environmental Protection Agency. (2002). *Chlorine emissions from activated sea-salt aerosols and their potential impact on ozone*. Sacramento, CA: California Environmental Protection Agency.

California Environmental Protection Agency (CalEPA). (1999). *Air toxics hot spots program risk assessment guidelines: Part III. Technical support document for the determination of noncancer chronic reference exposure levels. SRP draft*. Berkeley, CA: Office of Environmental Health Hazard Assessment.

Chanda, R. S., & Bhattacharjee, P. K. (1998). A reliability approach to transmission expansion planning using fuzzy fault-tree model. *Electric Power Systems Research, 45*(2), 101–108.

Chang, C. T., Liu, T. H., & Jeng, F. T. (2004). Atmospheric concentrations of the Cl atom, ClO radical, and HO radical in the coastal marine boundary layer. *Environmental Research, 94*(1), 67–74.

Dash, A. K., Pradhan, M. K., & Singh, R. (2018). Application of computational analysis for risk assessment of chlorine gas from tank in chlorine production unit: A case study. In *Advances in fire and process safety* (pp. 13–20). Singapore: Springer. https://doi.org/10.1007/978-981-10-7281-9_2.

Davy, H. (1811). VIII. On a combination of oxymuriatic gas and oxygene gas. *Philosophical Transactions of the Royal Society of London, 101*, 155–162.

Di Cristo, C., Leopardi, A., Quintiliani, C., & De Marinis, G. (2015). Drinking water vulnerability assessment after disinfection through chlorine. *Procedia Engineering, 119*(1), 389–397.

Doytchev, D. E., & Szwillus, G. (2009). Combining task analysis and fault tree analysis for accident and incident analysis: A case study from Bulgaria. *Accident Analysis & Prevention, 41*(6), 1172–1179.

Falakh, F., & Setiani, O. (2018). Hazard identification and risk assessment in water treatment plant considering environmental health and safety practice. *E3S Web of Conferences, 31*, 06011. EDP Sciences.

Ferdous, R., Khan, F., Veitch, B., & Amyotte, P. R. (2009). Methodology for computer aided fuzzy fault tree analysis. *Process Safety and Environmental Protection, 87*(4), 217–226.

Fukayama, M. Y., Tan, H., Wheeler, W. B., & Wei, C. I. (1986). Reactions of aqueous chlorine and chlorine dioxide with model food compounds. *Environmental Health Perspectives, 69*, 267–274.

Gao, Y. Q., Zhang, J., Li, C., Tian, F. X., & Gao, N. Y. (2020). Comparative evaluation of metoprolol degradation by UV/chlorine and UV/H_2O_2 processes. *Chemosphere, 243*, 125325.

Hazardous Substances Data Bank. (1993). *National toxicology information program*. Bethesda, MD: National Library of Medicine.

HazDat. (2007). *Chlorine. HazDat database: ATSDR's hazardous substance release and health effects database*. Atlanta, GA: Agency for Toxic Substances and Disease Registry.

Horton, D. K., Berkowitz, Z., & Kaye, W. E. (2002). The public health consequences from acute chlorine releases, 1993–2000. *Journal of Occupational and Environmental Medicine, 44*(10), 906–913.

Ila, P. (2006). Chlorine & its consequences. In *Medical geology/geochemistry lecture notes*.

Kendrick, M. A. (2018). Halogens in seawater, marine sediments and the altered oceanic lithosphere. In D. E. Harlov, & L. Y. Aranovich (Eds.), *The role of halogens in terrestrial and extraterrestrial processes*. Berlin: Springer.

Khan, F. I., & Abbasi, S. A. (2000). Analytical simulation and PROFAT II: A new methodology and a computer automated tool for fault tree analysis in chemical process industries. *Journal of Hazardous Materials, 75*(1), 1–27.

Knipping, E. M., & Dabdub, D. (2003). Impact of chlorine emissions from sea-salt aerosol on coastal urban ozone. *Environmental Science and Technology*, *37*(2), 275–284.

Kong, X., Wang, L., Wu, Z., Zeng, F., Sun, H., Guo, K., et al. (2020). Solar irradiation combined with chlorine can detoxify herbicides. *Water Research*, 115784.

Li, H., Peng, J., Long, H., Li, S., & Zhang, L. (2020). Cleaner process: Efficacy of chlorine in the recycling of gold from gold-containing tailings. *Journal of Cleaner Production, 125066.*

Liu, Z., Xu, B., Lin, Y. L., Zhang, T. Y., Ye, T., Hu, C. Y., et al. (2020). Mechanistic study on chlorine/nitrogen transformation and disinfection by-product generation in a UV-activated mixed chlorine/chloramines system. *Water Research*, *184*, 116116.

Lodders, K. (2003). Solar system abundances and condensation temperatures of the elements. *The Astrophysical Journal*, *591*(2), 1220.

Martinez, T. T., & Long, C. (1995). Explosion risk from swimming pool chlorinators and review of chlorine toxicity. *Journal of Toxicology: Clinical Toxicology*, *33*(4), 349–354.

Michael, P. J., & Cornell, W. C. (1998). Influence of spreading rate and magma supply on crystallization and assimilation beneath mid-ocean ridges: Evidence from chlorine and major element chemistry of mid-ocean ridge basalts. *Journal of Geophysical Research: Solid Earth*, *103*(B8), 18325–18356.

Mink, F. L., Coleman, W. E., Munch, J. W., Kaylor, W. H., & Ringhand, H. P. (1983). In vivo formation of halogenated reaction products following peroral sodium hypochlorite. *Bulletin of Environmental Contamination and Toxicology*, *30*(1), 394–399.

Nodelman, V., & Ultman, J. S. (1999a). Longitudinal distribution of chlorine absorption in human airways: Comparison of nasal and oral quiet breathing. *Journal of Applied Physiology*, *86*(6), 1984–1993.

Nodelman, V., & Ultman, J. S. (1999b). Longitudinal distribution of chlorine absorption in human airways: A comparison to ozone absorption. *Journal of Applied Physiology*, *87*(6), 2073–2080.

NTSB. (1998). Railroad accident brief report. LAX 96 FR 010. Derailment and hazardous materials release with fatality. In *Montano Rail Lin, Alberton, Montana. April 11, 1996*. Washington, DC: National Transportation Safety Board.

NTSB. (2005). *Collision of Norfolk Southern freight train 152 with standing Norfolk Southern local train P22 with subsequent hazardous materials release at Graniteville, South Carolina, January 06, 2005*. Washington, DC: National Transportation Safety Board.

NTSB. (2006). *National Transportation Safety Board railroad accident report: Collision of union Pacific railroad train MHOTU-23 with BNSF railway company train MEAP-TUL-126-D with subsequent derailment and hazardous materials release in Macdona, Texas, June 28* (p. 2004).

Occidental Chemical Corporation (OxyChem). (2012). *Chlorine handbook*. Available from https://www.oxy.com/ourbusinesses/chemicals/products/documents/chlorine/chlorhb1.pdf.

Pandya, N., Gabas, N., & Marsden, E. (2012). Sensitivity analysis of phast's atmospheric dispersion model for three toxic materials (nitric oxide, ammonia, chlorine). *Journal of Loss Prevention in the Process Industries*, *25*(1), 20–32.

Renjith, V. R., Madhu, G., Nayagam, V. L. G., & Bhasi, A. B. (2010). Two-dimensional fuzzy fault tree analysis for chlorine release from a chlor-alkali industry using expert elicitation. *Journal of Hazardous Materials*, *183*(1–3), 103–110.

Roy, P. K., Bhatt, A., & Rajagopal, C. (2003). Quantitative risk assessment for accidental release of titanium tetrachloride in a titanium sponge production plant. *Journal of Hazardous Materials*, *102*(2–3), 167–186.

Rudnick, R. L., & Gao, S. (2003). Composition of the continental crust. *The Crust*, *3*, 1–64.

Scenna, N. J., & Santa Cruz, A. S. (2005). Road risk analysis due to the transportation of chlorine in Rosario city. *Reliability Engineering & System Safety*, *90*(1), 83–90.

Schilling, J. G., Unni, C. K., & Bender, M. L. (1978). Origin of chlorine and bromine in the oceans. *Nature*, *273*(5664), 631–636.

Selinus, O., et al. (2005). *Essentials of medical geology*. Boston, MA: Elsevier Academic Press.
Shimizu, K., Saal, A. E., Myers, C. E., Nagle, A. N., Hauri, E. H., Forsyth, D. W., et al. (2016). Two-component mantle melting-mixing model for the generation of mid-ocean ridge basalts: Implications for the volatile content of the Pacific upper mantle. *Geochimica et Cosmochimica Acta, 176*, 44–80.
Shu, M. H., Cheng, C. H., & Chang, J. R. (2006). Using intuitionistic fuzzy sets for fault-tree analysis on printed circuit board assembly. *Microelectronics Reliability, 46*(12), 2139–2148.
Spicer, T., & Miller, D. (2018). Quantifying the mass discharge rate of flashing two phase releases through simple holes to the atmosphere. *Process Safety Progress, 37*(3), 382–396.
Spicer, T. O., & Tickle, G. (2020). Simplified source description for atmospheric dispersion model comparison of the Jack rabbit II chlorine field experiments. *Atmospheric Environment, 244*, 117866.
US EPA. (1992). *Drinking water criteria document for chlorine, hypochlorous acid and hypochlorite ion. (external review draft.) environmental criteria and assessment office*. Cincinnati, OH: Office of Health and Environmental Assessment.
US EPA. (1999). *Integrated risk information system (IRIS) on chlorine. National center for environmental assessment*. Washington, DC: Office of Research and Development.
Wang, J., Winskog, C., Edston, E., & Walther, S. M. (2005). Inhaled and intravenous corticosteroids both attenuate chlorine gas-induced lung injury in pigs. *Acta Anaesthesiologica Scandinavica, 49*(2), 183–190.
Wang, J., Zhang, L., & Walther, S. M. (2002). Inhaled budesonide in experimental chlorine gas lung injury: Influence of time interval between injury and treatment. *Intensive Care Medicine, 28*(3), 352–357.
Wang, J., Zhang, L., & Walther, S. M. (2004). Administration of aerosolized terbutaline and budesonide reduces chlorine gas–induced acute lung injury. *Journal of Trauma and Acute Care Surgery, 56*(4), 850–862.
Zhou, D., & Feng, W. (2016, August). Risk assessment on chlorine leakage dispersion. In *Vol. 1. 2016 8th International Conference on Intelligent Human-Machine Systems and Cybernetics (IHMSC)* (pp. 472–475). IEEE.

CHAPTER

10

Cyanogen: Risk assessment, environmental, and health hazard

Lakshami Prashad Purohit[a] *and Sonu Kumar*[b]

[a]Department of Physics, Gurukula Kangri (Deemed to be University), Haridwar, Uttarakhand, India

[b]Department of Chemistry, Gurukula Kangri (Deemed to be University), Haridwar, Uttarakhand, India

Abbreviations

AEGLs acute exposure guideline values
BCF bioconcentration factor
EPA Environmental Protection Authority
NRC National Research Council
TEL tolerable exposure limit

10.1 Introduction

The dinitrile compound of oxalic acid is known as cyanogen. This compound is an extremely poisonous gas having the general formula of C_2N_2. J. L. Gay-Lussac first prepared it in its purest form in 1815 (Brotherton & Lynn, 1959). In the laboratory, it is prepared by the oxidation of nitrile ion in an aqueous solution, and in the industry usually synthesized by the gas phase oxidation of hydrogen cyanide (HCN). It is a colorless gas with a pungent, penetrating, and almond-like odor. Cyanogen is also a very toxic and extremely flammable gas having melting point 245 K and boiling point 252 K. Due to their low melting and boiling point cyanogen readily react explosively with powerful oxidizing agents. Some industries use cyanogen as fuel gas in welding, where on the other hand NASA considers it as rocket fuel. In cyanogen two

https://doi.org/10.1016/B978-0-323-89857-7.00029-3

cyano groups bond together through their carbon atoms by a single bond that is, NC–CN. While some of its other isomers were discovered in 2008 (Ringer, Sherrill, King, & Crawford, 2008). There are certain derivatives of cyanogen that are also called cyanogen. These derivatives of cyanogen contain only cyano (CN) group for example, cyanogen chloride (CNCl), cyanogen bromide (CNBr), etc. (Hartman & Dreger, 1931).

It is one of the simpler carbonitrile compounds, which is a class of molecules whose common feature is a strong carbon-nitrogen (C≡N) triple bond (Fang et al., 2019). Cyanogen also can be polymerized into an electrically conducting material due to its relatively high energy. Cyanide (CN) radical is known as pseudo halogen because it behaves like a halogen atom chemically and forms cyanogen (NC-CN). It is thermodynamically unstable and readily decomposes into their elements, easily reacts with water, and gets oxidized by oxygen. Cyanogen (NC-CN) is a much-studied species due to its kinetic stability. Most of the alkyl nitriles and halonitriles give aromatic 1,3,5-trisubstituted triazines at ambient and high static pressure after trimerize symmetrically. The trimerizations and product yield for these compounds are affected by specific solvents, catalysts, and substituents that are strongly attached to the nitrile moiety. There is no report on the trimerization of cyanogen while it polymerizes at high temperatures. Cyanogen has a linear structure with a short C–C distance, which specifies considerable electron delocalization in the molecule. It burns in the air with a very hot, violet flame and looks a lot like the halogen. It is hydrolyzed by alkali and undergoes thermal dissociation to CN radical at high temperatures (Eqs. 10.1 and 10.2).

$$C_2N_2 + O_2 \rightarrow 2CO_2 + N_2 \tag{10.1}$$

$$C_2N_2 + 2OH^- \rightarrow OCN^- + CN^- + H_2O \tag{10.2}$$

10.2 Occurrence

Cyanogen can be obtained from the anhydride of oxamide (Eq. 10.3).

$$H_2N - C(O) - C(O) - NH_2 \rightarrow NC - CN + 2H_2O \tag{10.3}$$

Oxamide can be prepared by the hydrolysis of cyanogen (Greenwood & Earnshaw, 1997) (Eq. 10.4).

$$NC - CN + 2H_2O \rightarrow H_2N - C(O) - C(O) - NH_2 \tag{10.4}$$

Cyanogen can be made either by simply heating $Hg(CN)_2$ or from a combination of KCN and $CuSO_4$ solution. The latter method produces spectroscopically large amounts of hydrogen cyanide, a substance that has many dominant absorption bands. An insignificant fraction also exists in the gas from $Hg(CN)_2$, but a threefold distillation over mercuric oxide appropriately eliminate all traces of the contaminants (Eq. 10.5).

$$2Hg(CN)_2 \rightarrow (CN)_2 + Hg_2(CN)_2 \quad (10.5)$$

Alternatively, a combination of copper (II) salt solutions and cyanides gives an unstable copper (II) cyanide that rapidly decomposes into copper (I) cyanide and cyanogen, Eq. (10.6) (Brotherton & Lynn, 1959).

$$2CuSO_4 + 4KCN \rightarrow (CN)_2 + 2CuCN + 2K_2SO_4 \quad (10.6)$$

Industrially, at large-scale cyanogen is manufactured by oxidizing hydrogen cyanide in presence of catalysts such as chlorine supported SiO_2 or NO_2 over a salt of copper. It is also synthesized by the reaction of N_2 on acetylene along with electrical discharge or spark (Breneman, 1889).

Moreover, one of the best methods to prepare paracyanogen is by heating mercuric cyanide. Heating of silver cyanate, cyanogen iodide, silver cyanide, or cyanuric iodide also results in the formation of paracyanogen (Bircumshaw, Tayler, & Whiffen, 1954). Cyanogen at 300–500°C (572–932°F) in presence of trace impurities polymerizes to paracyanogen, which transforms back to cyanogen when heated up to 800°C (1470°F). Paracyanogen has a double-bonded system of sp2 carbon atoms with π conjugation (Maya, 1993).

10.3 Historical background

Joshap Louis Gay-Lussac was the first who synthesized, determined the empirical formula, and named the cyanogen in 1815. Gay-Lussac invented the word "Cyanogen" from the Greek words, "kyanos" which means "Blue" and "Genneo" means "I create" because Swedish chemist was the first who isolated the cyanide from the pigment Prussian blue (Gay-Lussac, 1815). Cyanogen soap was used by photographers to remove silver stains from their hands (William, 1859). In the late nineteenth century, cyanogen accomplished significance with the evolution of the fertilizer industry and stayed as a significant intermediate in the manufacturing of different fertilizers.

A spectroscopic analysis of Halley's Comet originates cyanogen in the comet tail in 1910, which led to the public fear that the Earth would be poisoned if it passed through the comet tail. There was no effect when the comet passed through the tail because of the diffuse nature of the tail. Cyanogen is an important photochemically active species, which is found in the Titan's atmosphere and comets, and that place would be a source of CN radicals whose production is noticeable in the cometary tail. In addition, this comparatively simple, linear tetra-atomic molecule has become an exemplary system for photochemical studies.

10.4 Physicochemical properties

Cyanogen is a colorless gas that has an odor like almonds. It freezes at −28°C and boils at 20.7°C. Cyanogen produces the second-hottest-known natural flame (after carbon subnitride) with a temperature of over 4525°C (8177°F) when it burns in oxygen (Conway, Wilson Jr, & Grosse, 1953; Thomas, Gaydon, & Brewer, 1952). This gas is heavier than air and its flame can travel back to the source of the leak very easily. There are some common properties of cyanogen are listed below in Table 10.1.

10.5 Physiological role

Cyanogen shows a toxic mechanism similar to that of hydrogen cyanide (Refer chapter 15) (Kopras, 2012). Hydrogen cyanide is a common poison that has a direct impact on the focal sensory central nervous system in living systems. It suppresses the cytochrome oxidase, which causes the hindering in cellular respiration and restricts the electron allocation to

TABLE 10.1 Physical properties of cyanogen.

Property	Value
Molecular weight	52.036 g/mol
Color/form	Colorless gas; burns with a purple-tinged flame
Odor	Almond-like odor, which is acrid and pungent in lethal concentrations
Boiling point	−6.1°F at 760 mmHg (United States Coast Guard, 1999)
Melting point	−18.2°F (United States Coast Guard, 1999)
Solubility	Soluble in Ethanol, Ethyl ether; Solubility at 20°C: water dissolves 4.5 times its volume of cyanogen; diethyl ether, 5.0 times; and ethanol, 23.0 times (Lide, 2007)
Density	0.954 at −5.8°F (United States Coast Guard, 1999)
Vapor density	1.8 (AIR = 1) (Quincy, 1991)
Vapor pressure	5.1 atm at 70°F
Octanol/water partition coefficient	0.07 (Log *P*)
Heat of combustion	258.3 kcal/gmole @ 20°C (gas) (Weast, 1988)
Heat of vaporization	5.778 kcal/mole (liquid) (O'Neil, 2006)
Ionization potential	13.57 eV

oxygen (Rieders, 1965). The oxygen concentration growth in tissue significantly improves with pressure and decreases delivery for oxyhemoglobin. Thus, it can't encounter the metabolic necessities at a stage because of moderate oxidative metabolism. That is prevalently serious in the brainstem nuclei, somewhere inadequacy of an energy source, consequences to discontinue the central respiratory and death.

Many of the enzymes predominantly that comprise iron or copper active metal center are obstructed by the cyanide and some enzymes for example, cytochrome oxidase seems to be the most sensitive (Rieders, 1965). Cyanides also produce the momentary period of hyperpnea by encouraging the chemoreceptors of carotid and aortic bodies resulting in cardiac irregularities; however, death or brain injury may also occur in certain cases (National Research Council (NRC), 2002).

10.6 Uses

Cyanogen may be encountered in the circumstances in which there is the heating of nitrogen comprising carbon bonds like in blast furnace gases. Cyanogen is used as a fumigant. In organic synthesis, cyanogen is used as an intermediate, fuel gas for welding and cutting heat-resistant metals and in rocket and missile as a propellant (with ozone or fluorine) (Brotherton & Lynn, 1959; Conway et al., 1953). It is also used as a stabilizer in the manufacturing of nitrocellulose.

10.7 Sources of contaminants

Cassava, the staple crop in drought-prone areas of East Africa contains chemicals called cyanogenic glycosides, which release cyanide in the body when consumed and can be fatal (Teles, 2002). Mainly two toxicological substances such as hydrogen cyanide (about 0.5%) and carbon dioxide (about 2%) concern cyanogen contamination. For detailed sources of contamination see Chapter 15; Section 15.7 and Chapter 7; Section 7.7. The low concentration of these gases provide high acute toxicity and high concentration of cyanogen activity due to the greater toxicity of them.

10.8 Risk assessment

Risk accompanying with the experience of cyanogen bystander or public has been assessed by utilizing two different ways: First, short-term exposure from 10 min to 8 h and second, repeated exposure over a longer time. The second stage was applied to people residing in adjacent vicinity to a fumigation facility in which experience intervals could be of a further constant nature.

The NRC compared acute exposure guideline values (AEGLs) for the short-term experience of cyanogen with the modeled introduction levels as recognized by authorities to protect the general population of variable compassions. The short-term exposure risk to bystanders is deliberated negligible, as the maximum-modeled experience was found to be 0.858 ppm and the AEGL threshold concern value ranged from 2.5 ppm to 1.0 ppm. The demonstrated exposure levels for the long-term experience of cyanogen was measured from the perspective of the EPA staff-derived tolerable exposure limit (TEL) of 0.034 ppm. The danger to observers where there is a numerous log heap exposure source is about 8 × over the TEL even a good way off of 120 m at the modeled exposure levels after all the uncertainty adjustment factors are incorporated is 0.266 ppm (Environmental Protection Authority (EPA), 2018).

10.9 Environmental effects

In organic synthesis, cyanogen is used as an intermediate, as a propellant in a rocket and missile, and as a fuel gas for welding and cutting heating resistant metals from where it could be consequently discharged to the atmosphere through various unwanted tributaries. It is directly released into the environment by its previous use as a fumigant. In the atmosphere, it exists especially in the gas phase indicated by the vapor pressure of the air around 4300 mm-Hg at 25°C. Cyanogen produces hydroxyl radicals photochemically and successively degrade in the environment and the predictable half-life for this reaction is found to be greater than 530 days. It is not projected to be susceptible to the direct photochemical reaction in sunlight because it does not contain chromophores that absorb the light energy at wavelengths greater than 290 nm. When cyanogen is released into the soil, then it is predictable to have very high mobility on the basis of measured Koc (the ratio of the adsorbed organic analyte to the dissolved) of 8.3 (Environmental Protection Authority (EPA), 2018). The removal process of cyanogen via volatilization from the surface of moist soil is predictable to be a significant process, which is based upon Henry's Law constant having the value of 3.54×10^{-4} atm-cu m/mole. The volatilization of cyanogen from the surface of dry soil is based upon the vapor pressure of it. Cyanogen is not expected to get adsorbed by the suspended solids and sediment in the water.

The removal process of cyanogen via volatilization from the surface of the water is predictable to be an important process. The predictable half-lives of the volatilization process of cyanogen are 2 h and 3.4 days for a model river and model lake, respectively. The potential for the bio-concentration in marine creatures is very low, as suggested by the BCF of 3.2. Since cyanogen hydrolyzes very slowly in an aqueous solution, consequently the hydrolysis process is expected to be an important providence process.

10.10 Hazardousness

1. A milder consequence of cyanogen in humans may be specious at the concentrations of 0.5–1.0 μg/mL and higher concentrations such as 2.5 μg/mL or greater than 2.5 μg/mL are associated with convulsions, unconsciousness, and death.
2. Metabolic acidosis through an inexplicable greater anion gap (that is the dimensions of lactic acid are not accessible) or metabolic acidosis through a greater concentration of lactic acid (known as lactic acidosis) could exist. Anaerobic metabolism replaces aerobic metabolism due to the fabrication of lactic acid; it may happen as oxygen cannot be utilized during this process. Some other disease circumstances may be reflected by lactic acidosis; however, cyanide poisoning could not be specified by lactic acidosis.
3. Cyanogen causes larger oxygen content in venous blood than normal oxygen. It also causes a disaster of cells to abstract oxygen from the arterial blood through the poisoning of the intramitochondrial respiratory sequence.

10.11 Treatment

Before knowing chemical methods, people of villages of southern Tanzania used to expose cassava to sunlight for the reduction of cyanogen content from it. Boron nitride nanosheet (doped also) acts as a suitable adsorbent for cyanogen and hence can be used for removal of cyanogen gas from the environment (Noei & Arjmand, 2015).

Biological methods to remove cyanide from wastewater are becoming popular these days as they are more environment friendly and advantageous than other chemical methods. The microorganisms chosen for investigation for cyanide removal must be able to survive in wastewater containing heavy metals and other toxic chemical compounds and also synthesize enzymes for cyanide removal. All over the world, up to 20% of cyanogen wastes originate from electroplating industries. Group of researchers have made an attempt to degrade cyanogen content from wastewater originated from metal plating, electroplating, and goldmining industries by using *Klebsiella pneumoniae* and *Klebsiella oxytoca* species (Avcioglu & Bilkay, 2019).

10.12 Risk management

The EPA staff consider that the recommended controls would deal with some of the recognized dangers to people and nature. Notwithstanding, it is suggested that the following prerequisites or controls included in a Safe Work Instrument or under Section 77 and Section 77A of the HSNO Act are suggested to sufficiently deal with the risks to human wellbeing and the Earth (Environmental Protection Authority (EPA), 2018):

- PPE ought to be worn by laborers when they are within 10 m of a single log stack or within 20 m of a process in which various log stacks are being fumigated.
- The concentration of cyanogen under the tarpaulin ought to be $\leq$700 ppm prior to the expulsion of the tarpaulin from a fumigated log heap.
- Hydrogen cyanide: 1% *v*/v (maximum) is the limit set for toxicologically relevant important impurities in the active ingredient cyanogen used to fabricate this substance.
- Maximum application rate for this substance is 150 g/m^3.
- Cyanogen must not be applied into or onto the water.
- Use and a label control stating, "Atmospheric conditions should be monitored and should not cyanogen be vented under very low wind speed conditions (less than 5 km/h) or reversal conditions".
- Fumigations directed at port areas must be attempted distinctly at areas where waterbird colonies are not known to exist.
- A permission control, which would enable the EPA to demand a site-specific risk assessment from the clients, including the area of, and species present of, close by bird colonies and their distance to the treatment site, and how potential risk will be overseen.

10.13 Current research

The reaction of cyanogen with metals to make co-ordination complexes are drawing the attention of scientists these days. Recently, Fang et al prepared cerium and neodymium isocyanides by the reaction of cyanogen with Ce and Nd atoms in argon matrices (Fang et al., 2019). Similarly, the preparation of cyanogen complexes of metals has been reported by Chen et al with Infrared spectroscopic and theoretical study on reaction products of cyanogen and metals (Pt, Pd, and Ir). They prepared complexes [Pt(NCCN)], [Pd(NCCN)], and [Ir(NCCN)] by the spontaneous reaction between metals in excess Ar at 4 K which decompose on UV-illumination with no proof of the formation of either isocyanide or cyanide that is common for early transition, lanthanide, and actinide metals (Chen, Xiong, Andrews, & Gong, 2020).

Nag et al. probed the short-lived anion states (resonances) in cyanogen. They utilized electron energy loss spectroscopy to uncover the resonance positions by observing excitation functions for vibrationally inelastic electron scattering and four resonances were observed in spectra at 7.3 eV, 5.3 eV, 4.1 eV, and 0.36 eV (Nag et al., 2020).

10.14 Safety measures

- Splash-proof chemical goggles, plastic face shields should be used when handling large amount of cyanogen or when the potential for splashing is greater.
- When working in the laboratory with a small amount of cyanogen, reusable nitrile lab gloves and double gloving should be used. Examine the gloves repeatedly. Whenever gloves are punctured, damaged, or contaminated, change it frequently and immediately. Teflon gloves should be used for the handling of a larger amount of material.
- Once gloves are removed, instantly wash the hands.
- If the regular laboratory coat has not been polluted with cyanogen, then it could be used again before laundering.
- If functioning through greater quantities of cyanogen solution or there is a greater potential for the splashing throughout the process, then a chemical-resistant appeal with committed jackets should be worn.
- In the laboratory closed-toed, leather shoes should be used, and otherwise, boots made by polyvinyl alcohol may damage.

10.15 Handling and storage

- Cyanogen releases highly toxic, flammable, corrosive, and explosive gases after reacting violently with strong acids such as sulfuric acid, nitric acid, and hydrochloric acid. Thus, it should be preserved away from flame, sparks, hot surfaces, or heat.
- Cyanogen can decompose and produce highly toxic gases such as hydrogen cyanide and hydrogen bromide if it is exposed to heat, water, or moist air. It can polymerize aggressively on persistent storage at the ambient temperature. Cyanogen should be kept in tightly sealed containers in a refrigerator and stored under the nitrogen environment in a secondary container.
- After working with cyanogen, hands should be washed carefully.

- All equipment used when handling the product must be grounded. Fully encapsulating, vapor-protective clothing should be worn for spills and leaks with no fire.
- Use water spray to reduce vapors or divert vapor cloud drift. Avoid water runoff to contact spilled material.
- While using cyanogen always wear double nitrile gloves, a lab coat, and safety glasses, and always use it in a chemical fume hood.

10.16 Precautions

- Cyanogen containers should be preserved away from flame and its sources.
- Cyanogen should always be used in a well-ventilated region or outside regions.
- During the handling of cyanogen, face and eye protection should be taken care of.
- Avoid smoking, drinking, and eating at the time of using cyanogen.
- Avoid cyanogen dispersal into the atmosphere.
- During the usage of cyanogen, wear full protective fire-fighting apparel having apparatus with the autonomous breath.
- Use safety spectacles with face protection and take an emergency eye shower at the workplace of cyanogen.

References

Avcioglu, N. H., & Bilkay, I. S. (2019). Cyanide removal in electroplating, metal plating and gold mining industries' wastewaters by using Klebsiella pneumoniae and Klebsiella oxytoca species. *European Journal of Biology, 78*(1), 5–10.

Bircumshaw, L. L., Tayler, F. M., & Whiffen, D. H. (1954). Paracyanogen: Its formation and properties. Part I. *Journal of the Chemical Society*, 931–935.

Breneman, A. A. (1889). The fixation of atmospheric, issue 2. *Journal of the American Chemical Society, 11*(1), 2–27.

Brotherton, T. K., & Lynn, J. W. (1959). The synthesis and chemistry of cyanogen. *Chemical Reviews, 59*(5), 841–883. https://doi.org/10.1021/cr50029a003.

Chen, X., Xiong, Z., Andrews, L., & Gong, Y. (2020). End-on cyanogen complexes of iridium, palladium, and platinum. *Inorganic Chemistry, 59*(9), 6489–6495. https://doi.org/10.1021/acs.inorgchem.0c00582.

Conway, J. B., Wilson, R. H., Jr., & Grosse, A. V. (1953). The temperature of the cyanogen-oxygen flame. *Journal of the American Chemical Society, 75*(2), 499.

Environmental Protection Authority (EPA) (2018). *Science memo for application to import or manufacture EDN for release (APP202804)*: (pp. 1–121). .

Fang, Z., Vasiliu, M., Chen, X., Gong, Y., Andrews, L., & Dixon, D. A. (2019). Formation of cerium and neodymium isocyanides in the reactions of cyanogen with Ce and Nd atoms in argon matrices. *The Journal of Physical Chemistry A, 123*(38), 8208–8219.

Gay-Lussac, J. L. (1815). Recherches sur l'acide prussique. *Annales de Chimie, 95*, 136–231.

Greenwood, N. N., & Earnshaw, A. (1997). *Chemistry of the elements*: (pp. 320–321) (2nd ed.). Elsevier.

Hartman, W. W., & Dreger, E. E. (1931). Working with hazardous chemical. *Organic Syntheses 11*(30)https://doi.org/10.15227/orgsyn.011.0030.

Kopras, E. J. (2012). Cyanides and nitriles. In E. Bingham, & B. Cohrssen (Eds.), *Patty's toxicology* (pp. 1–52). John Wiley & Sons, Inc. https://doi.org/10.1002/0471435139.tox061.pub2.

Lide, D. R. (2007). *CRC handbook of chemistry and physics: A ready-reference book of chemical and physical data.* CRC Press.

Maya, L. (1993). Paracyanogen reexamined. *Journal of Polymer Science Part A: Polymer Chemistry, 31*(10), 2595–2600.

Nag, P., Čurík, R., Tarana, M., Polášek, M., Ehara, M., Sommerfeld, T. , et al. (2020). Resonant states in cyanogen NCCN. *Physical Chemistry Chemical Physics, 22*(40), 23141–23147. https://doi.org/10.1039/D0CP03333B.

National Research Council (NRC) (2002). Nineteenth interim report of the committee on acute exposure guideline levels: Part A. *Acute Exposure Guideline Levels for Selected Airborne Chemicals* (pp. 211–276). *17*(pp. 211–276). National Research Council (NRC).

Noei, M., & Arjmand, M. (2015). Removal of cyanogen toxic gas from environmental systems by using BN Nanosheet. *Indian Journal of Fundamental and Applied Life Sciences, 5*(S1), 5074–5080.

O'Neil, M. J. (2006). *The Merck index an encyclopedia of chemicals, drugs, and biologicals*: (p. 450). Whitehouse Station, NJ: Merck and Co., Inc.

Quincy, M. A. (1991). National fire protection guide. In *Fire protection guide on hazardous materials*.

Rieders, F. (1965). Noxious gases and vapors 1 carbon monoxide, cyanides, methemoglobin and sulfhemoglobin. In *Drill's pharmacology in medicine* (p. 932). New York: McGraw-Hill.

Ringer, A. L., Sherrill, C. D., King, R. A., & Crawford, T. D. (2008). Low lying singlet excited states of isocyanogen. *International Journal of Quantum Chemistry, 108*(6), 1137–1140. https://doi.org/10.1002/qua.21586.

Teles, F. F. F. (2002). Chronic poisoning by hydrogen cyanide in cassava and its prevention in Africa and Latin America. *Food and Nutrition Bulletin, 23*(4), 407–412.

Thomas, N., Gaydon, A. G., & Brewer, L. (1952). Cyanogen flames and the dissociation energy of N_2. *The Journal of Chemical Physics, 20*(3), 369–374.

United States Coast Guard (1999). *Chemical hazard response information system (CHRIS). In Hazardous Chemical Data. Commandant Instruction 16465. 12C.* Washington, D.C., US: Government Printing Office(1999). *https://www.hsdl.org/?view&did=24079.* Accessed 11 September 2020.

Weast, R. C. (Ed.). (1988). *Handbook of chemistry and physics.* In Vol. 68. Boca Raton, FL: CRC Press.

The photographic news. William, C. (Ed.). (1859). *A weekly record of the progress of photography*In Vol. 1, .

CHAPTER

11

Cyanogen chloride: Risk assessment, environmental, and health hazard

Rajeev Jain[a] *and Sonu Kumar*[b]

[a]Pondicherry University, Puducherry, India

[b]Department of Chemistry, Gurukula Kangri (Deemed to be University), Haridwar, Uttarakhand, India

Abbreviations

CK cyanogen chloride
DBP disinfectant byproduct
MOF metal organic framework
NIOSH National Institute for Occupational Safety and Health
NOAEL no-observed-adverse-effect level
OSHA Occupational Safety and Health Administration
TDI tolerable daily intake
TEDA triethylenediamine

11.1 Introduction

Cyanogen chloride (CK) is a chemical compound with the formula CNCl (Lide, 2007). It is a colorless gas or liquid (National Institute for Occupational Safety and Health (NIOSH), 2010) with a linear triatomic pseudohalogen that can be easily condensed into a liquid. It is also a highly toxic lachrymatory chemical that finds its use as tear gas. It boils at about 13°C, freezes at −6°C with a vapor density of 2.14. If impure, it readily polymerizes to cyanuric chloride but is comparatively stable in pure form. It has a pungent peppery smell, with a background of bitter almonds, and is extremely irritating to breathe. It is soluble in most organic solvents and up to 8% in water at 8°C, in which it is fairly stable at neutral or acidic reactions at room temperature (Aldridge, 1951). CK

Hazardous Gases
https://doi.org/10.1016/B978-0-323-89857-7.00024-4

and HCN are also used as blood agents, which inhibit the cytochromes oxidase and other enzymes, interrupting cellular respiration. The resulting metabolic acidosis and tissue hypoxia lead to convulsions and cardiorespiratory arrest. CK is utilized as an antecedent to the sulfonyl cyanides and chlorosulfonyl isocyanate, a helpful reagent in organic synthesis.

11.2 Occurrence

CK is a toxic disinfectant byproduct (DBP) found in chlorinated and chloraminated drinking water as well as in wastewater. The formation of CK in natural waters and the corresponding role of water quality parameters are not well understood. Model solution experiments, for example, have shown that CK is produced after the chlorination and chloramination of amino acids and formaldehyde.

CK is generated by the oxidation of sodium cyanide with chlorine (Eq. 11.1).

$$NaCN + Cl_2 \rightarrow CNCl + NaCl \tag{11.1}$$

CK also forms when monochloramine and formaldehyde react at concentrations representative of drinking water treatment (Pedersen, Urbansky, Marinas, & Margerum, 1999).

11.3 Historical background

Cyanide poisoning was first reported with the effects of an extract of bitter almonds; then cyanide was identified and isolated from cherry laurel. CK was first prepared in 1787 by the action of chlorine upon hydrocyanic acid (prussic acid) and was called "oxidized prussic acid". The correct formula for CK was first established in 1815. CK was also used in World War I in 1916 (Abdollahi & Hosseini, 2014).

By 1945, the U.S. Armed force's Chemical Warfare Service created chemical warfare fighting rockets planned for the new M9 and M9A1 Bazookas. An M26 gas rocket was adjusted to fire CK-filled warheads from these rocket launchers. As it was fit for infiltrating the defensive channel obstructions in certain gas masks, which can be proved to be an effective agent against Japanese powers (especially against those who were hiding in bunkers or caves) (Smart, 1997). Since their standard issues, gas protections did not have the boundaries that would give security against CK the US added the weapon to its munitions stockpile; however, the CNCl rocket was never sent or given to the battle workforce (Smart, 1997).

11.4 Physiochemical properties

CK is delivered as a liquid restricted under its vapor pressure. It is a profoundly harmful lachrymator with fumes heavier than air. Prolonged exposure of the repository to fire or extreme warmth may cause brutal cracking and soaring. It exists merely as a gas in air, confirmed based on vapor pressure 1220 mmHg at 25°C during the introduction to the environment (Daubert & Danner, 1989). In an alkaline solution, it is rapidly hydrolyzed to cyanic acid and HCI, as shown in Eq. (11.2) (Aldridge, 1951). See Table 11.1 for more details.

$$CNCl + H_2O \rightarrow HCNO + HCl \tag{11.2}$$

Furthermore, this slowly converts to CO_2 and ammonium chloride (Eq. 11.3):

$$HCNO + HCl + H_2O \rightarrow NH_4Cl + CO_2 \tag{11.3}$$

Cyanate and chloride ions are released by gradual hydrolysis of CK by water (Eq. 11.4).

$$CNCl + H_2O \rightarrow NCO^- + Cl^- + 2H^+ \tag{11.4}$$

TABLE 11.1 Physical properties of cyanogen chloride.

Property	Value
Color	Colorless volatile liquid or gas (Haynes, 2010)
Odor	Acrid, choking odor (United States Coast Guard (USCG), 1984); Lachrimating and irritating odor (Lewis, 2001); Strong acid/pungent odor
Boiling point	55.6°F at 760 mmHg
Melting point	20°F or −6.55°C (Haynes, 2010)
Solubility	Soluble in water, ethanol, ethyl ether (Haynes, 2010); Soluble in oxygenated solvents (Ashford, 1994)
Density	1.222 at 32°F
Vapor density	2.1 (Relative to Air)
Vapor pressure	704.36 mmHg at 50°F
Stability	Tends to form polymers on storage.

11.5 Physiological role

CK reacts with hemoglobin and glutathione eventually liberating the CN ion. Cyanide is distributed to many organs and the blood. At lethal or nearly lethal doses of cyanide, relatively high concentrations were found in the liver, lungs, kidneys, brain, and blood. Various biotransformation pathways have been identified for cyanide. The most important way is the formation of thiocyanate by the acceptance of a sulphane-sulfur of thiosulphate or other sulphane-sulfur containing compounds (trans-sulphurization), the key enzyme being rhodanese.

The rate-limiting factor of this pathway is the lack of sulphane-sulfur sources in the body. Cyanide is largely eliminated from the body via the urinary excretion of thiocyanates in the case of high exposure levels. Other minor elimination routes include the exhalation of carbon dioxide and traces of hydrogen cyanide. For more detailed pathways refer to Chapter 15. The relative importance of various biotransformation and elimination routes are unknown for lower, clearly sublethal exposure levels (Health Council, 2002).

CK was stated to be detoxified at rates of 0.03–0.06 and 0.02–0.04 mg/kg/min by rabbits and dogs, respectively, depending on the injection rate. For man, the detoxification rate has been recorded between 0.02 and 0.1 mg/kg/min (Moore & Gates, 1946).

11.6 Uses

Because of the warning characteristics of CK, formerly it was much used as a pesticide. Nowadays it has been used in organic synthesis, such as the preparation of methane tetracarbonitrile and tetracyanomethane by heating silver tricyanomethanide in liquid CK. It is also used in the Lonza process in the preparation of extremely pure malononitrile. It is used as a metal cleaner (in ore refining or production of synthetic rubber), as well as for electroplating and photography. CK is widely used in industry in fumigating ships and warehouses and in ore-extracting processes. It is used as a warning agent in fumigant gases and pesticides (American Conference of Governmental Industrial Hygienists (ACGIH), 1991; U.S. National Library of Medicine (NLM), 2003).

11.7 Sources of contaminants

CK is a hazardous compound and can be generated in significant concentration in the environment by chloramination of nitrogen-containing organic compounds such as amino acids and some other compounds such as L-arginine, L-histidine, L-serine, glycine, catechin, humic acid, uric acid, benzene, etc., causing an adverse effect in living beings (Hirose, Maeda, Ohya, Nojima, & Kanno, 1988; Li & Blatchley, 2007; Zheng, Dzombak, & Luthy, 2004). CK formed by photolysis of chlorinated compounds with guanidine structure such as acetamide, acetamidine, guanidine, guanidine acetic acid, and compounds with imidazole compounds such as 1-methylimidazole, imidazole, 4-methylimidazole were investigated by Weng and Blatchley and it was concluded that this gas results in UV irradiation of chlorinated L-arginine and L-histidine (Weng & Blatchley, 2013). But the generation of CK by this is relatively less than that from L-glycine, an amino acid which is known as an efficient precursor of CK (generation of CK by chlorination of L-glycine) (Lee, Na, Ramirez, & Olson, 2006). Human body fluids such as sweat and urine contain L-arginine and L-histidine along with glycine in higher concentrations in urine (Wentworth, 1982). The reaction between free chlorine and glycine is quick, the chlorinated intermediate (chloroglycine) quickly change over to CK owing to the availability of free chlorine in the framework. However, with the aid of UV light, the improvement of CK formation was seen without the addition of free chlorine. Acceleration of N–Cl bond cleavage is responsible for the improvement of CK generation. UV light (254 nm wavelength) irradiation improves CNCl development from glycine by promoting the rate-limiting process (dechlorination) exhibited that arrangement of CK and NCl_3 from chlorination of uric acid is supported at pH 6.0 as compared to higher pH values (Lian, E, Li, & Blatchley, 2014).

Low concentration of cyanide in crude waters changed over to CK by chlorination and may likewise be generated during the production of chloramines in situ as a leftover disinfectant to keep up the sterile state of the distribution system. An occupational introduction to CK may happen through inward breath and dermal contact with this compound at working environments where CK is created or utilized. Observing information demonstrates that most of the population is expected to be exposed to CK through the uptake of drinking water. Exposure may happen to people interacting with the compound in areas where it is utilized as a chemical warfare agent.

11.8 Risk assessment

CK has the same general type of toxicity and mode of action as hydrogen cyanide (refer Chapter 15), but it is considered to be much more irritating to the respiratory tract and the eyes, even at very low concentrations. According to data, the lowest concentration inducing irritation in humans was found to be 1 ppm (2.6 mg/m^3) at a 10-min exposure period. Other symptoms observed in humans exposed to unknown concentrations include muscular weakness, lung congestion, skin irritation, hoarseness, conjunctivitis, edema of the eyelids, cold sensation, weight loss, chronic headache, and burning urine (Greim, 1997). In addition, after long-term exposure to CK severe symptoms could appear such as coma, seizures, and dilated pupils (mydriasis). Some cardiovascular effects include shock, abnormal, or disordered heart rhythms (dysrhythmias), critically low blood pressure, and cardiac arrest; respiratory effects: abnormally rapid, followed by abnormally slow respirations; accumulation of fluid in the lungs (pulmonary edema); and respiratory arrest; and eye effects: dilated pupils, inflammation of the surface of the eye, and temporary blindness (National Institute for Occupational Safety and Health (NIOSH), 2010).

CK produces systematic possessions in the whole body at the cellular level through interfering in oxygen utilization with a profound central nervous system, respirational, and cardiovascular effects. Cyanosis is the main characteristic of the diminished oxygen supply in the tissues that is confirmed by the bluish discoloration of the skin. Cyanosis generally could appear during the course of poisoning if circulatory failure and breathing interruption perceive. Severe spasmodic blinking and tears in the eyes mainly occur due to the strong irritation of CK (National Institute for Occupational Safety and Health (NIOSH), 2010).

11.9 Environmental effects

CK generation and use in different processes lead to its delivery to the environment through waste streams. CK is not prone to photolysis by sunlight as it does not contain chromophores that absorb at wavelength >290nm (Lyman, Reehl, & Rosenblatt, 1990). The ecological destiny of CK is generally site-explicit in the biotic and abiotic qualities of soil and water enormously impacting the capacity of this compound to absorb, persist, dissociate, degrade, and complex (Boening & Chew, 1999). CK is incredibly unstable and hydrolyzes quickly in water with a half-life of 1.45 min to 10 h at 5°C in ambient water (Boening & Chew, 1999), hence considered as a nonpersistent agent (Munro et al., 1999).

11.10 Hazardousness

According to the US Department of Human Services CNCl is a colorless gas with a sharp and penetrating smell and its exposure limit has been set to 0.3 ppm as chloramines (National Institute for Occupational Safety and Health (NIOSH), 2007).

Various symptoms that could be accompanied by a low dose of CK introduction are as follows: (1) Bronchorrhea, (2) Lacrimation, and (3) Rhinorrhea. (Abdollahi & Hosseini, 2014).

A moderate dose of CK exposure could be accompanied by the following symptoms:

(i) Feeling of anxiety and apprehension
(ii) Onset possibly takes several minutes, except for immediate irritant effects
(iii) Nausea and vomiting
(iv) Vertigo
(v) Seizures
(vi) Prolonged prodrome before the loss of consciousness
(vii) Cardiac arrest
(viii) Bradypnea followed by apnea

The symptoms that could be accompanied by a low dose of CK introduction are given as follows:

(i) Transient hyperpnea
(ii) Seizures 15–30 s later
(iii) Apnea 2–3 min later
(iv) Cardiac arrest 6–8 min post introduction

Although patients may suffer from any of the above symptoms, physical findings are generally nonspecific and similar to those of severe hypoxemia. The respirational distress, apnea, and convulsions are due to the severe continued introduction of CK. Due to the induced level of cyanide or associated carbon monoxide poisoning, a lack of oxygen extraction at tissue level has been observed that consequently shows pink or cherry red skin of the exposed personnel.

11.11 Treatment

However, there is limited data accessible on the removal of CK but processes involving chloramination result in its generation. So, it is necessary to apply a treatment to keep down the formation of CK while keeping up sufficient chloramine residuals where chloramination is practiced. Notwithstanding, CK is known to be unstable in chlorinated water (Na & Olson, 2004) and can be removed by chemical reducing agents, for example, sodium sulfite, sodium thiosulfate, and sodium disulfite.

The removal of CK from air streams can be achieved by use of carbon impregnant (such as TEDA and basic Cu^{+2} individually or combinedly) in aqueous and non-aqueous (acetonitrile) media wherein humid condition TEDA act as OH^- source and acid-base reaction occur and result into unstable cyanic acid which further hydrolyzed to CO_2 and NH_3 or NH_4^+ with other unanticipated products of side reactions. Whereas, in absence of water, CK may substitute TEDA directly resulting in various monomeric, dimeric, and trimeric products. The combination of TEDA and Cu^+ removes CK more effectively in aqueous media due to the formation of the soluble complex $[Cu(TEDA)_2(OH)(H_2O)]^-$ in the carbon-adsorbed water layer and show synergism (Mahle et al., 2010). CK can also be removed more efficiently by using $Zr(OH)_4$ impregnated with TEDA (Peterson, Wagner, Keller, & Rossin, 2010). TEDA-modified Zr-based MOF (T-MOF-808) can also be used for the effective removal of CK (Cha et al., 2020).

11.12 Risk management

Management of oral exposure is symptomatic and supportive and should not induce vomiting but should immediately administer 100% oxygen. In the case of ingestion of large amounts, the activated charcoal can be considered. In the eye and dermal exposure, remove the patient from the source of exposure and then wash the eye with large amounts of tepid water for at least 15 min. In skin exposure, remove the clothes and wash the skin with water. In inhalation exposure, immediately remove the patient from the source of exposure and administer 100% oxygen. Airway management is needed in patients who are comatose or have altered mental status. In each of the types of exposure, a cyanide antidote, either hydroxocobalamin or the sodium nitrite/sodium thiosulfate kit, should be administered. The cyanide antidote kit (nitrites and thiosulfate) in clinical use consists of amyl nitrite, sodium nitrite, and sodium thiosulfate. Enhanced elimination can be performed with antidotes; however, the role of hemodialysis is uncertain (Abdollahi & Hosseini, 2014).

11.13 Current research

There has been tremendous research on CK since its discovery but recent studies molded the motive, detection methods, and application strategies, etc. Recently, theories of detection of CK by changing the semiconductor's electrical conductivity is proposed with the ways for diminishing troublesome effects. Nanosensing of organic and inorganic materials is becoming an interesting and potential direction for researchers nowadays. Since the last few decades, after the revelation of tremendous properties of graphene and other carbonaceous materials such as fullerene, nanotubes, etc., different detection and sensing techniques using graphene also came into light in which, some are strongly effective and appropriate. Nano sensing of CK by using doped graphene was studied by Morteza Rouhani using Density functional theory (DFT), where doping of B, Al, Si, Ga on graphene was done to increase reactivity (Rouhani, 2019). DFT is also used by Movlarooy and Fadradi explaining the adsorption of CK gas on the surface of Al- and Ga-doped boron nitride nanotube surface (Movlarooy & Fadradi, 2018). C_{20} fullerene has been used as a sensor to detect CK more efficiently than graphene. According to the studies, the adsorption process and chemical reactions are possible in this case, where CK molecules first get adsorbed on the fullerene and then these two go through a [2+2] cycloaddition process, a reversible chemical reaction (Pakravan & Siadati, 2017). Phosphorene nanosheets have also been used as a sensing medium for CNCl molecules these days (Jyothi, Nagarajan, & Chandiramouli, 2020). Blue phosphorene nanosheet has been used for fast probing of CK along with improvements with the substitution of Al (Nagarajan & Chandiramouli, 2019).

11.14 Cyanogen chloride toxicity

A recent case study presented by Graeme W. Norval shows the chronic impact on the lungs of laboratory workers on exposure to chloramines and CNCl fumes. The incident occurred in a biological research laboratory where the researcher needed to use the laboratory fume hood for an entire week for a specific task but he started to suffer an upper respiratory tract injury in the form of vocal cord dysfunction due to exposure to the resultant chloramine and CK gases (Norval, 2020).

In the First World War, CNCl was utilized as a war gas. The concentration of 2.5 mg/m^3 caused irritation and a concentration of 120 mg/m^3 was found deadly to human beings (NAS, 1977). As, it tends to convert to cyanide in the body readily. However, there isn't a lot of information available on the oral poisonousness of CNCl. Hence, the guidelines are based on the toxicology of cyanide (WHO, 2007). Research showed that CNCl metabolizes to CN^- ion by glutathione and hemoglobin in an in-vitro investigation with rodent blood (Aldridge, 1951).

In lethality tests, a concentration of 100 mg/m^3 was reported to be deadly to cats for 18 min, 120 mg/m^3 for 6 h deadly to dogs, 5 mg/m^3 for 2 min to goats, and a subcutaneous dose of 20 mg/kg of body weight to rabbits (Flury & Zernik, 1951).

The NOAEL in a later rodent subchronic study in which animals were exposed through the drinking-water was reported to be 5.4 mg/kg of body weight every day for minor changes in the testis (National Toxicology Program (NTP), 1993). This examination was chosen as the basic investigation to base the guideline's value. Taking into account the minor nature of the changes observed and the NOAEL in a past ongoing examination, it isn't considered as important to incorporate an additional uncertainty factor to consider the length of the investigation. The use of an uncertainty factor of 100 to the NOAEL of 5.4 mg/kg of body weight gives a tolerable daily intake (TDI) of 0.054 mg/kg of body weight. Considering 60-kg grown-up drinking 2 l of water every day and permitting 20% of the TDI to originate from water because of the potential for exposure to cyanogenic glycosides in food (WHO, 2007), the guideline value for long-term exposure was recorded as 0.3 mg/l (rounded value).

11.15 Safety measures

It is necessary for employees to be trained for each hazard and when and how to use protective equipment.

11.15.1 Clothing

- Always wear protective clothes and gloves and elude skin contact with CK. The manufacturers or suppliers of safety equipment should deliver endorsements on the most protective clothing material and gloves for the process.
- Each day, the available protecting clothing such as gloves, footwear, suits, etc. should be cleaned and put on before work.
- The manufacturers or suppliers of safety equipment should use chlorinated polyethylene as a protecting material.

11.15.2 Eye protection

- During working with CK vapors or fumes, always use non-vented, impact-resistant goggles.
- If working with liquid CK, one should use impact and splash-resistant goggles.
- Always wear a face shield alongside goggles at the time of working with CK.
- When working with CK contact lenses should be avoided.

11.15.3 Respiratory protection

- Improper use of respirators is dangerous. Such equipment should only be used if the employer has a written program that takes account of workplace conditions, requirements for worker training, respirator fit testing, and medical exams, as described in OSHA 1910.134.
- Where the potential exists for exposure over 0.3 ppm, use a NIOSH-approved supplied-air respirator with a full facepiece operated in a

pressure-demand or other positive-pressure mode. For increased protection use an auxiliary self-contained breathing apparatus operated in a pressure-demand or other positive-pressure mode.

11.16 Handling and storage

Fine and systematic handling and storage of CK is needed due to its reactive and toxic nature (Hazardous Substance Fact Sheet, 2004). These are some of the ways to manage CK carefully-

- Before working with CK, one should be trained on its proper handling and storage.
- CK reacts slowly with water to form hydrogen cyanide and heat.
- CK is not compatible with strong acids (such as hydrochloric, sulfuric, and nitric); carbon dioxide; nitrate salts; alkalis; ammonia; and alcohol.
- Store in tightly closed containers in a cool, well-ventilated area away from copper, brass, and moisture.
- Use a stabilizer to prevent polymerization.

11.17 Precautions

- CK should be preserved away from fires, sparks, and other causes of ignition.
- Avoid breathing CK vapors.
- Avoid handling a broken package of CK without particular protecting equipment.
- Wash skin instantly in the case of contamination with liquid CK.
- Significantly wet clothes with liquid CK should be removed instantly.
- During working with CK one should not drink, eat, or smoke.
- Contaminated protecting clothing by CK should be separated in such a way that there is no direct individual contact of the body of the operator.

References

Abdollahi, M., & Hosseini, A. (2014). Cyanogen chloride. In P. Wexler (Ed.), *Encyclopedia of toxicology* (pp. 1096–1099). (3rd ed.). Elsevier Inc., Academic Press. https://doi.org/10.1016/B978-0-12-386454-3.00484-X.

Aldridge, W. N. (1951). The conversion of cyanogen chloride to cyanide in the presence of blood proteins and sulphydryl compounds. *Biochemical Journal, 48*(3), 271–276.

American Conference of Governmental Industrial Hygienists (ACGIH) (1991). *Documentation of the threshold limit values and biological exposure indices.* .

Ashford, R. D. (1994). *Ashford's dictionary of industrial chemicals*: (p. 246). London, England: Wavelength Publications Ltd..

Boening, D. W., & Chew, C. M. (1999). A critical review: General toxicity and environmental fate of three aqueous cyanide ions and associated ligands. *Water, Air, and Soil Pollution, 109* (1–4), 67–79.

Cha, G. Y., Chun, H., Hong, D. Y., Kim, J., Cho, K. H., Lee, U. H. , et al. (2020). Unique design of superior metal-organic framework for removal of toxic chemicals in humid environment via direct functionalization of the metal nodes. *Journal of Hazardous Materials, 398,* 122857.

Daubert, T. E., & Danner, R. P. (1989). *Physical and thermodynamic properties of pure chemicals: A data compilation.* Washington, DC: Taylor & Francis.

Flury, F., & Zernik, F. (1951). HARMFUL GASES, Berlin, 1931, quoted by S. Wilska. *Acta Chemica Scandinavica, 5,* 1359.

Chlorcyan. Greim, H. (Ed.). (1997). *Gesundheidsschädliche Arbeitsstoffe. Toxikologischarbeitsmedizinische Begründungen von MAK-werten (Maximale Arbeitsplatzkonzentrationen).* (25st ed.). Weinheim, FRG: VCH Verlagsgesellschaft mbH.

Haynes, W. M. (Ed.). (2010). *CRC Handbook of Chemistry and Physics* (pp. 4–56). (91st ed.). Boca Raton, FL: CRC Press Inc..

Hazardous Substance Fact Sheet (2004). *CAS NO-506-77-4.* New Jersey Department of Health and Senior Services.

Health Council (2002). *Hydrogen cyanide, sodium cyanide, and potassium cyanide. Health-based recommended occupational exposure limits.* The Hague, the Netherlands: Health Council of the Netherlands.

Hirose, Y., Maeda, N., Ohya, T., Nojima, K., & Kanno, S. (1988). Formation of cyanogen chloride by the reaction of amino acids with hypochlorous acid in the presence of ammonium ion. *Chemosphere, 17*(5), 865–873.

Jyothi, M. S., Nagarajan, V., & Chandiramouli, R. (2020). Investigation on adsorption properties of HCN and ClCN blood agents on θ–phosphorene nanosheets—A first–principles insight. *Chemical Physics, 538,* 110896.

Lee, J. H., Na, C., Ramirez, R. L., & Olson, T. M. (2006). Cyanogen chloride precursor analysis in chlorinated river water. *Environmental Science & Technology, 40*(5), 1478–1484.

Lewis, R. J. (2001). Sax's dangerous properties of industrial materials. 10th Edition. Vol. 1–3, *Journal of the American Chemical Society,* 123(23), 5618. https://doi.org/10.1021/ja015211y.

Li, J., & Blatchley, E. R. (2007). Volatile disinfection byproduct formation resulting from chlorination of organic – nitrogen precursors in swimming pools. *Environmental Science and Technology, 41*(19), 6732–6739.

Lian, L., E, Y., Li, J., & Blatchley, E. R. (2014). Volatile disinfection byproducts resulting from chlorination of uric acid: Implications for swimming pools. *Environmental Science and Technology, 48*(6), 3210–3217. https://doi.org/10.1021/es405402r.

Lide, D. R. (2007). CRC handbook of chemistry and physics–87th edition. *Journal of the American Chemical Society, 129*(3), 724. https://doi.org/10.1021/ja069813z.

Lyman, W. J., Reehl, W. F., & Rosenblatt, D. H. (1990). *Handbook of chemical property estimation methods: Environmental behavior of organic compounds.* United States: N.p. Web.

Mahle, J. J., Peterson, G. W., Schindler, B. J., Smith, P. B., Rossin, J. A., & Wagner, G. W. (2010). Role of TEDA as an activated carbon impregnant for the removal of cyanogen chloride from air streams: Synergistic effect with Cu (II). *The Journal of Physical Chemistry C, 114* (47), 20083–20090.

Moore, S., & Gates, M. (1946). Hydrogen cyanide and cyanogen chloride. *Summary technical report of Division 9: Chemical warfare agents and related chemical problems* (pp. 7–16). Vol. 1 (pp. 7–16). Washington, DC: National Defense Research Council.

Movlarooy, T., & Fadradi, M. A. (2018). Adsorption of cyanogen chloride on the surface of boron nitride nanotubes for CNCl sensing. *Chemical Physics Letters, 700,* 7–14.

Munro, N. B., Talmage, S. S., Griffin, G. D., Waters, L. C., Watson, A. P., King, J. F. , et al. (1999). The sources, fate, and toxicity of chemical warfare agent degradation products. *Environmental Health Perspectives, 107*(12), 933–974.

Na, C., & Olson, T. M. (2004). Stability of cyanogen chloride in the presence of free chlorine and monochloramine. *Environmental Science and Technology, 38*(22), 6037–6043. https://doi.org/10.1021/es0498235.

Nagarajan, V., & Chandiramouli, R. (2019). Probing cyanogen chloride gas molecules using blue phosphorene nanosheets based on adsorption properties: A first-principles study. *Computational and Theoretical Chemistry, 1150*, 63–70.

NAS (1977). *Drinking water and health.* Washington, DC: National Academy of Sciences.

National Institute for Occupational Safety & Health (NIOSH) (2007). *NIOSH Pocket Guide to Chemical Hazards.* Department of Health and Human Services, US Department of Labour. https://www.cdc.gov/niosh/docs/2005-149/pdfs/2005-149.pdf. Accessed 10 April 2020.

National Institute for Occupational Safety & Health (NIOSH) (2010). *NIOSH pocket guide to chemical hazards.* Department of Health & Human Services, Centers for Disease Control & Prevention. Publication No. 2010-168.

National Toxicology Program (NTP) (1993). *Sodium cyanide administered in drinking water to F344/N rats and B6C3F1 mice. National Toxicology Program (Toxicology Report Series No 37)* Research Triangle Park, NC: National Institutes of Health.

Norval, G. W. (2020). Chronic lung impact on laboratory worker exposed to chloramines and cyanogen chloride. *ACS Chemical Health and Safety, 27*(2), 129–132.

Pakravan, P., & Siadati, S. A. (2017). The possibility of using C20 fullerene and graphene as semiconductor segments for detection, and destruction of cyanogen-chloride chemical agent. *Journal of Molecular Graphics and Modelling, 75*, 80–84.

Pedersen, E. J., Urbansky, E. T., Marinas, B. J., & Margerum, D. W. (1999). Formation of cyanogen chloride from the reaction of monochloramine with formaldehyde. *Environmental Science and Technology, 33*(23), 4239–4249. https://doi.org/10.1021/es990153q.

Peterson, G. W., Wagner, G. W., Keller, J. H., & Rossin, J. A. (2010). Enhanced cyanogen chloride removal by the reactive zirconium hydroxide substrate. *Industrial and Engineering Chemistry Research, 49*(22), 11182–11187.

Rouhani, M. (2019). DFT study on adsorbing and detecting possibility of cyanogen chloride by pristine, B, Al, Ga, Si and Ge doped graphene. *Journal of Molecular Structure, 1181*, 518–535.

Smart, J. K. (1997). History of chemical and biological warfare: An American perspective. In *Medical Aspects of Chemical and Biological Warfare* (pp. 09–86). Washington, DC: Office of the Surgeon General.

U.S. National Library of Medicine (NLM) (2003). *Cyanogen chloride.* https://pubchem.ncbi.nlm.nih.gov/compound/Cyanogen-chloride. Accessed 10 June 2020.

United States Coast Guard (USCG) (1984). *CHRIS—Hazardous chemical data.* (Vol. II). Washington, D.C: U.S. Government Printing Office.

Weng, S., & Blatchley, E. R. (2013). Ultraviolet-induced effects on chloramine and cyanogen chloride formation from chlorination of amino acids. *Environmental Science and Technology, 47*(9), 4269–4276. https://doi.org/10.1021/es400273w.

Wentworth, J. (1982). In C. Lentner (Ed.), *Geigy scientific tables, vol. 1: Units of measurement, body fluids, composition of the body, nutrition* (p. 295). West Caldwell, NJ: Medical education division, Ciba-Geigy Corp..

WHO (2007). *Cyanide in drinking-water. Background document for development of WHO guidelines for drinking-water quality.* WHO SDE/WSH/0701/2.

Zheng, A., Dzombak, D. A., & Luthy, R. G. (2004). Formation of free cyanide and cyanogen chloride from chloramination of publicly owned treatment works secondary effluent: Laboratory study with model compounds. *Water Environment Research, 76*(2), 113–120.

CHAPTER

12

Diazomethane: Risk assessment, environmental, and health hazard

Sonu Kumar

Department of Chemistry, Gurukula Kangri (Deemed to be University), Haridwar, Uttarakhand, India

Abbreviations

EPA	Environmental Protection Agency
IARC	International Agency for Research on Cancer
NIOSH	National Institute for Occupational Safety and Health
OSHA PEL	Occupational Safety and Health Administration Permissible Exposure Level
TWA	time-weighted average

12.1 Introduction

Diazomethane is an explosive, yellow-colored gas with a musty odor, whose diazo structure (Fig. 12.1) was developed by von Pechmann in 1894. Diazomethane is also called azimethylene or diazirine and is the most versatile reagent in synthetic organic chemistry. It is mainly used for the fabrication of carbon–heteroatom and carbon–carbon bonds. It is a low molecular weight compound, and a reaction using diazomethane shows good atomic efficiency with nitrogen as a sole by-product. However, especially in industry, the difference between minimal application and its convenience is enormous due to its associated hazards. It is an extremely poisonous compound and works similarly to various strong alkylation agents.

The time-weighted average concentration of diazomethane is 0.2ppm, which is suggested by the Occupational Safety and Health Administration Permissible Exposure Level (OSHA PEL). This concentration (0.2ppm) of diazomethane makes exposure monitoring very challenging.

Hazardous Gases
https://doi.org/10.1016/B978-0-323-89857-7.00018-9

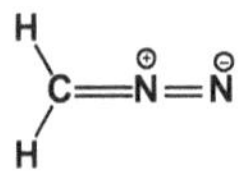

FIG. 12.1 Structure of diazomethane.

Additionally, diazomethane is outstandingly sensitive to explosion through heat, shock, and light in both its gaseous and liquid forms. Consequently, diazomethane is solely used in solution form in diethyl ether or a diluted mixture in inert gas. This makes handling and safe preparation difficult in batch mode and it requires specific precautions, safety measures, and dedicated equipment at the laboratory scale during its handling.

12.2 Occurrence

The decomposition of derivatives of *N*-methyl-*N*-nitrosamine gives diazomethane. Among various techniques for the preparation of diazomethane, the most common and frequently used is the consumption of 1-methyl-3-nitro-nitroguanidine (MNNG, 2), *N*-methyl-*N*-nitroso-*p*-toluene sulfonamide (Diazald, **1**), or *N*-methyl-*N* nitrosourea (**3**) as in Scheme 12.1.

The fabrication of diazomethane through the original technique elaborates the utilization of *N*-methyl-*N*-nitrosourea where analogous techniques are still in practice today.

12.3 Historical background

Diazomethane was discovered by von Pechmann in 1894 and is confirmed as a yellowish gas having different practical applications when experiments were carried out with a minimum amount at ordinary

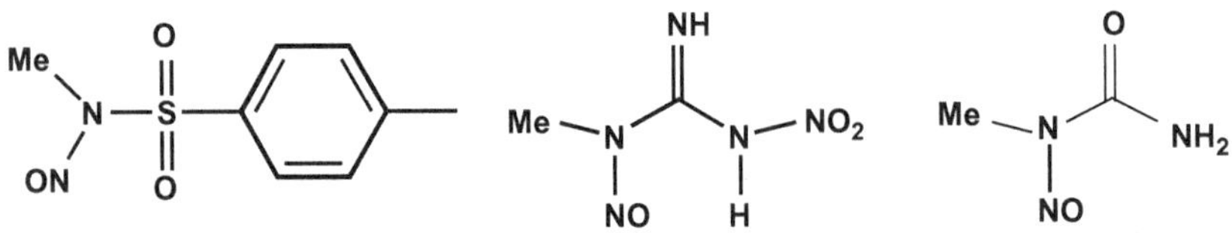

(1) (2) (3)

SCHEME 12.1 Various precursor compounds for the synthesis of diazomethane.

temperature. Its suitability as a reagent for microsynthesis was estimated in 1895 (Roper & Ma, 1957). A batch process for the production of gaseous diazomethane, "A new method for the preparation of diazomethane," was disclosed (De Boer & Backer, 1954) as the homebrew method using *N*-methyl-*N*-nitrosourea as the precursor. The process comprises the introduction of a solution of potassium hydroxide in a mixture of carbitol-water to *p*-tolysulfonylmethylnitrosamide in anisole.

A "large-scale" preparation was disclosed by Acevedo et al. in US Pat. No. 5,459,243, "Apparatus and processes for the large scale generation and transfer of diazomethane," issued October 17, 1995. The reactions disclosed were performed on the 100 millimole scale, which generated a dilute solution of diazomethane in dichloromethane.

12.4 Physicochemical properties

Diazomethane (molecular weight: 42.04 g/mol) (Budavari, 1989) is a poisonous and explosive yellow gas in nature (Hazardous Substances Data Bank, 1993; World Health Organization (WHO), 2001; Budavari, 1989; Clayton & Clayton, 1981). It is a gas with a musty odor and the threshold of the odor has not been established (Hazardous Substances Data Bank, 1993). Nitrosomethylurethane releases diazomethane after it is treated with alkali. Upon heating to 100°C or on the surface of an irregular glass surface, diazomethane tends to explode. It is soluble in dioxane or ethers, and at room temperature diazomethane solution decomposes gradually in alcohol and water. Thus, it is generally used in the form of a solution in the presence of diethyl ether. The physical properties of diazomethane are given in Table 12.1.

12.5 Physiological role

DNA methylation can be sensed (Friedman, Mahapatra, Dash, & Stevenson, 1965; Kriek & Emmelot, 1964) when it is protected through diazomethane in vitro; however, the methyl carbenium ion $[CH_3]^+$ is a very effective alkylating agent. Macromolecules such as proteins and RNA methylation are not excluded. Methylation processes are responsible for the hazardousness effects of diazomethane; such effects are irritative and carcinogenic. It has been suggested that the strong irritation caused by diazomethane results in the intracellular formation of formaldehyde (Flury & Zernik, 2013). Diazomethane gives methanol and nitrogen when it gradually reacts with water (Sunderman, 1970; Sunderman, Connor, & Fields, 1938).

TABLE 12.1 Physical properties of diazomethane.

Property name	Value
Color	Yellow
Odor	Musty odor
Boiling point	−23°C or −9°F at 760 mmHg (National Institute for Occupational Safety and Health (NIOSH), 2016)
Melting point	−145°C or −229°F (National Institute for Occupational Safety and Health (NIOSH), 2016)
Flashpoint	Flammable
Solubility	Reacts with water (National Institute for Occupational Safety and Health (NIOSH), 2016), is sparingly soluble in benzene and alcohol, and strongly soluble in ether and dioxane
Density	1.45 (water = 1)
Vapor density	1.4 (air = 1)
Vapor pressure	760 mmHg (National Institute for Occupational Safety and Health (NIOSH), 2016)
Main hazards	Toxic and explosive
Explosion	100°C
Ionization potential	9.0 eV
Dipole moment	1.50
Standard molar enthalpy	249.9 kJ/mol K (at 298 K)
Heat capacity	52.5 J/mol K (at constant pressure and 298 K)

12.6 Uses

Diazomethane is used in organic synthesis widely. It is an exceptionally versatile C1 building block in organic synthesis and is used in carbene reactions, methylations, homologations, and cycloaddition reactions (Yang, Martin, & Schenkel, 2018). It is used as a methylating agent for acidic compounds such as carboxylic acids, phenols, and enols (Budavari, 1989; Sittig & Marshall, 1985) and to convert carboxylic acids to their methyl esters. The reaction is supposed to progress through proton migration from carboxylic acid to diazomethane to give methyl diazonium cation that quickly gives methyl ester and nitrogen gas after reacting with the carboxylate ion. Diazomethane and its homologs are used in

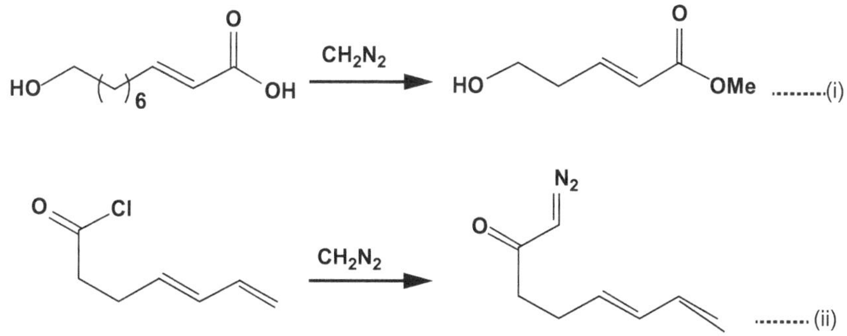

SCHEME 12.2 (i) Arndt–Eistert synthesis and (ii) α-diazoketone synthesis using diazomethane.

Arndt–Eistert synthesis and the Buchner–Curtius–Schlotterbeck reaction (Buchner & Curtius, 1885). Arndt–Eistert synthesis and α-diazoketone preparation using diazomethane are shown in Scheme 12.2.

12.7 Risk assessment

The concentrate solution of gaseous and liquid diazomethane has an excessive tendency to explode if it comes in contact with metals or the rough surface of glass. It is a strong irritant and in high concentration causes lung congestion, difficulty in breathing, cyanosis, and pulmonary edema. After 7h of diazomethane exposure in humans, different symptoms such as pharyngitis, bronchitis, dyspnea, and laryngitis can be seen. Additionally, short-term effects by inhalation may initiate asthmatic symptoms, coughing, wheezing, eye irritation, pulmonary edema, weakness, dizziness, chest pains, and headache. In rats and mice, diazomethane exposure by skin and by inhalation causes increased incidences of lung tumors. At very low concentrations, repetitive exposure may cause respiratory complaints. The US Environmental Protection Agency has not classified diazomethane irrespective of its carcinogenic potential.

Human death could be a consequence of fluid in the lungs (pulmonary edema) and oxygen deficiency. Asthma and lung edema symptoms do not develop until a few hours have elapsed.

12.8 Environmental effects

The production and uses of diazomethane in the laboratory as a methylating agent for phenols, carboxylic acids, enols, etc., cause an increase in its concentration in the environment through several disposal methods.

In the atmosphere, gaseous phase diazomethane undergoes hydrolysis with atmospheric water vapor (National Center for Biotechnology Information, 2020). Diazomethane does not absorb natural light (wavelength > 290 nm) because it does not comprise any chromophore; consequently, direct photolysis of diazomethane is not possible in sunlight. The absorption of diazomethane by sediments or soils, bioconcentration, and biodegradation is also not possible due to prompt reactivity with water under atmospheric conditions. Therefore, it does not demonstrate any significant effects on environments (National Center for Biotechnology Information, 2020).

12.9 Hazardousness

Acute effects: Acute toxication in humans may be in the form of irritation of the eyes, asthmatic symptoms, pneumonia, cough, dizziness, wheezing, weakness, headache, fever, chest pains, liver enlargement, tremors, malaise, moderate cyanosis, hypersensitivity, and shock. In animals, diazomethane may cause severe respiratory area irritation, bronchopneumonia, hemorrhaging, pulmonary edema, and emphysema. Acute toxicity through inhalation has also been revealed by an experiment in cats (Registry of Toxic Effects of Chemical Substances (RTECS), 1993).

Chronic effects (noncancer): In humans and animals, no evidence has been reported on the chronic effects of diazomethane. Any reference concentration and a reference quantity for diazomethane have not been recognized by the Environmental Protection Agency (United States Environmental Protection Agency, 1999).

Reproductive/developmental effects: No evidence has been reported on the propagative or improvement effects of diazomethane in humans and animals.

Cancer risk: No evidence has been reported on the carcinogenic effects of diazomethane in humans, but in rodents, increased incidences of pulmonary adenomas have been observed when rodents were exposed through inhalation. It has not been categorized based on its perspective of carcinogenicity by the Environmental Protection Agency (Braun, 1951; United States Environmental Protection Agency, 1999). Diazomethane has been categorized as group 3 by the International Agency for Research on Cancer and is not classifiable based on its carcinogenicity to humans.

12.10 Treatment

Basic treatment: If required, create a patent airway (oropharyngeal or nasopharyngeal airway). Provide suction if necessary. Watch for signs of respiratory insufficiency and assist ventilation if needed. Administer oxygen by a nonrebreather mask at 10–15 L/min. Monitor for pulmonary edema, shock, or seizures and treat if necessary. For eye contamination, flush eyes immediately with water. Irrigate each eye continuously with 0.9% saline during transport. Do not use emetics. For ingestion, rinse mouth and administer 5 mL/kg up to 200 mL of water for dilution if the patient can swallow, has a strong gag reflex, and does not drool. Cover skin burns with dry sterile dressings after decontamination (Currance, Clements, & Bronstein, 2005).

Advanced treatment: Consider orotracheal or nasotracheal intubation for airway control in the unconscious patient with severe pulmonary edema or severe respiratory distress. Positive-pressure ventilation techniques with a bag valve mask device may be beneficial. Consider drug therapy for pulmonary edema. Consider administering a β-agonist such as albuterol for severe bronchospasm. Monitor cardiac rhythm and treat arrhythmias as necessary. Start IV administration of D5W/SRP to maintain minimal blood flow rate. Use 0.9% saline (normal saline) or lactated Ringer's solution if signs of hypovolemia are present. For hypotension with signs of hypovolemia, administer fluid cautiously. Watch for signs of fluid overload. Treat seizures with diazepam or lorazepam. Use proparacaine hydrochloride to assist eye irrigation (Currance et al., 2005).

12.11 Risk management

Public Institute of Occupational Safety and Health's promptly hazardous to life or wellbeing limit (NIOSH IDLH) and The National Institute for Occupational Safety and Health (NIOSH) have defined exposure limit to ensure that a worker can escape from an exposure condition that is expected to cause death or instantaneous or delayed everlasting adverse health effects or prevent escape from the environment. The NIOSH-suggested exposure limit is 8 or 10 h weighted average exposure and/or ceiling. OSHA PEL (Occupational Safety and Health Administration (OSHA), 1998) is expressed as a time-weighted average; the concentration of a substance to which most employees can be exposed without adverse impact is averaged over a typical 8-h workday or a 40-h workweek.

In the event of a spill or leak involving diazomethane, persons not wearing protective equipment and clothing should be restricted from entering contaminated areas until cleanup has been completed. The following steps should be undertaken following a spill or leak:

- Notify safety personnel.
- Remove all sources of heat and ignition.
- Ventilate potentially explosive atmospheres.

- Provide the use of fully encapsulating, vapor-protective clothing and equipment for cleanup personnel.
- If possible, without risk, stop the flow of gas.
- If the source of the leak is a cylinder and cannot be stopped in place, remove the leaking cylinder to a safe place in the open air, and repair the leak or allow the cylinder to empty.
- If the leak is in liquid form, allow diazomethane to evaporate.

A diazomethane-exposed area of the body (i.e., flush affected) should be washed immediately with a large quantity of water for at least 15–20 min; medical attention should then be sought. The eyes may experience pain and redness of the tissues; consequently, they should be washed immediately with water for 20 min. Medical observation and rest are necessary. Prompt medical consideration is compulsory in all cases of overexposure. Conscious persons should be given access to fresh and unpolluted air.

12.12 Current research

For constructing C–C and C–heteroatom bonds in organic synthesis, diazomethane is one of the most versatile but extremely poisonous and potent carcinogenic reagents. Therefore, it has always been a topic of interest for researchers. Recently, the continuous synthesis of diazomethane followed by in situ methyl esterification has been proposed where complete conversion of diazomethane to a useful product has been achieved (Duan et al., 2020), which is an important practice for lab-scale synthesis as well as for industrial production. Due to its explosive nature, its generation and purification are risky by distillation. Therefore, a protocol has been suggested (Dallinger & Kappe, 2017) to build a configurationally simple tube-in-flask reactor for the in situ on-demand generation of anhydrous diazomethane using membrane separation technology; hence, avoiding distillation methods. A theoretical study was made based on the simultaneous formation and decomposition of diazomethane (Pipim, Tia, & Adei, 2020), where (3+2) cycloaddition of a diazomethane derivative and 7-isopropylidenebenzonorbornadiene was proposed. This periselective addition results in an *exo*-cycloadduct as the major product. Similarly, catalytic cyclopropanation of diazomethane with spiro[2,4] hepta-4,6-diene has also been observed in the presence of Pd or Cu compounds resulting in mono- and dicyclopropanation products: spiro[bicyclo[3,1,0]hex-3-ene-2,1′-cyclopropane] and spiro[cyclopropane-1,5′-tricyclo[4.1.0.0]heptane]. Here, the situation of simultaneous formation and decomposition of diazomethane of Pd compound as catalysts allow cyclopropanation to be performed (Shulishov, Pantyukh, Menchikov, & Tomilov, 2019).

12.13 Diazomethane toxicity: Case study

A 25-year-old chemistry student worked repeatedly with diazomethane at long intervals for half a year. After another half a year, he was exposed again to diazomethane, which resulted in breathlessness during physical activity, attacks of palpitations, coughing with yellowish sputum, dyspnea on exertion, tachycardia, lack of appetite with weight loss (height 175 cm, weight 51 kg), and severe cyanosis and polyglobulism. In addition to bronchitis and bronchiolitis processes, bronchopneumonia developed in both lungs. The symptoms, however, cleared up completely within 10 days after treatment (Hanusch, Schäfer, & Hanusch, 1966).

12.14 Safety measures

- The following personal protective equipment should be used when working with diazomethane: lab coat, covered shoes, safety glasses, latex gloves, long pants, etc.
- A well-ventilated fume cupboard should always be used for working with diazomethane and the use of a further blast shield is intensely suggested.
- When a spill of diazomethane inside a chemical fume hood occurs, initiate the emergency exhaust, instantaneously close the sashes, and let the fumes evaporate. If there is a larger spill (>100 mL) outside the chemical fume hood, then vacate the lab. If there is a possibility that vacating the lab sets off the fire alarms, then all staff should leave the building and the chemical fume hood sashes should be opened.
- Even if the amount of spill is small (<100 mL), it is still advisable to evacuate the lab.
- Neutralization of diazomethane after exposure with skin could be effected by simply adding soap and allowing it to react with a substrate.

12.15 Handling and storage

- If the existence work existence requires irregular quantities of diazomethane then it should be securely kept in a freezer in a stoppered bottle, further sealed by parafilm.
- Diazomethane explodes in its crystallization form if stored below its melting point (−145 °C). Hence, diazomethane should not be stored at very low temperatures. Avoid the cryogenic storage of diazomethane with the use of dry ice or a freezer.

- Diazomethane is very explosive and extremely toxic in nature; hence, it should be stored in cool conditions for minor periods (for 1–2 weeks).
- Before storing diazomethane, it should be assured that the bottle is not damaged, and potassium hydroxide pellets should be added as a storage drying agent.
- Diazomethane reacts with certain metal ions that cause explosions; hence, do not use any other drying agent.
- Be sure to indicate the contents on the container label.
- Precursors such as Diazald and *N*-nitosomethylurea are used for the synthesis of diazomethane and may be stored in a refrigerator at temperatures that diminish the degradation of reagent and increase the superiority of reagent.

12.16 Precautions

- Diazomethane is attractive as a methylating agent in organic synthesis and for phenols, carboxylic acids due to its larger reactivity and greater efficiency with the production of the only N_2 as a by-product. Therefore, it is essential that any glassware used in handling diazomethane be fire polished and not contain any scratches or ground glass joints.
- As diazomethane reacts, its natural yellow tone gives a spontaneous indication of reaction evolution. Consequently, diazomethane is incredibly harmful; it should be produced and utilized uniquely in a properly functioning smoke hood.
- Diazomethane should be handled behind a safety shield, and other defensive equipment should be utilized because of its explosive nature with certain metals or ground glass of any sort (joints, syringes, stoppers, stopcocks).
- Diazomethane should be handled in ether (solvent) due to its low boiling point (about 23 °C).
- Diazomethane vapors or solutions should never be dried with drierite because it explodes on reaction with calcium sulfate.
- Despite all of these hazards, diazomethane may be used safely if appropriate precautions are observed.

References

Braun, H. (1951). On diazomethane poisoning, with particular reference to a case observed in the medical policlinic of the University of Munich. *Medizinische Monatsschrift*, 5(4), 284.

Buchner, E., & Curtius, T. (1885). Synthesis of ketone ethers from aldehydes and diazoacetic acid ethers. *Reports of the German Chemical Society*, *18*(2), 2371–2377.

Budavari, S. (1989). *The merck index. An encyclopedia of chemicals, drugs, and biologicals* (11th ed.). Rahway, NJ, USA: Merck, ISBN:9780911910285.

Clayton, G. D., & Clayton, F. E. (1981). Patty's industrial hygiene and toxicology. In *Vol. 2A. Toxicology*. Chichester, Sussex: John Wiley & Sons, Inc.

Currance, P. L., Clements, B. W., & Bronstein, A. C. (2005). *Emergency care for hazardous materials exposure* (3rd ed.). St. Louis: Elsevier Mosby.

Dallinger, D., & Kappe, C. O. (2017). Lab-scale production of anhydrous diazomethane using membrane separation technology. *Nature Protocols, 12*(10), 2138–2147.

De Boer, T. J., & Backer, H. J. (1954). A new method for the preparation of diazomethane. *Recueil des Travaux Chimiques des Pays-Bas, 73*(3), 229–234. https://doi.org/10.1002/recl.19540730308.

Duan, X., Huang, Z., Qian, G., He, K., Zhang, J., Mleczko, L., et al. (2020). Unprecedented yield of methyl-esterification with in-situ generated diazomethane in a microchannel reactor with methanol as solvent. *Chemical Engineering Science, 213*, 115397.

Flury, F., & Zernik, F. (2013). *Harmful gases: Vapors, mists, fumes and dust*. Springer-Verlag.

Friedman, O. M., Mahapatra, G. N., Dash, B., & Stevenson, R. (1965). Studies on the action of diazomethane on deoxyribonucleic acid: The action of diazomethane on deoxyribonucleosides. *Biochimica et Biophysica Acta (BBA)-Nucleic Acids and Protein Synthesis, 103*(2), 286–297.

Hanusch, A. W., Schäfer, H., & Hanusch, A. (1966). Diazomethan-Intoxikation. *Zentralbl Arbeitsmed Arbeitsschutz, 16*, 261–266.

Hazardous Substances Data Bank. (1993). *U.S. Department of health and human services*. Bethesda, MD: National Toxicology Information Program, National Library of Medicine.

Kriek, E., & Emmelot, P. (1964). Methylation of deoxyribonucleic acid by diazomethane. *Biochimica et Biophysica Acta (BBA)-Specialized Section on Nucleic Acids and Related Subjects, 91*(1), 59–66.

National Center for Biotechnology Information. (2020). *PubChem compound summary for CID 9550, diazomethane*. https://pubchem.ncbi.nlm.nih.gov/compound/Diazomethane. (Accessed 11 September 2020).

National Institute for Occupational Safety and Health (NIOSH). (2016). *Diazomethane*. https://www.cdc.gov/niosh/npg/npgd0182.html. (Accessed 5 March 2020).

Occupational Safety and Health Administration (OSHA). (1998). *Occupational safety and health standards, toxic and hazardous substances. Code of federal regulations*. 29 CFR 1910.1000.

Pipim, G. B., Tia, R., & Adei, E. (2020). (3+2)cycloaddition reaction of 7-isopropylidenebenzonorbornadiene and diazomethane derivatives: A theoretical study. *Journal of Molecular Graphics and Modelling, 101*, 107713.

Registry of Toxic Effects of Chemical Substances (RTECS). (1993). *U.S. Department of health and human services*. Bethesda, MD: National Toxicology Information Program, National Library of Medicine.

Roper, R., & Ma, T. S. (1957). Diazomethane as a reagent for microsynthesis. *Microchemical Journal, 1*(2), 245–260. https://doi.org/10.1016/0026-265X(57)90070-X.

Shulishov, E. V., Pantyukh, O. A., Menchikov, L. G., & Tomilov, Y. V. (2019). Catalytic cyclopropanation of spiro[2.4]hepta-4,6-diene with diazomethane. *Tetrahedron Letters, 60*(31), 2043–2045.

Sittig, & Marshall. (1985). *Handbook of toxic and hazardous chemicals and carcinogens* (2nd ed.). Park Ridge, NJ: Noyes Publications.

Sunderman, F. W. (1970). *Diazomethane poisoning. Laboratory diagnosis of diseases caused by toxic agents St. Louis* (pp. 292–295). MO: Warren H. Green, Inc.

Sunderman, F. W., Connor, R., & Fields, H. (1938). Diazomethane poisoning. First clinical case report. *The American Journal of the Medical Sciences, 195*(4), 469–472.

United States Environmental Protection Agency. (1999). *Integrated Risk Information System (IRIS) on Diazomethane*. Washington, DC: National Center for Environmental Assessment, Office of Research and Development.

World Health Organization (WHO). (2001). *IARC monographs on the evaluation of the carcinogenic risk of human. Vol. 79.* Lyon, France https://monographs.iarc.fr/wp-content/uploads/2018/06/mono79.pdf. (Accessed 11 September 2020).

Yang, H., Martin, B., & Schenkel, B. (2018). On-demand generation and consumption of diazomethane in multistep continuous flow systems. *Organic Process Research & Development, 22*(4), 446–456.

CHAPTER

13

Fluorine: Risk assessment, environmental, and health hazard

Preeti Das and Pooja

Department of Chemistry, Gurukula Kangri (Deemed to be University), Haridwar, Uttarakhand, India

Abbreviations

ATRP	atom transfer radical polymerization
ER/AR	estrogen receptor to androgen receptor ratios
FSH	follicle stimulating hormone
HQ	hazardous quotient
ITP	iodine transfer polymerization
LH	luteinizing hormone
OTC	over-the-counter
RAFT	reversible addition-fragmentation chain transfer
SPADNS	sodium 2-(parasulfophenyl largo)-1,8-dihydroxy-3,6-naphthalene disulfonate

13.1 Introduction

The 13th most abundant element, Fluorine is a member of the halogen family and is found naturally in water, soil, animals, and plants (World Health Organization, 1996). It exists in trace amount in all tissues of the body in the form of enamel, dentin, and bone as well as in many enzymatic reactions (Ullah, Zafar, & Shahani, 2017). Water is an important medium for fluorine distribution, in addition to beverages, like tea. Fluorine is either present naturally or added during water fluoridation (Harrison, 2005; Tahir & Rasheed, 2012).

It is present far and wide in our environment, constituting from 0.06% to 0.09% by weight in the upper layers of the lithosphere. Fluorine (atomic number 9; Symbol F) is a highly toxic, pale yellow, diatomic, lightest

Hazardous Gases
https://doi.org/10.1016/B978-0-323-89857-7.00027-X

halogen that exists as a gas at standard conditions. The name "fluorine" was coined by the French chemist Amperé as "le fluor" after its ore fluorspar. Due to its reactive nature, it is essentially found only in mineral compounds. The main commercial source of fluorine is fluorite. In a gaseous state, it is used as a potential oxidizing agent, which practically reacts with organic and inorganic materials (Jorgensen, Shariati, Shields, Durr, & Proskin, 1989).

Fluorine is usually found in mineral forms such as fluorspar, cryolite, and fluorapatite. Phosphate rocks (apatite) also contain a great amount of fluorine, often in form of various fluoride compounds, such as silicon fluoride that are presently uneconomical to extract.

Fluorocarbon compounds often persist in the atmosphere due to the strength of carbon-fluorine bonds. Fluorine is not known to have any metabolic role in mammals; although some plants and sea sponges are found to synthesize fluorine poisons (mostly often monofluorosates) that help to avoid detachment (Lee et al., 2014).

13.2 Occurrence

A major part of fluorine originates during coal combustion and emission in the form of CaF_2, MgF_2, and AlF_3. The occurrence mode of fluorine in bottom ash and slag is inorganic villiaumite (e.g., soluble NaF, KF, and insoluble CaF_2), which is difficult to break down even at high temperatures. The highest content of fluorine is found in fly ash from the coal-fired power plant as it is physically adsorbed, whereas the insoluble fluorine is discharged from the plant in the environment as gaseous products (HF, SiF_4).

The natural organofluorines such as fluoroacetate have been found in microorganisms and plants responsible for defense against herbivores in many parts of the world. Some more examples include terminally fluorinated fatty acids, fluoroacetone, and 2-fluorocitrate.

13.3 Historical background

The synthesis of elemental fluorine has been pursued for approximately 175 years by many chemists, including Davy, Fermy, Moissan, and Otto Ruff. But all their attempts failed. It was firstly prepared in the form of gas in a laboratory by Henri Moissan, who was awarded Nobel Prize for his work in 1906. He successfully isolated fluorine in 1886, after being poisoned several times in his detection. The isolation was carried out by the process of electrolysis and fusion of anhydrous potassium hydrogen fluoride (KHF_2) and anhydrous hydrofluoric acid (HF) in a copper cell. The copper cell was used as a cathode. However, the limitation of cell that is, erratic current flocculation and increased melting point of the electrolytes, may deplete in hydrofluoric acid.

On the commercial scale, an enlargement of nuclear devices during World War II accelerated the development of fluorine chemistry as a result of the vital for elementary fluorine; usually for the manufacture of uranium hexafluoride (UF_6) for the isotopic enrichment of the uranium and chloride trifluoride (ClF_3) both in the United States and Germany.

The present methods for the production of fluorine are basically similar. To a great extent, progress has been made in quantity manufacture of fluorine, and in the modern era, fluorine has conceded from being a laboratory curiosity to an item of industrial production and an object of investigation for rocket propulsion (Forbes & Splinter, 2003).

13.4 Physicochemical properties

Fluorine is a pale-yellow gas with a density of 1.695 g per liter, that is about 1.3 times as dense as air. It changes from a gas to a liquid at a temperature of −188.13°C (−306.5°F) and from a liquid to a solid at −219.61°C (−363.30°F). Fluorine has a strong and characteristic odor that can be detected in very small amounts, as low as 20 parts per billion; this property is helpful for those who deal with fluorine. It means that the gas can be detected and avoided in case it leaks into a work area.

Fluorine is the most reactive element as it combines easily with every other element except helium, neon, and argon. It reacts with most compounds, often violently. For example, when mixed with water, it reacts explosively. For these reasons, it must be handled with extreme care in the laboratory. Some of the properties of fluorine are discussed in Table 13.1.

13.5 Physiological role

Owing to its highly electronegative nature, it forms HF in acidic conditions. Additionally, its permeability has been demonstrated in airway epithelial cells (Wilk-Blaszczak, French, & Man, 1992), but reports (Gofa & Davidson, 1996) also suggest the potential activity in osteoblastic cells. The activity of potassium and calcium selective ions channel; pH gradient-dependent, carrier-mediated mechanisms mediate many of the early events in fluoride-induced cell activation. Either HF transportation or diffusion completes by proton gradient of cells (Gutknecht & Walter, 1981; Şireli & Bülbül, 2004). Based on the pH dependence of

TABLE 13.1 Physicochemical properties of fluorine.

Symbol	F
Atomic number	9
Atomic mass	18.998403 u
Block	p-block
Electronic configuration	[He] $2s^2$ $2p^5$
Phase	Gas
Melting point	−219.6°C
Boiling point	−188.1°C
Electronegativity	3.98
Density (at STP)	1.696 g/L
Critical point	144.41 K, 5.1724 MPa
Oxidation states	−1
Covalent radius	64 pm
Van der Waals radius	147 pm
Crystal structure	Cubic
Magnetic ordering	Diamagnetic
Appearance/Color	Colorless

F transport, several investigators have suggested that F^- crosses cell membranes mainly in the nonionic form, that is, HF.

The fluoride in the blood gets deposited on calcium-rich areas such as bone and teeth (dentine and enamel), whereas about 80%–90% of it is retained by infants and 60% by adults (ATSDR, 2003). The enlargement in HF concentration primarily is a reason for excretion via urine, which is affected by urine pH, thus, affecting the fluorine clearance and retention (ATSDR, 2003). Fig. 13.1 demonstrates the summary of inhalation, metabolism, and excretion of fluoride.

13.6 Uses

Before World War II, fluorine was used for welding and frosting glass but was not commercially used. Fluorine-containing polymers or fluorinated polymers with polar covalent bonds between carbon and fluorine atoms are one of the important dielectric materials used in our daily lives.

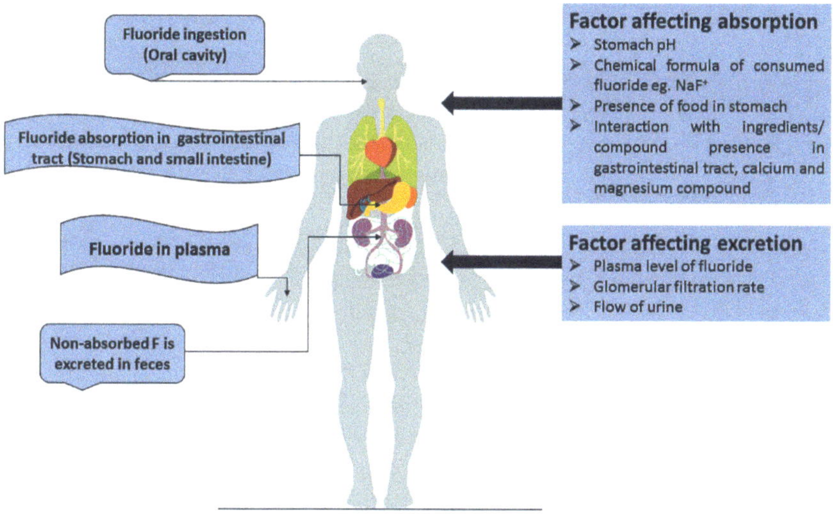

FIG. 13.1 Summary of absorption, metabolism, and excretion of fluoride following oral intake. *Credit: Pooja*

Fluorinated polymers take the advantage of inertness to chemical corrosion, prominent weather resistance, low flammability, thermal stability, and easy processing. They are widely used in microelectronics, optics, aerospace, transducers, etc. Therefore, numerous fluorine-containing polymers have been prepared to fulfill the requirements owing to its versatile nature (Shen, 2018).

Fluorine has several uses like in the preparation of uranium hexafluoride (needed by the nuclear power industry to separate uranium isotopes); in the manufacturing of sulfur hexafluoride (the insulating gas for high-power electricity transformers); used in many fluorochemicals (including solvents and high-temperature plastics), such as Teflon (also known as polytetrafluoroethene, PTFE); CFCs (chloro-fluoro-carbons) that is used as aerosol propellants, refrigerants, and for "blowing" expanded polystyrene. Some of the uses are discussed in the given section.

(1) Fluorochemical industry: Globally fluorine is extracted from fluorite mining. The mined fluorite is separated by froth flotationinto two main parts; one 60–85% pure met spar used in iron smelting and the second 97% pure acidspar converted to hydrogen fluoride.
(2) Dental Care: There are theories stating that fluorine aids enamel growth in small dental caries. This could be due to the presence of fluorine content in drinking water.
(3) Pharmaceuticals: Major amount of modern pharmaceuticals contain fluorine such as the cholesterol-reducer atorvastatin (Lipitor),

anti-asthmatic Seretide. The reason for usage may be delaying inactivation and lengthen dosage periods because of a stable C—F bond. Fluorination also enhances lipophilicity often helping in cell membrane penetration and hence bioavailability.

(4) Positron emission tomography scanning: The computer-assisted tomography of glucose requiring tissues use fluorodeoxyglucose as a tracer.

(5) Semiconductor manufacturing: Atomic as well as molecular fluorine are often used for plasma etching in semiconductor manufacturing.

13.7 Sources of contaminants

Some of the native sources of fluorine include volcanic activity, groundwater, streams, rivers, lakes, and some wells exposed to fluoride-containing rocks, industrial emissions, or community water fluoridation (National Research Council, 2006). The presence of the perfluorinated compounds in the ground and surface water has led scientists from across the world to sign the Madrid Statement calling for government and manufacturer action (Blum et al., 2015; Domingo, 2012).

Owing to human activity, particularly, the use of pesticides, foodstuff is also being presented in majority by significant levels of fluorine. There are reports depicting fluorine content in cow's milk (Erdal & Buchanan, 2004) as well as processed chicken. Some types of cookware or oil are also responsible for fluorine contamination of food for oil/water-resistant packaging.

The tea and commercial drinks, such as juices and soft drinks, alcoholic beverages, especially alcohol and beer, have recorded significant levels of fluorine (Goschorska, Gutowska, Baranowska-Bosiacka, Rać, & Chlubek, 2016; Warnakulasuriya, Harris, Gelbier, Keating, & Peters, 2002).

Anthropogenic sources of atmospheric fluoride are from coal combustion or refineries, metal ore smelters, aluminum production plants, phosphate fertilizer plants, chemical production facilities, steel mills, magnesium plants, brick, and structural clay manufacturers as well as copper and nickel producers, phosphate ore processors, glass producers, and ceramics constructors.

Other sources include the addition of fluoride to the toothpaste (Bralić, Buljac, Prkić, Buzuk, & Brinić, 2015), dental floss (Flatt, Warren-Morris, Turner, & Chan, 2008), mouthwashes (Parashar, 2015), fluoridated toothpicks, and interdental brushes (Särner, 2008), topical fluoride gel and foam, fluorine vanish. NaF is more toxic as it is more soluble and releases more amounts of fluoride compared to calcium fluoride (CaF) (Ullah et al., 2017).

Other examples of sources include dental material for fillings such as glass ionomers cement, resin-modified glass ionomers cement, giomers, polyacid-modified composites (Shimazu, Ogata, & Karibe, 2013; Wiegand, Buchalla, & Attin, 2007), dental mercury amalgams (Franklin Garcia-Godoy & Chan, 1991; Tveit & Gjerdet, 1981), dental materials for pits and resin-based such as dissecting sealant (Poggio et al., *2016;* Salmerón-Valdés, Scougall-Vilchis, Alanis-Tavira, & Morales-Luckie, 2016), and dental material for tooth sensitivity/erosion treatment, such as silver diimine fluoride. Concerns have been raised about the use of fluoride-containing toothpaste for children (Basch & Kernan, 2016).

About 20%–30% of pharmaceutical/prescription compounds contain a differing quantity of fluorine (Walker et al., 2013) for example Prozac, Lipitor, and Ciprobay (ciprofloxacin) (Müller, Faeh, & Diederich, 2007), as well as the rest of the fluoroquinolone family (gemifloxacin, levofloxacin, moxifloxacin, norfloxacin, and ofloxacin). Products prepared from perforated compounds include carpets and fabrics, paints, cosmetics, non-stick coatings and paper coatings for oil and moisture resistance, as well as protective coatings for paper, and cardboard, etc.

Household dust may contain poly- and perfluoroalkyl substances (PFAs) and significant levels of fluoride have also been associated with heavy smokers. Synergistic exposure to fluoride and aluminum can occur through water, tea, food residues, baby formula, aluminum-containing antacids, or through drugs, deodorants, cosmetics, and glassware. Fluorine gas is used to make uranium hexafluoride, which separates isotopes of uranium in nuclear reactors and weapons.

13.8 Risk assessment

Usually, health risk assessment consists of four stages, such as risk identification; response assessment; exposure evaluation; and characterization of hazard, as recommended by the US-EPA (United States-Environmental Protection Agency). In accordance, human beings are largely exposed to contaminants through three pathways: intake of water, inhalation, and straight dermal contact. Ingestions of contaminated drinking water and fluorinated beverages are considered as the most remarkable and chief pathway (Radfard et al., 2018).

In order to estimate the probability of adverse health effects, it is necessary to assess risks associated with human beings. Risk due to exposure to fluorine through the intake of drinking water, date, vegetables, and fish can be assessed using the methodology provided by US-EPA (United States-Environmental Protection Agency), which estimates the hazardous quotient (HQ) by computing the ratio of the estimated dose of a contaminant to the reference dose. A fluoride ion-selective electrode (ISE) can be utilized to measure the fluoride content of food samples. Sodium 2-(parasulfophenyl largo)-1,8-dihydroxy-3,6-naphthalene disulfonate colorimetric method (SPADNS) has been used for the determination of fluoride concentrations in water samples (Keshavarz, Ebrahimi, & Nikaeen, 2015).

Furthermore, risk assessment due to exposure, through the consumption of contaminated groundwater was used to estimate the noncarcinogenic health risk of fluoride in three groups of age such as adults, children, and infants, in rural areas of Gonabad Bajestan, Iran (Qasemi, Afsharnia, Zarei, Farhang, & Allahdadi, 2018).

On the other hand, consumption of tea can be a potential source of fluoride exposure; the health risk assessment was carried out using the method Monte Carlo simulation. For the assessment related to the exposure, chronic daily intake (CDI) and hazard quotient (HQ) has been used as a model and followed in a different age group in men and women as well as children (men and women (21–72 years and children 0–12 years) in Iran (Karami et al., 2019).

13.9 Environmental effects

When fluorine from the air ends up in the water it settles into the sediment where it is strongly attached to soil particles. Fluorine is not destroyed but is converted into other forms. Fluorine present in soils may tend to accumulate in plants. The amount of uptake by plants depend upon the type of plant, soil, and concentration of the type of fluorine compound found in the soil. With plants that are sensitive to fluorine exposure, even low concentrations of fluoride can cause damage leading to a decline in growth. Too much fluoride, whether taken from the soil through roots or absorbed from the atmosphere by the leaves, retards the growth of plants and reduces crop yield.

Animals that eat fluorine-containing plants may accumulate large amounts of fluoride in their bodies. Consequently, animals that are exposed to high concentrations of fluorine suffer from dental decay and bone degradation.

Fluorine gas in the form of chlorofluorocarbons (CFCs), hydrochlorofluorocarbons (HCFCs) perfluorocarbons (PFCs), sulfur hexafluoride (SF_6), and nitrogen trifluoride (NF_3) are the substance found to deplete the ozone layer. However, most F-gases have a great potential for global warming, and thus are known as so-called greenhouse gases (GHG), and some of which are nearly inert to removal by a chemical process.

Airborne fluorides are readily absorbed by plant leaves, and it is by this means that phytotoxic (plant poisoning) damage is introduced into the plant. After absorption, the fluoride is transported to the apical (tips) or margin portions of the leaves. Once the fluoride has been incorporated into the foliar (leaf) tissue, translocation to other parts of the plant generally does not occur.

The introduction of airborne fluorides into the environment has a known secondary effect on foraging animals. When pastures are fumigated with fluorides, the forage crops absorb and store the fluorides in their leaves. Relatively high concentrations of hydrogen fluoride, a few ppm, may reach pastures in the event of an accidental release of liquid fluorine.

While these exposures would not be high enough to cause acute damage to animals, it must be considered whether any chronic long-term damage would result. This secondary effect, called fluorosis (fluoride poisoning), strikes primarily cattle and is observed in areas that have industries that daily introduce airborne fluorides into the environment. Since cattle are the most susceptible animals to fluorosis and most of the economic damage has occurred to the cattle industry, many studies have been done on chronic fluoride poisoning in cattle.

13.10 Hazardousness

Effect on reproductive health: One of the growing concerns is infertility caused due to fluoride intake. In accordance with studies, too much consumption of fluoride causes the problem of low birth weight (LBW) during pregnancy and also affects the development of the fetus (Gupta, Sharma, & Gupta, 2001). A study review has also concluded the lessening of total fertility rate with rising fluoride levels in both male and female individuals. The reason may be the increased exposure to fluoride (F^-) associated with the increased levels of follicle-stimulating hormone (FSH) and luteinizing hormone (LH) (Ortiz-Pérez et al., 2003; Savchenko, 1977), decreased estrogen levels (Jiang, Fan, Cheng, & Cui, 2005), reduced testosterone levels and potent metabolite conversion (Ortiz-Pérez et al., 2003; Savchenko, 1977), reduced thyroid hormones (Bachinskiĭ, Gutsalenko, Naryzhniuk, Sidora, & Shliakhta, 1985; Bobek, Kahl, & Ewy, 1976; Wang et al., 2009), disturbed androgen to estrogen ratios and estrogen receptor to androgen receptor ratios (ER/AR) (Ando et al., 2001; Palmero, Prati, Bolla, & Fugassa, 1995). Some of the major reported changes in males include changes in the structure and functional activities of spermatozoa, disruption of spermatogenesis, and instability of multiple hormone systems decreased circulating testosterone concentration. There are reports showing a lowering of pregnancy rate and a number of implantation due to high fluoride content (Al-Hiyasat, Elbetieha, & Darmani, 2000). Such disturbances in reproductive and endocrine systems probably contribute to more reproductive health problems in the present day.

Effect on neuronal health: Fluoride may cause neurotoxicity (Choi, Sun, Zhang, & Grandjean, 2012) including effects on learning and memory (Grandjean & Landrigan, 2006; Li, 2003; Shivaprakash, Ohri, & Noorani, 2011).

Effect on skeletal and dental health: The higher quantity of fluoride has been accompanied by dental fluorosis, weakening of bones, which may be misdiagnosed as rheumatoid arthritisor ankylosing spondylitis if reported at in early age (National Research Council, 2006; Wong et al., 2010). The symptoms showing dental fluorosis include intrinsic tooth discoloration or faint white lines or specks or white mottled patches, while severe fluorosis is characterized by brown discoloration and brittle, pitted, and rough enamel.

Fluoride may also indirectly alter the action of protease via a decrease in the availability of free calcium ions in the mineralization environment (Fejerskov & Kidd, 2009). Liquid fluorine spills may result in unpredictable short-term exposure of the general public to hydrogen fluoride. Unpredictable short-term exposures may occur when pollutants are released in an uncontrolled manner, at unpredictable times and places as the result of accidents such as damage to transportation equipment, fire in a chemical storage facility, or the rupture of a propellant tank. Even for such circumstances, provisions should be in place to predict the possibilities with fair accuracy leading to such an accident and hence to study the corresponding exposure parameters.

13.11 Treatment

The emergency medical services personnel must place patients with known digestion of fluoride on a cardiac monitor and establish intravenous access to calcium (Bayless & Tinanoff, 1985).

13.12 Risk management

Specific risk management measures are not required beyond good industrial hygiene and safety procedures (Safety Data Sheet, 2011). The concept that formerly controlled fluoride looked after the development of wounds that were essentially considered systemic and pre-eruption in action, altering the enamel structure, but well established by the late 1980s. It was noted that this effect was mainly a topical effect. Therefore, fluoride must be available at the right place (biofilm fluid, saliva) in small concentrations (ppm) and at the right time (sugar exposure) to interfere with demineralization and remineralization events.

13.13 Current research

The fluoropolymers can be prepared by radical polymerization in a controlled way, for example, Iodine Transfer Polymerization (ITP), Atom Transfer Radical Polymerization (ATRP), Reversible Addition-Fragmentation Chain Transfer (RAFT), and so on. Fluorinated acrylates or methacrylates have been extensively synthesized by these methods (Shen, 2018).

Recently, Fluorine containing graphene quantum dotes (F-GQDs) with a high singlet oxygen generation (SOG) has been synthesized for photodynamic therapy (PDT). GQDS is extensively applied in biomedical areas such as bio-imaging, bio-sensing, and photo-thermal therapy due to their superior optical and physicochemical properties (Li et al., 2020).

Another achievement is the use of fluorine in the field of medicine to explain the mechanism of action of alumino fluoride compress (AlF) as a messenger to detect protein biosynthesis and their role in Alzheimer's disease.

13.14 Fluorine toxicity: Case study

A 65-year-old skilled worker was in the process of cleaning iron pipes associated with a cooling apparatus by using a high-pressure stream of water, in which HF gas was liquefied into HF acid. He began working alone, which was unusual, while his co-worker went downstairs to get his rubber gloves. Approximately 5 min later, the co-worker heard him crying for help and ran up to him. The co-worker found him crouching and his face turned gray in color. He complained of chest pain. At that time he was wearing a helmet, goggles, rubber gloves, and a raincoat. It was not clear whether during the accident he had been wearing the plastic face shield that was found nearby. The co-worker and the field overseer then removed his contaminated clothing. However, they neither irrigated with water nor applied a calcium gluconate gel to the affected area, the efficacies of which have been reported by previous investigators, when an ambulance arrived approximately 30 min after the accident, he was in a state of cardiopulmonary arrest. Despite aggressive cardiopulmonary resuscitation efforts and the intravenous administration of adrenaline, atropine, and sodium bicarbonate at a nearby hospital, he was pronounced dead approximately 1.5 h after the accident. Initial biochemical tests in the hospital revealed that the serum calcium level was lower (4.9 mg/dL) than the normal range (8.4–10.2 mg/dL) (Takase et al., 2004).

13.15 Safety measures

- Must wear ventless goggles, lab coat, hydrofluoric acid-resistant gloves, long pants, and closed-toe shoes when working with fluorine.
- Keep valves and fittings free from oil and grease.
- Wear protective gloves/protective clothing/eye protection/face protection. Keep away from combustible materials.
- Keep suitable chemically resistant protective clothing readily available for emergency use.
- Rescue mechanisms, particularly for the body, must be selected based on the task and the risks involved.

13.16 Handling and storage

- Fluorine should be used in a ventilated gas cabinet or fume hood.
- Fluorine can be stored in passivated steel cylinders or nickel or Monel metal cylinders at temperatures below 200°C (392°F).
- Store in a well-ventilated place with temperature below 50°C.
- Observe all regulations and local requirements regarding the storage of containers.
- Cylinders should be stored in the vertical position and properly secured to prevent falling over.
- Stored containers should be periodically checked for general condition and leakage. Container valve guards or caps should be in place.
- Store containers in a location that is free from the risk of fire and away from sources of heat and ignition.
- Containers should not be stored in conditions likely to encourage corrosion.

13.17 Precautions

- Do not breathe gas and vapors.
- Keep away from combustible materials.
- Protect eyes, face, and skin from contact with the product.

References

Al-Hiyasat, A. S., Elbetieha, A. M., & Darmani, H. (2000). Reproductive toxic effects of ingestion of sodium fluoride in female rats. *Fluoride, 33*(2), 79–84.

Ando, S., Sirianni, R., Forastieri, P., Casaburi, I., Lanzino, M., Rago, V., et al. (2001). Aromatase expression in prepuberal sertoli cells: Effect of thyroid hormone. *Molecular and Cellular Endocrinology, 178*(1–2), 11–21.

ATSDR (Agency for Toxic Substances and Disease Registry). (2003). *Toxicological profile for fluorides, hydrogen fluoride, and fluorine.* Atlanta, GA: U.S. Department of Health and Human Services, Public Health Service.

Bachinskiĭ, P. P., Gutsalenko, O. A., Naryzhniuk, N. D., Sidora, V. D., & Shliakhta, A. I. (1985). Action of the body fluorine of healthy persons and thyroidopathy patients on the function of hypophyseal-thyroid the system. *Problemy endokrinologii, 31*(6), 25–29.

Basch, C. H., & Kernan, W. D. (2016). Ingredients in children's fluoridated toothpaste: A literature review. *Global Journal of Health Science, 9*(3), 1.

Bayless, J. M., & Tinanoff, N. (1985). Diagnosis and treatment of acute fluoride toxicity. *The Journal of the American Dental Association, 110*(2), 209–211. https://doi.org/10.14219/jada.archive.1985.0246.

Blum, A., Balan, S. A., Scheringer, M., Trier, X., Goldenman, G., Cousins, I. T., et al. (2015). The Madrid statement on poly-and perfluoroalkyl substances (PFASs). *Environmental Health Perspectives, 123*(5), A107–A111.

Bobek, S., Kahl, S., & Ewy, Z. (1976). Effect of long-term fluoride administration on thyroid hormones level blood in rats. *Endocrinologia Experimentalis, 10*(4), 289–295.

Bralić, M., Buljac, M., Prkić, A., Buzuk, M., & Brinić, S. (2015). Determination fluoride in products for oral hygiene using flow-injection (FIA) and continuous analysis (CA) with homemade FISE. *International Journal of Electrochemical Science, 10*, 2253–2264.

Choi, A. L., Sun, G., Zhang, Y., & Grandjean, P. (2012). Developmental fluoride neurotoxicity: A systematic review and meta-analysis. *Environmental Health Perspectives, 120*(10), 1362–1368.

Domingo, J. L. (2012). Health risks of dietary exposure to perfluorinated compounds. *Environment International, 40*, 187–195.

Erdal, S., & Buchanan, S. N. (2004). A quantitative look at fluorosis, fluoride exposure, and intake in children using a health risk assessment approach. *Environmental Health Perspectives, 113*(1), 111–117.

Fejerskov, O., & Kidd, E. (Eds.). (2009). *Dental caries: The disease and its clinical management* John Wiley & Sons.

Flatt, C. C., Warren-Morris, D., Turner, S. D., & Chan, J. T. (2008). Effects of a stannous fluoride-impregnated dental floss on in vivo salivary fluoride levels. *The Journal of the American Dental Hygienists' Association, 82*(2), 19.

Forbes, F. S., & Splinter, P. A. V. (2003). Liquid rocket propellants. In R. A. Meyers (Ed.), *Encyclopedia of physical science and technology* (3rd ed., pp. 741–777). Academic Press Inc.

Franklin Garcia-Godoy, D. D. S., & Chan, D. C. (1991). Ionomer-lined amalgam restorations. *American Journal of Dentistry, 4*(5).

Gofa, A., & Davidson, R. M. (1996). NaF potentiates a K+-selective ion channel in G292 osteoblastic cells. *The Journal of Membrane Biology, 149*(3), 211–219.

Goschorska, M., Gutowska, I., Baranowska-Bosiacka, I., Rać, M. E., & Chlubek, D. (2016). Fluoride content in alcoholic drinks. *Biological Trace Element Research, 171*(2), 468–471.

Grandjean, P., & Landrigan, P. J. (2006). Developmental neurotoxicity of industrial chemicals. *The Lancet, 368*(9553), 2167–2178.

Gupta, A., Sharma, U., & Gupta, S. K. (2001). Increased incidence of low birth weight babies in high fluoride areas. *Journal of Obstetrics and Gynaecology of India, 51*, 95–98.

Gutknecht, J., & Walter, A. (1981). Hydrofluoric and nitric acid transport through lipid bilayer membranes. *Biochimica et Biophysica Acta (BBA) - Biomembranes, 644*(1), 153–156.

Harrison, P. T. (2005). Fluoride in water: A UK perspective. *Journal of Fluorine Chemistry*, *126*(11 – 12), 1448–1456. https://doi.org/10.1016/j.jfluchem.2005.09.009.
Jiang, C. X., Fan, Q. T., Cheng, X. M., & Cui, L. X. (2005). Relationship between spermatogenic cell apoptosis and serum estradiol level in rats exposed to fluoride. *Wei Sheng Yanjiu Journal of Hygiene Research*, *34*(1), 32–34.
Jorgensen, J., Shariati, M., Shields, C. P., Durr, D. P., & Proskin, H. M. (1989). Fluoride uptake into demineralized primary enamel from fluoride-impregnated dental floss in vitro. *Pediatric Dental Journal*, *11*(1), 17–20.
Karami, M. A., Fakhri, Y., Rezania, S., Alinejad, A. A., Mohammadi, A. A., Yousefi, M., et al. (2019). Non-carcinogenic health risk assessment due to fluoride exposure from tea consumption in Iran using monte carlo simulation. *International Journal of Environmental Research and Public Health*, *16*(21), 4261.
Keshavarz, S., Ebrahimi, A., & Nikaeen, M. (2015). Fluoride exposure and its health risk assessment in drinking water and staple food in the population of Dayyer, Iran, in 2013. *Journal of Education Health Promotion*, *4*(72). https://doi.org/10.4103/2277-9531.171785.
Lee, S. T., Cook, D., Pfister, J. A., Allen, J. G., Colegate, S. M., Riet-Correa, F., et al. (2014). Monofluoroacetate containing plants that are potentially toxic to livestock. *Journal of Agricultural and Food Chemistry*, *62*(30), 7345–7354. https://doi.org/10.1021/jf500563h.
Li, L. (2003). The biochemistry and physiology of metallic fluoride: Action, mechanism, and implications. *Critical Reviews in Oral Biology and Medicine*, *14*(2), 100–114.
Li, Z., Wang, D., Xu, M., Wang, J., Hu, X., Anwar, S., et al. (2020). Fluorine-containing graphene quantum dots with a high singlet oxygen generation applied for photodynamic therapy. *Journal of Materials Chemistry B*, *8*(13), 2598–2606.
Müller, K., Faeh, C., & Diederich, F. (2007). Fluorine in pharmaceuticals: Looking beyond intuition. *Science*, *317*(5846), 1881–1886.
National Research Council. (2006). *Fluoride in drinking water: A scientific review of EPA's standards*. Washington, DC: The National Academies Press.
Ortiz-Pérez, D., Rodrı´guez-Martı´nez, M., Martı´nez, F., Borja-Aburto, V. H., Castelo, J., Grimaldo, J. I., et al. (2003). Fluoride-induced disruption of reproductive hormones in men. *Environmental Research*, *93*(1), 20–30.
Palmero, S., Prati, M., Bolla, F., & Fugassa, E. (1995). Tri-iodothyronine directly affects rat sertoli cell proliferation and differentiation. *Journal of Endocrinology*, *145*(2), 355–362.
Parashar, A. (2015). Mouthwashes and their use in different oral conditions. *Scholars Journal of Dental Sciences*, *2*(2B), 186–191.
Poggio, C., Andenna, G., Ceci, M., Beltrami, R., Colombo, M., & Cucca, L. (2016). Fluoride release and uptake abilities of different fissure sealants. *Journal of Clinical and Experimental Dentistry*, *8*(3), 284.
Qasemi, M., Afsharnia, M., Zarei, A., Farhang, M., & Allahdadi, M. (2018). Non-carcinogenic risk assessment to human health due to intake of fluoride in the groundwater in rural areas of Gonabad and Bajestan, Iran: A case study. *Human and Ecological Risk Assessment: An International Journal*, *25*(5), 1222–1233.
Radfard, M., Rahmatinia, M., Akbari, H., Hashemzadeh, B., Akbari, H., & Adibzadeh, A. (2018). Data on health risk assessment of fluoride in water distribution network of Iranshahr, Iran. *Data in Brief*, *20*, 1446–1452. https://doi.org/10.1016/j.dib.2018.08.184.
Safety Data Sheet. (2011). *Fluorine, Version: 1.2*. SDS No.: 057, 1-6.
Salmerón-Valdés, E. N., Scougall-Vilchis, R. J., Alanis-Tavira, J., & Morales-Luckie, R. A. (2016). Comparative study of fluoride released and recharged from conventional pit and fissure sealants versus surface prereacted glass ionomer technology. *Journal of Conservative Dentistry: JCD*, *19*(1), 41.
Särner, B. (2008). *On approximal caries prevention using fluoridated toothpicks, dental floss and interdental brushes*. Institute of Odontology. Department of Cariology.

Savchenko, O. N. (1977). Effect of inorganic fluorine compounds on the functional state of the pituitary-testis system. *PRO, 23*(4), 104–107.

Shen, Q.-D. (2018). Preparation, structure and properties of fluorine-containing polymers. In *Dielectric Polymer Materials for High-Density Energy Storage* (pp. 59–102). https://doi.org/10.1016/b978-0-12-813215-9.00003-8.

Shimazu, K., Ogata, K., & Karibe, H. (2013). Evaluation of the caries-preventive effect of three orthodontic band cements in terms of fluoride release, retentiveness, and microleakage. *Dental Materials Journal, 32*(3), 376–380.

Shivaprakash, P. K., Ohri, K., & Noorani, H. (2011). Relation between dental fluorosis and intelligence quotient in school children of Bagalkot district. *Journal of the Indian Society of Pedodontics and Preventive Dentistry, 29*(2), 117.

Şireli, M., & Bülbül, A. (2004). The effect of acute fluoride poisoning on nitric oxide and methemoglobin formation in the Guinea pig. *Turkish Journal of Veterinary and Animal Sciences, 28*(3), 591–595.

Tahir, M. A., & Rasheed, H. (2012). Fluoride in the drinking water of Pakistan and the possible risk of crippling fluorosis. *Drinking Water Engineering and Science Discussions, 5*, 495–514. https://doi.org/10.5194/dwesd-5-495-2012.

Takase, I., Kono, K., Tamura, A., Nishio, H., Dote, T., & Suzuki, K. (2004). Fatality due to acute fluoride poisoning in the workplace. *Legal Medicine, 6*(3), 197–200. https://doi.org/10.1016/j.legalmed.2004.04.003.

Tveit, A. B., & Gjerdet, N. R. (1981). Fluoride release from a fluoride-containing amalgam, a glass ionomer cement and a silicate cement in artificial saliva. *Journal of Oral Rehabilitation, 8*(3), 237–241.

Ullah, R., Zafar, M. S., & Shahani, N. (2017). Potential fluoride toxicity from oral medicaments: A review. *Iranian Journal of Basic Medical Sciences, 20*(8), 841–848. https://doi.org/10.22038/IJBMS.2017.9104.

Walker, M. C., Thuronyi, B. W., Charkoudian, L. K., Lowry, B., Khosla, C., & Chang, M. C. (2013). Expanding the fluorine chemistry of living systems using engineered polyketide synthase pathways. *Science, 341*(6150), 1089–1094.

Wang, H., Yang, Z., Zhou, B., Gao, H., Yan, X., & Wang, J. (2009). Fluoride-induced thyroid dysfunction in rats: Roles of dietary protein and calcium level. *Toxicology and Industrial Health, 25*(1), 49–57.

Warnakulasuriya, S., Harris, C., Gelbier, S., Keating, J., & Peters, T. (2002). Fluoride content of alcoholic beverages. *Clinica Chimica Acta, 320*(1–2), 1–4.

Wiegand, A., Buchalla, W., & Attin, T. (2007). Review on fluoride-releasing restorative materials—Fluoride release and uptake characteristics, antibacterial activity and influence on caries formation. *Dental Materials, 23*(3), 343–362.

Wilk-Blaszczak, M. A., French, A. S., & Man, S. P. (1992). Halide permeation through 10 pS and 20 pS anion channels in human airway epithelial cells. *Biochimica et Biophysica Acta (BBA) - Biomembranes, 1104*(1), 160–166.

Wong, M. C., Glenny, A. M., Tsang, B. W., Lo, E. C., Worthington, H. V., & Marinho, V. C. (2010). Topical fluoride as a cause of dental fluorosis in children. *Cochrane Database of Systematic Reviews, 1*.

World Health Organization. (1996). *Trace elements in human nutrition and health*. Geneva: World Health Organization.

CHAPTER

14

Formaldehyde: Risk assessment, environmental, and health hazard

Santosh Kumar Dubey[a] *and Preeti Das*[b]

[a]Department of Chemistry, Kurukshetra University, Kurukshetra, Haryana, India

[b]Department of Chemistry, Gurukula Kangri (Deemed to be University), Haridwar, Uttarakhand, India

Abbreviations

DHHS	Department of Health and Human Services
EPA	Environmental Protection Agency
IDLH	immediately dangerous to life or health
NIOSH	National Institute for Occupational Safety and Health
NOAEL	no observed adverse effect level
NTP	National Toxicology Program
OSHA	Occupational Safety, Health Administration
PEL	permissible exposure limit
TOXNET	toxicology data network

14.1 Introduction

Formaldehyde is a highly flammable, colorless, irritating, and poisonous low molecular weight gas that readily undergoes polymerization. It has a relative molecular mass of 30.03 g/mol and a pungent odor at normal room temperature and pressure. The gas is fairly soluble in water, alcohol, and diethyl ether. It is used in the liquid phase as well as in polymerized form. However, it frequently photo-oxidizes in the presence of sunlight to carbon dioxide under atmospheric conditions. Formaldehyde rapidly reacts with trace substances and pollutants present in the air being affected by sunlight. The half-life of formaldehyde is nearly 50 min in

Hazardous Gases
https://doi.org/10.1016/B978-0-323-89857-7.00026-8

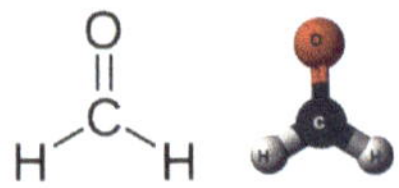

FIG. 14.1 Structure of formaldehyde.

the absence of nitrogen dioxide during day time. However, when nitrogen dioxide is present, it reduces to 35 min. Formaldehyde tends to react with hydrogen chloride or other inorganic chlorides under specific conditions to form bis(chloromethyl) ether.

Due to its reactivity, it is used for the polymerization process in industrial applications. Formaldehyde is used in the preparation of plastics, paints, fabrics, and especially in permanent press fabrics, curtains, pillows, and textiles, explosives, pulpy paper products, four-wheel vehicle parts, farraginous wood products, and many more (King, 2019).

The shape of formaldehyde is a trigonal planar and resembles two triangles when joined to a point, as shown in Fig. 14.1.

14.2 Occurrence

Formaldehyde being an organic compound originates from nonliving sources such as tobacco smoke, automobile fumes, and wildfires. These may consist of methane and other organic compounds in the atmosphere that react with sunlight. This gas can happen in places where no life exists in the universe. Formaldehyde has also been observed by astrophysicists in the space where CO_2 with H_2 is formed from ice. It is also commonly present in the human body. In the normal process of natural metabolism, a person of 68 kg produces 68 g of formaldehyde each day. Formaldehyde was reported as a high bulk industrial chemical with production reaching 52 million tons in 2017.

Ninety percent of formaldehyde is used for atmospheric processes. It is produced by the reaction of sunlight and oxygen on methane and other hydrocarbons in the atmosphere; in this instance, it becomes an important constituent of smog.

Formaldehyde and its acids are present in living organisms. It builds up in the metabolism of endogenous amino acids and is found in the bloodstream of humans at concentrations of about 0.1 mL mol (National Research Council, 2014). Formaldehyde does not exist primarily in the environment, as it is fragmented in a few hours by bacteria present in soil or water or sunlight. Furthermore, it is readily metabolized in humans into formic acid in the body. Formaldehyde is formed endogenously in mammals, fruits, certain foods, carpets, paints, and humans, resulting in oxidative metabolism.

14.3 Historical background

Russian chemist Aleksandra Butlerov (1828–86) in 1859 (Butlerow, 1859) reported the presence of formaldehyde. It was finally recognized in 1869 by August Wilhelm von Hofmann (Read, 1935) who was a professor of chemistry and director of the laboratory of the University of Berlin. Formaldehyde was not identified until 1868, when von Hofmann set out to establish both its structure and identity. The method that was used by Hofmann to identify formaldehyde established the base for the latest formaldehyde manufacturing processes.

Commercially in the 1880s, formaldehyde production started in Germany. Later on, it also started in Belgium, France, and the United States by the beginning of the 1900s. At present, formaldehyde is used as a preserving agent in the medical field. However, these uses are only 1% of total formaldehyde sales nowadays. Initially, it was manufactured in quite small amounts (5–20 kg) for laboratory usage in universities and factories. Later on, due to improved manufacturing processes, it was produced on a larger scale. Over the past century, demand for formaldehyde has increased rapidly because of improvement in science and technology. A large variety of applications of formaldehyde have been discovered. It is worth mentioning that commercial-grade particleboard was prepared at a factory in Bremen, Germany, in the 1940s and later became widely used in the furniture and construction industries.

14.4 Physicochemical properties

Formaldehyde is a highly flammable, colorless gas at room temperature having a pungent smell and suffocating odor (Budavari, O'Neil, Smith, Heckelman, & Kinneary, 1989). Its odor thresholds range in between 0.5 to 1.0 (ATSDR, 1999) and 0.06 to 0.5 ppm (Gerberich, Seaman, & Hoechst-Celanese Corporation, 1994). One of the important features of formaldehyde is that it reacts rapidly with a lot of substances and polymerizes easily. Because of this, it is the world's most exigent industrial chemical (Gerberich et al., 1994). See Table 14.1 for more physical properties of formaldehyde.

14.5 Physiological role

Due to water solubility, formaldehyde is rapidly absorbed and metabolized by the respiratory and gastrointestinal tract. These metabolites may enter human skin, stimulating allergies and inflammations, though

TABLE 14.1 Physical properties of formaldehyde.

Chemical formula	CH_2O
IUPAC name	Formaldehyde, formalin, methylene oxide, formic aldehyde, methyl aldehyde, methanal
Appearance	Colorless gas
Molecular weight	30.03
Melting point	−92°C
Boiling point	−21°C
Density at −20°C	0.815 g/mL
Odor	Pungent, suffocating odor; highly irritating
Solubility	Up to 55% in freshwater at 20°C, ether, alcohol, acetone, benzene
Geometry	Trigonal planar
Vapor pressure at 25°C	>1 atm
Polymerization	Readily polymerizes in water
Photolysis	Half-life (in sunlight) 1.6–19 h producing H_2 and CO or H^+ and HCO^-
Explosive limits	7%–73%
Autoignition temperature	300°C
Flashpoint	60°C
Flammability limits at 25°C	7%–73%
Incompatibilities	Reacts with alkalis, acids, and oxidizers

dermal absorption is very small. Due to its quick metabolism, inhaled formaldehyde does not change the concentration of formaldehyde in the bloodstream, which is approximately 2–3 mg/L (Fig. 14.2).

Formaldehyde forms methylol derivatives almost directly by reacting with primary and secondary amines, thiols, and hydroxyls. It normally behaves like an electrophile and reacts with deoxyribonucleic acid, ribonucleic acid, and proteins such as macromolecules to form reversible joints or irreversible cross-links. The formaldehyde that is absorbed can be oxidized in three different ways to form and emit CO_2 or may be converted into biological macromolecules via a tetrahydrofolate-dependent one-carbon biosynthetic path. In a small amount, it is produced as a normal metabolite as

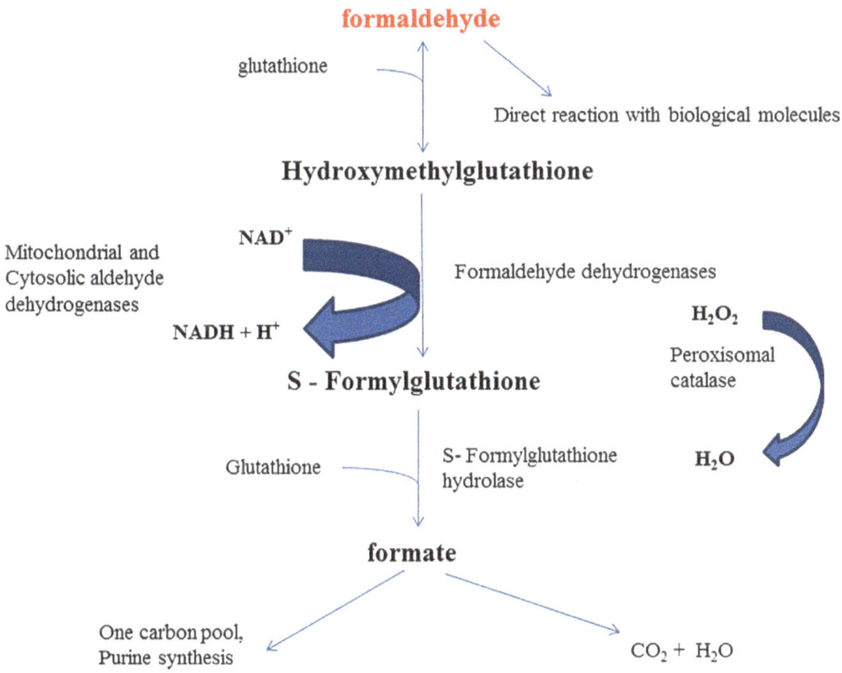

FIG. 14.2 Schematic representation showing the physiology of formaldehyde. *Credit: Preeti Das.*

well as in the oxidative demethylation of xenobiotics in the human body. It can also be found in the liver.

It takes about 1–1.5min to disappear from plasma; some of it is converted into CO_2 and passed out through the lungs. A very small amount may be excreted in the form of salts and a lot of other metabolites are excreted in the urine.

14.6 Uses

Formaldehyde is used in the preparation of resins, molding compounds, photographic films, tissue preservatives, and bactericides (Naya & Nakanishi, 2005). Generally, formaldehyde is a type of chemical that can be used in the preparation of many things required for the upgradation of daily life.

In building and construction work: Formaldehyde-based resins may be utilized for the preparation of wood products that comprise countertops, furniture, shelving, moldings, stair systems, flooring, wall sheathing, support beams, and a lot of other similar items and structures. Glues that use formaldehyde as a building block deliver high-quality performance that is extremely economical. The wood products industry uses formaldehyde-based resins in a wide range of panel and board products, enabling sustainable utilization of forestry resources and minimizing waste. For example, composite wood panels are typically made from recovered wood waste that would otherwise be burned or disposed of in a landfill.

In personal care and consumer products: Many personal care items direct utility of formaldehyde-based chemistry. These products contain formaldehyde-releasing ingredients that behave like preservative and also kill and prevent growth of bacteria and other microbes.

In health care applications: Formaldehyde also finds its use in the preparation of vaccines, antiinfective medicine, and hard-gel capsules.

In automobiles: The technology associated with formaldehyde helps make vehicles lighter and energy efficient. Formaldehyde resins may be used to produce internal molded components and under-the-hood components that need to withstand high-temperature. These resins are also used in the production of highly durable exterior primers, clear coat paints, tire-cord adhesives, brake pads, and fuel system components.

14.7 Sources of contaminants

Formaldehyde is a pollutant gas produced by human activities and natural sources. The noncommercial sources of exposure to formaldehyde include processes in the automobile industry, power plants, incinerators, refineries, stoves, and kerosene emissions. It is also present in tobacco smoke. Half of the formaldehyde present in the environment is removed from the air by photochemical processes, precipitation, and bioradiation (International Agency for Research on Cancer, 2012).

A small amount of formaldehyde is produced naturally by plants, animals, and humans. Formaldehyde is also found in rainwater and safe water released from manufacturing processes and industries.

14.8 Risk assessment

In April 1987, the Environmental Protection Agency (EPA) released "Assessment of Health Risks for Garment Workers and Some Domestic Habitats for Formaldehyde From Expellers," which stated that formaldehyde is a carcinogenic agent, and a material that irritates the eyes and human respiratory system. According to the draft, published in 1991, which includes data related to the epidemiology and toxicology of the formal assessment that EPA has received since completion of the earlier assessment, examines its effect on estimates of airborne formaldehyde (Hernandez et al., 1994). According to the details in the EPA guidelines, a study by Kerns et al. presented an extrapolation model for formaldehyde, which was different from the linearized multistage procedure suggested by the EPA's guidelines for carcinogen risk assessment (Wiltse & Dellarco, 1996). Under the Guidelines for Carcinogen Risk Assessment, the EPA has classified formaldehyde as a "potential human carcinogen," which is based on:

- Finite evidence for carcinogenesis in humans and animals.
- Supplementary supporting evidence (i.e., structure-activity, mutation, and mechanistic considerations).

Hazard identification: This technique has been put through the entrenched evaluation procedures of the Organic Methods Evaluation Branch. Because of the 1988 reduction of the formaldehyde permissible exposure limit (PEL), auxiliary evaluation data were also collected:

- The OSHA Fact Sheet (April 2011), which gives information regarding the harmful effects of formaldehyde on workers.
- The Immediately Hazardous to Life or Health document provided by the National Institute of Occupational Safety and Health (May 1994), which contains critical toxicity data for formaldehyde.
- TOXNET for formaldehyde, which is the National Library of Medicine Hazardous Substance Database.
- The US Department of Health and Human Services, which identifies and discusses substances, mixtures, or exposure conditions in the National Toxicology Program, and which may pose health hazards because of their carcinogenicity. The list of substances that are in the Report onCarcinogens indicates only one potential hazard. They do not establish at-risk conditions that may pose a cancer risk for individuals.

Risk analysis: Due to the ubiquitous sources of formaldehyde, the overall magnitude of relative contributions from various sources of exposure presented here is expected to be reasonably representative of those in other parts of the world. To determine the relative contributions from various media, estimates of the total daily intake of formaldehyde by six age groups of the general population of Canada were developed, which indicated that the daily intake of formaldehyde via inhalation was consistently less than that estimated for the ingestion of foodstuffs. However, the critical effects regarding the exposure to formaldehyde occurred primarily at the site of the first contact. The respiratory tract following inhalation and aerodigestive tract, including oral and gastrointestinal mucosa, following ingestion are related to the concentration of formaldehyde in media to which humans are exposed, rather than the total intake of this substance. Therefore, the effects of exposure by inhalation and ingestion are addressed separately. The principal focus of the assessment is airborne exposure because of the primary limitations of available data as a basis for the characterization of exposure via ingestion. The less representative assessment for ingestion comprises the comparison of the concentration of formaldehyde in a limited number of food products with bearable concentration (ingestion).

The methods used for determining exposure limits for workers are concentrated on proof concerning nasal/nasopharyngeal carcinogenicity and regarded it as a significant part of developing a more nonsegregated assessment in the general population where very low concentration exposure occurs. On the basis of several factors, an extreme workroom concentration of 0.3ppm was set up by the German MAK Commission in 2006 (Andersen et al., 2019):

- Very high levels of exposure were required due to DNA-protein crosslink and cell proliferation in rats.
- According to the biologically based dose-response model, the risk limits were between 0.3 and 15ppm for 6h.
- An evaluation showed that for 0.3ppm practically no observed adverse effect level for nasal and/or eye sensory irritation in workers existed.

Risk evaluation: With various stages placed on different information streams, e.g., toxins and epidemiological and biomarker studies in humans or mechanism of action studies in animals, current hazard classification and significant risk estimates for formaldehyde are different among official bodies and regulatory agencies. Measurable risks approximated that exposures to 1 ppb of formaldehyde ranged from 2.3×10^{-10} to 1×10^{-4}, i.e., over 6 orders of magnitude. These differences arose from large-scale decisions in the same way that formaldehyde caused cancer in mice. Various official bodies came to the conclusion that formaldehyde is a known human carcinogen for nasopharyngeal cancer and leukemia (International Agency for Research on Cancer, 2012). The effect is unlikely to occur beyond those tissues, such as the portal of entry, where cytotoxicity has been reported.

14.9 Environmental effects

Formaldehyde is a natural component of the human body and environment; however, it is not bioaccumulative. It easily biodegrades in air, water, and soil under both aerobic and anaerobic conditions. In air, formaldehyde breaks down in sunlight to form carbon monoxide and formic acid. It is rarely found in drinking water and occurs naturally in minimum quantities in foods such as cheese and grains, and is also added to kill pathogens.

14.10 Hazardousness

The consumption of formalin by humans may cause loss of perception, specific failure of blood circulation, pneumonia, inflammation of the kidney, and abortion and can injure the larynx and trachea (Bohmer, 1934; Kline, 1925). Using formaldehyde to decontaminate dental pulp may produce paresthesia, soft-tissue necrosis, and bone appropriation (Grossman, 1978; Heling, Ram, & Heling, 1977; Montgomery, 1976). Filters saturated with melamine-formaldehyde resin are associated with an outbreak of hemolytic anemia in hemodialysis patients (Orringer & Mattern, 1976).

Skin: Skin contact can cause irritation, allergy (Glass, 1961; Pirilä & Kilpiö, 1949), and inflammation (Blejer & Miller, 1966; Epstein & Maibach, 1966; Jordan, Sherman, & King, 1979; Marzulli & Maibach, 1973; North American Contact Dermatitis Group, 1973; Odom & Maibach, 1977; Rudner et al., 1975). Formaldehyde (shampoos, clothes, etc.) may contain amounts below the threshold for sensitization induction and waxing, rather than being allergic to daily exposure or contacting infrequently to produce a response (Marzulli & Maibach, 1977).

Eyes: Formaldehyde can produce eye irritation in the range of 4–20 ppm (Barnes & Speicher, 1942; Bourne Jr & Seferian, 1959; Morrill Jr, 1961; Sim & Pattle, 1957; Walker, 1964; Weber-Tschopp, Jermini, & Grandjean, 1976).

Respiratory system: Formaldehyde may cause conditions like irritation, dryness of nose and throat, and fatigue. In eight textile plants, symptoms of smell, itching mucous membranes, shortness of breath, tearing, excessive thirst, and restless sleepiness have been experienced by the employees (Shipkovitz, 1968). Cough, chest tightness, and wheezing are shreds of evidence for lower airway irritation. There are reports of dyspnea and asthma after acute inhalation of formalin vapor (Zanni & Russo, 1957), pulmonary edema, actual respiratory distress after 15 h of exposure to formaldehyde (Porter, 1975), pneumonitis, and death (Fassett, 1963).

Nervous system: Among plywood and particleboard workers, affective fatigue with olfactory thresholds for rosemary, thymol, camphor, and tar have been observed (Yefremov, 1970). Several nonsymptoms related to nervous system reactions such as thirst, headache, fainting, laziness, and inability to concentrate have been noted (Helwig, 1977; Weger, 1927).

CNS: Unhealthiness, headache, irritability, sleep disturbances, and loss of memory, dexterity, and balance may proceed exposure to a single, high-level, formaldehyde.

Dermal: Skin irritation may occur due to exposure to formaldehyde vapor or formalin. In sensitized individuals, contact dermatitis can develop at very low-risk levels.

14.11 Treatment

The effects of formaldehyde can be reduced by treatment but cannot be eliminated because it has no antidote. Patients who are at serious risk (symptoms such as tearing eyes, runny nose, or severe or persistent cough) may require immediate hospitalization. Emergency treatment or long-term effects may occur when exposed to concentrated vapor or liquid or in the case of swallowing.

Inhalation of formaldehyde may cause bronchospasm, pulmonary edema, airway irritation, etc. The concentration of large amounts of formaldehyde by either method can cause severe systemic toxicity, which can lead to tissue and organ damage, metabolic acidosis, or coma.

14.12 Risk management

- While working with any form of formaldehyde, proper engineering controls must be utilized.
- Formaldehyde should be used in laboratories in a functioning chemical fume hood.
- Other forms of local exhaust ventilation (e.g., canopy hood, elephant trunk) should be used in those areas where a chemical fume hood is unavailable.
- Respirators must be worn on exposure to vapors in case local exhaust ventilation is also not feasible or insufficient to keep the concentration below PEL.
- Those who are using respirators should have a complete medical evaluation and fitness test by the Division of Safety and Compliance.
- Minimum standard laboratory attire should be worn including long pants, closed-toe shoes, safety glasses with splash goggles or side shields, lab coat, and gloves.
- An experienced observer should be present to ensure that workers are not overexposed.

14.13 Current research

There are several methods for the analysis and testing of formaldehyde, leading to a new idea of the relevance of its emission and loading ratio factors of wood-based panels under the joint action of impacting. The quantitative indicators of wood-based panel load in a specific space according to the formaldehyde emission requirements we determined, which provide technical precondition support to improve the interior decoration and furniture design (Zhang, Song, Tao, Zhang, & Shi, 2018).

14.14 Formaldehyde toxicity: Case study

On January 21, 2018, at 5:30 p.m., an employee was working in a dairy parlor close to a formalin intermediate bulk container. The employee ingested 120–150 mL of methanol and formaldehyde solution in order to poison himself. He needed to have his stomach pumped and intubated; he was also hospitalized.

At approximately 7:00 p.m. on March 19, 1999, an employee was writing a report in the main laboratory when a coworker started a fumigation procedure using formaldehyde in the tissue culture room. The employee suffered from overexposure to formaldehyde. She informed the owner, left the facility, and sought medical attention.

Inhaled formaldehyde is an effective sensory irritant in mice at a dose of 0.5 ppm for 3 h/day for 3 days (Kane & Alarie, 1977). Rabbits and guinea pigs were kept for 10 days for 40–70 ppm formaldehyde inhalation, in which lacrimation was observed. The application of a drop of 0.5 mL of formalin to the rabbit eye was observed to cause edema of the cornea (National Research Council, 2014). When 0.1%–20% formaldehyde was applied to the skin of guinea pigs, skin irritation was observed (Naya & Nakanishi, 2005).

14.15 Safety measures

- Use plastic-supported absorbent pads. Pads can be disposed of after the completion of tasks.
- Only handle formaldehyde powder and solutions in a chemical fume hood or other approved control or glove box.
- If the weight-reducing formaldehyde is solid, it cannot be placed in a fume hood or biosafety cabinet. Hence, the following method should be used: With a vacant container and an open lid, move to the hood, pour the solid into the container, and apply the lid tightly.
- Use as much of the fume hood as possible, especially when agitating or transferring solutions.
- Only use formaldehyde-containing chemicals in spill-proof, labeled, nonbreakable secondary containers.
- Employers must also ensure that contaminated sharps are disposed of in sharps disposal containers immediately or as soon as feasible after use. Disposable syringes or needle locking syringes are recommended for use. Remove the blade from the scalpel with a scalpel blade remover.

- Dispose of hazardous chemicals in a suitable waste container.
- Label waste as nonhazardous waste before disposal. Keep gross contaminated items in double bags for hazardous waste pickup such as paper towels, gloves, and absorbent pads.
- After the work is completed, clean the area using detergent water and wash hands properly with water and soap.

14.16 Handling and storage

- While handling formalin solutions, keep away from sources of ignition.
- When not in use, keep containers tightly closed.
- Store properly in a well-ventilated place in containers of more than 4 L.
- Store away from heat and flame.
- Ensure that formaldehyde is stored in labeled chemically compatible containers.
- Always keep large quantities of containers on a low and protected shelf to prevent them from being accidentally dropped or knocked off.
- Do not store formaldehyde bottles in areas where leaks may flow into drains.
- While handling formaldehyde, wear neoprene, medium or heavyweight nitrile, PVC, or natural rubber gloves.

14.17 Precautions

- Formaldehyde must always be used with sufficient ventilation.
- Always wear goggles or a face shield, lab coat, and shorts or open-toe shoes to deal with formaldehyde to reduce the risk of vapor or small splashes to the cornea.

References

Andersen, M. E., Gentry, P. R., Swenberg, J. A., Mundt, K. A., White, K. W., Thompson, C., et al. (2019). Considerations for refining the risk assessment process for formaldehyde: Results from an interdisciplinary workshop. *Regulatory Toxicology and Pharmacology, 106*, 210–223. https://doi.org/10.1016/j.yrtph.2019.04.015.

ATSDR (Agency for Toxic Substances and Disease Registry). (1999). *Toxicological profile for formaldehyde*. Atlanta, GA: U.S. Department of Health and Human Services, Public Health Services, Agency for Toxic Substances and Disease Registry. https://www.atsdr.cdc.gov/MHMI/mmg111.pdf. (Accessed 8 August 2020).

Barnes, E. C., & Speicher, H. W. (1942). The determination of formaldehyde in air. *The Journal of Industrial Hygiene and Toxicology, 24*, 10–17.

Blejer, W. P., & Miller, B. H. (1966). *Occupational health report of formaldehyde concentrations and effects on workers at the Bayly manufacturing company, Visalia, Calif. Study report number S–1806*. Los Angeles, state of California health and welfare agency, Dept. of public health, Bureau of Occupational Health.

Bohmer, K. (1934). Formalin poisoning. *Deutsche Zeitschrift für die Gesamte Gerichtliche Medizin, 23*, 7–18.

Bourne, H. G., Jr., & Seferian, S. (1959). Formaldehyde in wrinkle-proof apparel produces; tears for milady. *Industrial Medicine & Surgery, 28*(5), 232.

Budavari, S., O'Neil, M. J., Smith, A., Heckelman, P. E., & Kinneary, J. F. (1989). The Merck index. In *An encyclopedia of chemicals, drugs and biologicals* (11th ed.). Rahway NJ: Merck & Co.

Butlerow, A. (1859). Ueber einige derivate des Jodmethylens [On some derivatives of methylene iodide]. *Annalen der Chemie und Pharmacie., 111*, 242–252.

Epstein, E., & Maibach, H. I. (1966). Formaldehyde allergy: Incidence and patch test problems. *Archives of Dermatology, 94*(2), 186–190.

Fassett, D. W. (1963). Aldehydes and Acetals. In F. A. Patty (Ed.), *Industrial hygiene and toxicology* (2nd ed., pp. 1959–1989). New York: Interscience.

Gerberich, H. R., Seaman, G. C., & Hoechst-Celanese Corporation. (1994). Formic acid. In *Vol. 11. Kirk-Othmer encyclopedia of chemical technology* (4th ed., pp. 929–951). New York: Wiley.

Glass, V. I. (1961). An outbreak of formaldehyde dermatitis. *New Zealand Medical Journal, 60*, 423–427.

Grossman, L. I. (1978). Paresthesia from N_2 or N_2 substitute: Report of a case. *Oral Surgery, Oral Medicine, and Oral Pathology, 45*(1), 114–115.

Heling, B., Ram, Z., & Heling, I. (1977). The root treatment of teeth with Toxavit: Report of a case. *Oral Surgery, Oral Medicine, and Oral Pathology, 43*(2), 306–309.

Helwig, H. (1977). How harmless is formaldehyde? *Deutsche Medizinische Wochenschrift, 102*, 1612–1613.

Hernandez, O., Rhomberg, L., Hogan, K., Siegel-Scott, C., Lai, D., Grindstaff, G., et al. (1994). Risk assessment of formaldehyde. *Journal of Hazardous Materials, 39*(2), 161–172. https://doi.org/10.1016/0304-3894(94)00067-0.

International Agency for Research on Cancer. (2012). Formaldehyde. In *IARC monographs on the evaluation of carcinogenic risks to humans, No. 100F*. Available from https://www.ncbi.nlm.nih.gov/books/NBK304432/.

Jordan, W. P., Sherman, W. T., & King, S. E. (1979). Threshold responses in formaldehyde-sensitive subjects. *Journal of the American Academy of Dermatology, 1*(1), 44–48.

Kane, L. E., & Alarie, Y. (1977). Sensory irritation to formaldehyde and acrolein during single and repeated exposures in mice. *American Industrial Hygiene Association Journal, 38*(10), 509–522. https://doi.org/10.1080/0002889778507665.

King, H. (2019). *Does molecule remove formaldehyde from the air*. Available from https://molekule.science/does-molekule-remove-formaldehyde-from-the-air/.

Kline, B. S. (1925). Formaldehyde poisoning: With report of a fatal case. *Archives of Internal Medicine, 36*(2), 220–228.

Marzulli, F. N., & Maibach, H. I. (1973). Antimicrobials: Experimental contact sensitization in man. *Journal of Cosmetic Science, 24*, 399–421.

Marzulli, F. N., & Maibach, H. I. (1977). Contact allergy: Predictive testing in humans. In F. N. Marzulli, & H. I. Maibach (Eds.), *Advances in modern toxicology, dermatotoxicology and pharmacology* (pp. 353–372). Washington, DC: Hemisphere Publishing.

Montgomery, S. (1976). Paresthesia following endodontic treatment. *Journal of Endodontics, 2*(11), 345–347.

Morrill, E. E., Jr. (1961). Formaldehyde exposure from paper process solved by air sampling and current studies. *Heating, Ventilation and Air Conditioning, 58*(7), 94–95.

National Research Council. (2014). Review of the formaldehyde assessment. In *National toxicology program 12th report on carcinogens*. Washington, DC: The National Academies Press.

Naya, M., & Nakanishi, J. (2005). Risk assessment of formaldehyde for the general population in Japan. *Regulatory Toxicology and Pharmacology*, *43*(3), 232–248. https://doi.org/10.1016/j.yrtph.2005.08.002.

North American Contact Dermatitis Group. (1973). Epidemiology of contact dermatitis in North America: 1972. *Archives of Dermatology*, *108*, 537–540.

Odom, R. B., & Maibach, H. I. (1977). Contact urticaria: A different contact dermatitis. In F. N. Marzulli, & H. I. Maibach (Eds.), *Advances in modern toxicology. dermatotoxicology and pharmacology* (pp. 441–453). Washington, DC: Hemisphere Publishing.

Orringer, E. P., & Mattern, W. D. (1976). Formaldehyde-induced hemolysis during chronic hemodialysis. *New England Journal of Medicine*, *294*(26), 1416–1420.

Pirilä, V., & Kilpiö, O. (1949). On dermatitis caused by formaldehyde and its compounds. *Annales Medicinae Internae Fenniae*, *38*(1), 38–51.

Porter, J. H. (1975). Acute respiratory distress following formalin inhalation. *The Lancet*, *306*(7935), 603–604.

Read, J. (1935). *Text-book of organic chemistry*. London: G Bell & Sons.

Rudner, E. J., Clendenning, W. E., Epstein, E., Fisher, A. A., Jillson, O. F., Jordan, W. P., et al. (1975). The frequency of contact sensitivity in North America 1972–74 north American contact dermatitis group. *Contact Dermatitis*, *1*(5), 277–280.

Shipkovitz, H. D. (1968). Formaldehyde vapor emissions in the permanent-press fabrics industry. In *Report no. TR–52. Cincinnati: U.S. Dept. of health, education, and welfare* Public Health Service. Consumer Protection and Environmental Health Service. Environmental Control Administration.

Sim, V. M., & Pattle, R. E. (1957). Effect of possible smog irritants on human subjects. *Journal of the American Medical Association*, *165*(15), 1908–1913.

Walker, J. F. (1964). Formaldehyde. *American Chemical Society monograph series no.159* (3rd ed.). New York: Reinhold.

Weber-Tschopp, A., Jermini, C., & Grandjean, E. (1976). Air pollution and irritations due to cigarette smoke. *Sozial- und Präventivmedizin*, *21*(2–3), 101–106.

Weger, A. (1927). Thalamic symptom complex in formalin intoxication. *Journal of Neurology and Psychiatry*, *111*(1), 370–382.

Wiltse, J., & Dellarco, V. L. (1996). U.S. Environmental Protection Agency guidelines for carcinogen risk assessment: Past and future. *Mutation Research/Reviews in Genetic Toxicology*, *365*(1–3), 3–15. https://doi.org/10.1016/s0165-1110(96)90009-3.

Yefremov, G. G. (1970). The state of the upper respiratory tract in formaldehyde production employees. *Zhurnal Ushnykh, Nosovykh i Gorlovykh Bolezneĭ*, *30*, 11–15.

Zanni, D., & Russo, L. (1957). Long-standing lesions in the respiratory tract following acute poisoning with irritating gases. *Lavoro Umano*, *9*, 241–254.

Zhang, J., Song, F., Tao, J., Zhang, Z., & Shi, S. Q. (2018). Research progress on formaldehyde emission of wood-based panel. *International Journal of Polymer Science*, 1–8. https://doi.org/10.1155/2018/9349721.

CHAPTER

15

Hydrogen cyanide: Risk assessment, environmental, and health hazard

Manila and Payal Devi
Department of Chemistry, Gurukula Kangri (Deemed to be University), Haridwar, Uttarakhand, India

Abbreviations

ATSDR	Agency for Toxic Substances and Disease Registry
ATP	adenosine triphosphate
IDLH	immediately dangerous to life or health
NIOSH	National Institute for Occupational Safety and Health
OSHA	Occupational Safety and Health Administration
PPI	pixels per inch
POI	proof of identity
PVC	poly vinyl chloride
SCBA	self-contained breathing apparatus

15.1 Introduction

Hydrogen cyanide (HCN) is a colorless or pale blue (hydrocyanic acid) extremely volatile liquid that is a potential threat to life at room temperature. The odor of HCN has a faint resemblance to bitter almonds. Generally, it is produced commercially and further used as a reagent in several synthetic procedures ranging from polymer to pharmaceutical industries.

It is often used as a 96% solution in water. This volatilizable composite is fatal not only to humans but also to rodents and whales, the reason for which might be the complexation with ferric ion present in metalloenzymes. HCN is a starting material for nylon 66 which is known for its wide range of applications. It is also used as an intermediate for the production of acrylonitrile that finds its use in the synthesis of acrylic fiber, synthetic

https://doi.org/10.1016/B978-0-323-89857-7.00010-4

rubber, and plastics. It finds a major role as an insecticide for fumigation of enclosed areas.

The toxicity of HCN is responsible for the obstruction of the cellular oxidative processes. The normal human tolerance limit of cyanide is up to 50–60 ppm that can turn life-threatening when exposed to concentrations of 200–500 ppm for 15 min. In earlier times HCN was used as a method for giving capital punishment.

Because of its hazardousness, it is not preferred as a solvent although it is highly stable and polymerizes with ammonia or sodium cyanide. In electrolytic processes, this salt is used for the extraction of minerals and the handling of steel.

15.2 Occurrence

It has been assumed that a massive bombardment from the cascade of carbon asteroids resulted from the communication of Jupiter and Saturn, blasted the earth's surface on reacting with nitrogen in the air to form HCN. Minute concentrations of HCN have been found to be available in the stratosphere also. A few theories support that HCN played a crucial role in the origin of fauna and flora on this planet as it is capable of the formation of amino acids. It is present in human blood also in trace amounts. A few bacterial, algal, and fungal species are known to produce HCN. It can be obtained from pit fruits such as cherries, apricots, apples, and bitter almonds as pits contain cyanohydrins such as mandelonitrile and amygdalin (Vetter, 2000). Some millipedes also release HCN post activation of their defense mechanism.

HCN occurs in small amounts in plants, where it remains in a combined form with sugar. It is largely used in laboratories and commonly prepared by breaking formamide ($HCONH_2$), treating sodium cyanide with sulphuric acid, or oxidizing combination of methane-ammonia.

Some authors have reported the activation of opioid receptors by endogenous or exogenous opioids that results in the release of HCN by neurons. It also plays a vital role in neuronal production and neuronal cells. In addition, increased endogenous neuronal HCN production requires opioid analgesia, which was observed by scavengers of Oracle Partner Network. They considered endogenous HCN as a neuromodulator (Borowitz, Gunasekar, & Isom, 1997).

15.3 Historical background

Since the ancient days of Rome, cyanide and its derivatives have been used as a weapon (Sykes, 1981). Nero, the famous Roman emperor poisoned his family members and others who displeased him with cherry laurel water having cyanide as its chief toxin. Napoleon proposed the use of cyanides to enhance the effectiveness of his soldier's bayonets during the Franco-Prussian war. During World War II, the Nazis employed hydrocyanic acid adsorbed onto a dispersible pharmaceutical base (Zyklon B) to kill millions of civilians and enemy soldiers in the death camps. It was one of the gases employed in gas chambers used for high degree torture. In the 1980s, reports indicating the use of cyanide against residents of the Syrian city of Hama and its Kurdish city, and possibly Shahabad, Iran during the Iraq War in Iran also surfaced.

15.4 Physicochemical properties

HCN is a colorless liquid with a bitter almond-like odor, boiling point 78°F, density 5.7 lb./gal, flashpoint 0°F, and evaporates easily (or boils) at room temperature. Vapor is slightly lighter than air. Some properties are discussed in Table 15.1 (National Center for Biotechnology Information, 2020).

Since 1981, three groups have reported spectroscopic detections and measurements of HCN in the atmosphere. The HCN is linear with a triple bond between hydrogen and carbon. It is a weak acid and moderately ionized in water to give ionic cyanide. The cyanide anion salts are also known as cyanide.

TABLE 15.1 Physical Properties of hydrogen cyanide.

Chemical formula	HCN
IUPAC name	Formonitrile
Synonyms	Prussic acid, hydrocyanic acid, Cyclone B
Molecular weight	27.0253 g/mol
Form	Colorless gas and liquid
Density	$0.6876\,g\,cm^{-3}$
Boiling point	25.6°C
Melting point	−13.4°C
Odor	Faint bitter almonds
Solubility in water	Ethanol and ether
Vapor pressure	100 kPa (25°C)
Shape	Linear

15.5 Physiological role

HCN is produced by the combustion of materials prepared from nitrogen and carbon, such as fabric, silk, cotton, and paper, as well as gaseous forms of cyanide in polymers and synthetic materials. Combustion of these materials near fire sources can result in the rapid and fatal inefficiency of a victim (Traber, Herndon, Enkhbaatar, Maybauer, & Maybauer, 2012).

HCN may be generated by brain cells (Gunasekar et al., 2004; Gunasekar, Borowitz, Turek, Van Horn, & Isom, 2000) and neutrophils with little evidence for acting as neuromodulators (Cipollone & Visca, 2007). The probable route of cyanide metabolism is given in Fig. 15.1 (Asiah et al., 2014).

15.6 Uses

1. HCN as a precursor of sodium cyanide and potassium cyanide is commonly employed for electroplating and mining of metals such as silver and gold.
2. A wide variety of organic compounds are manufactured from HCN through the intermediate like cyanohydrins including various monomers, and chelating agents.
3. The adiponitrile, precursor of nylon-6, 6 can be prepared by the hydrocyanation method involving the addition of butadiene to HCN (Gail et al., 2011).

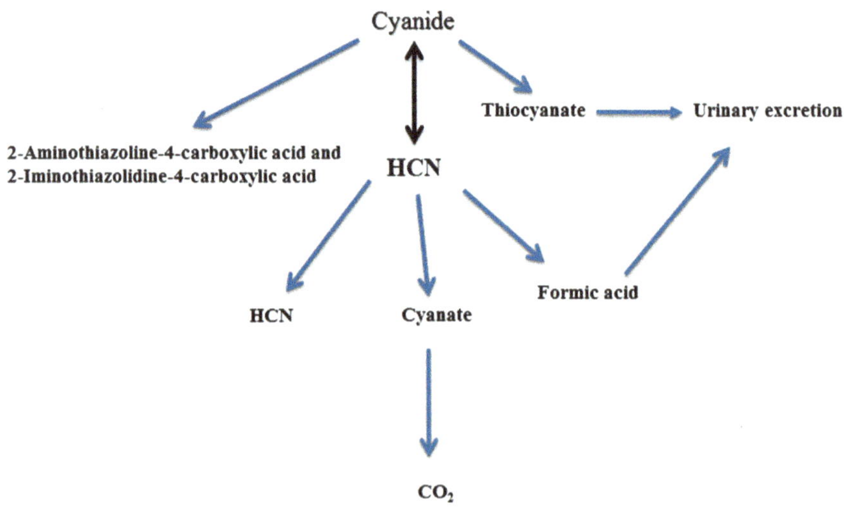

FIG. 15.1 The probable routes of cyanide metabolism. *Credit: Manila.*

4. HCN is taken as the precursor to amino acids and nucleic acids that plays an important role in the source of life (Matthews, 2000). Numerous ideas have been discovered in the differential regions for the study of organic compounds derived from HCN concentrations (Al-Azmi, Elassar, & Booth, 2003).
5. The presence of HCN in the interstellar medium and the atmosphere of carbon stars (Snyder & Buhl, 1971) has widened the examination of its formation and destructive pathways. Pure rotational transitions are seen from $J=1 \rightarrow 0$, $J=3 \rightarrow 2$, $J=4 \rightarrow 3$, and $J=10 \rightarrow 9$ (Bieging, Shaked, & Gensheimer, 2000; Schilke & Menten, 2003).

15.7 Sources of contaminants

Cyanide ions are mainly present in the form of HCN. The source of contamination involves the escapism of HCN from mines, firefighting, and metallurgical chemical and galvanic industries during the gold extraction process especially from low-grade ores; a solution of sodium cyanide tricked onto pads of crushed ore.

15.8 Risk assessment

HCN is an extremely toxic gas that has a high solubility in the air and can lead to the formation of explosive mixtures. The population at risk is mainly due to occupational exposure, which is the class of workers working in the manufacturing facilities involved either in the production of HCN or its usage for preparation of other products like, nylon, pesticides, intermediate reagents, etc. In order to avoid any mishap, a certain systematic procedure has to be followed which comes under this heading. Risk assessment involves the identification of the hazard, population at risk, its sources of contamination, and then techniques to avoid any accident leading to loss of lives or any destruction to the ecosystem as well.

Population at risk: The population at risk here is mainly the workers working in the industrial units as discussed earlier. Exposure might take place through pesticide fumigation also. The general population might be exposed to HCN either from certain foods or cigarette smoke.

Hazard identification: As far as hazard identification is concerned, HCN is well known for its toxicity. Although acute oral doses of cyanide cause cardiac, respiratory, and neurophysiological changes, the organs most susceptible to acute poisoning appear to be the central nervous system and heart. Several studies of the acute effects of cyanide in humans, following suicide attempts or accidental poisoning by oral and inhalation paths, have been followed, but a limited characterization of the risk has been identified (Newhouse, Chiu, Haber, Strawson, & Stern, 2010). HCN can enter the human body from different paths like inhalation, ingestion, absorption through the skin, or eye contact also.

Risk evaluation: HCN is a very useful product due to its broad spectrum of applications in various fields; therefore, its manufacturing should be done under controlled conditions barring all risk creating situations. All precautions must be followed without any relaxation by systematic procedures to make sure all safety measures are in place. The management team completely dedicated to ensuring the safety of the plant should be in place. Periodic assessment of all the equipment and workers processing HCN must be done to keep up the credibility of the risk assessment process.

15.9 Environmental effects

HCN is more likely to occur in the atmosphere as a result of industrial activities such as incidental release from drums or bottles and dust particles. Gaseous HCN cannot be easily removed from wind, rain, or snow as it has a half-life of approximately 1–3 years. The major amount of HCN evaporates from surface water but still has no significant harmful effects on humans.

Airborne cyanide is generally well below levels that could be a cause for concern. Some cyanide in water turns into less harmful chemicals by microorganisms (very small sized plants and animals) or forms a complex with metals such as iron (Singh & Chawla, 2018).

15.10 Hazardousness

Generally, a lower concentration of cyanide has a negligible effect; however, exposure to the concentration greater than 1000 μg/m^3 may show symptomatic toxicity and 7000 μg/m^3 may show signs of thyroid enlargement. In addition to the above, the exposure for at least 8.5 months or 24-h in a confined area of 27m^3 at a concentration of >17,000 μg/m^3 or > 8000 μg/m^3, respectively, may lead to death (Asiah et al., *2014*).

Some after exposure health effects are explained below:

Dermal effects: On the skin, exposure may cause irritation and burns.

Endocrine effects: Results in thyroid effect as it affects iodine uptake by the interference of thiocyanide, a metabolite of cyanide.

CNS effects: Higher-level exposure causes CNS exacerbation followed by depression, spasms, abolished deep reflexes and thin pupils, paralysis, or death.

Hematological effects: There are reports of increased hemoglobin and lymphocyte count on the exposure of 6.4–10.4 ppm of an unspecified cyanide form during electroplating (Houeto, Hoffman, & Imbert, 1995). One study observed an increase in neutrophil values and erythrocyte sedimentation speed, and a decrease in hemoglobin levels in workers exposed to unspecified concentrations of cyanide indefinitely during case hardening and electroplating (Ko & Wang, 2005).
Hepatic effect: An increase in cyanide alkaline phosphatase was noted in workers exposed to unspecified levels of cyanide; however, serum bilirubin was found to be within normal limits in these workers.
Cardiovascular effects: The common effects reported include palpitations and hypotension post incidental cyanide inhalation.
Renal effect: Occupational exposure of 200 ppm HCN may cause anuria followed by polyuria (El Ghawabi et al., 1975).

15.11 Treatment

Cyanide poisoning can be managed with specific antidotes and supportive medical care. The physiological pathway of toxicity includes ferric ion binding to mitochondrial cytochrome oxidase that hinder the transport of electrons in the cytochrome system and inhibit oxidative phosphorylation and ATP production. The inhibition of oxidative metabolism increases the demand for glycolysis, that results in lactic acid production causing a severe acid-base imbalance.

Besides, the bitter almond scent of HCN gas, the blood levels may be useful in documenting exposure. In all serious cases of exposure to cyanide, a successful outcome hinges on the immediate provision of basic life support and decontamination, and follow-up with excellent supportive care. Oxygen, decontamination, nitrites, and sodium thiosulfate are the main therapeutic modalities. A small amount of cyanide can also be detoxified by the vitamin B_{12} precursor, hydroxy cobalamin. This chelating agent binds cyanide to form nontoxic cyanocobalamin.

15.12 Risk management

OSHA determines cyanide levels and guidelines that are in the workplace air. The permitted exposure limit for HCN and most cyanide salts is 10 ppm or 11 mg cyanide per cubic meter (mg/m^3) on 8-h workdays and 40-h days. NIOSH also determines level that are harmful to human life and health. The IDLH level for most cyanide salts is 50 ppm for HCN or 25 mg/m^3 as cyanide.

15.13 Current research

HCN has been believed to be one of the first molecules present on early Earth that was brought down by some comet or asteroid. This theory has been a topic of continuous research overtime again by a number of researchers. One such research regarding the role of HCN in prebiotic chemistry has been formulated after a few experimental observations. Experiments performed under artificial conditions that might exist during that period of time indicate HCN to be a feedstock molecule responsible for the formation of biomolecules like sugars, ribonucleotides, amino acids, etc. This study has also assessed the relationship between the amount of HCN present in the comet or asteroid and its impact time along with its magnitude while striking the earth surface thousands to millions of years ago (Todd & Öberg, 2020).

Moreover, Lino et al. explained the observed HCN augmentation on pole and equator through the meridional point of view, which suggests accumulation and production of HCN at the downward branches of the two-cell meridional circulation model in the context of abundant availability of HCN in Neptune's stratosphere. The data was retrieved from the archived 2016 observation of the Atacama Large Millimeter/submillimeter Array. Cometry collisions or the tropospheric N_2 transportation was believed to be the reason behind the spatial distribution of HCN (Iino, Sagawa, Tsukagoshi, & Nozawa, 2020).

HCN is an industrially viable chemical that is manufactured from ammonia and natural gas. However, a separate chemical process is available for HCN purification from hydrogen and water along with recovery of leftover ammonia. Earlier solvent absorption technique was used for this purpose, which has been replaced by gas separation membrane technique. A study on the effectiveness of gas membrane as a potentially improved technique has been done in which HCN selectivity over two different polymeric membranes was examined by studying permeability of HCN. The permeability of Rubbery polymeric membranes showed a strong correlation with respect to the condensability of the gases, which was found to be $H_2O > H_2 \approx HCN > N_2$. Glassy polymeric membranes showed disparity as they corresponded their order of permeability with the kinetic diameter of the gases, giving the order as $H_2O > H_2 > HCN \approx N_2$. An attempt had also been made to develop a mechanism for the facilitation of transferring HCN within the two types of membrane mentioned earlier with the aid of metal chlorides. A complexation between HCN and metal ion not only facilitated the diffusion of HCN within the membranes but also enhanced the levels of HCN inside the polymeric membranes. An improved HCN/Zn selectivity was observed for the two membranes under study which was further supplemented by the presence of water vapor, as it increases the complexation between HCN and the metal ion, gaining an increase in its selectivity (Scholes, 2020).

15.14 Cyanide toxicity: Case studies

Fluor Daniels Employees Suffer, Unconfirmed Exposure to Cyanide: On 15 October 1994 an incident of HCN poisoning took place where Fluor Daniels (F/D) employees were removing an I-beam used as a monorail for a hoist in preparation for replacing a 100% HCN tank. One of the pipefitters reported about the 13-ppm cyanide reading on his monitox, which enhanced to 161 ppm within the time period of 20 min. The pipefitter took it back to the control room while the other F/D employees left the area. At this point, the pipefitter began feeling ill, coughing up blood, and experiencing difficulty breathing. The alarm did not rang because detector tubes and the perimeter monitoring system indicated nondetection during the incident. Overall, 34 employees from the four different nearby companies were affected. The weather that day consisted of heavy rain and 100% humidity. The monitoxes were sent to the manufacturer, who said that water was inside them and hence they could not find traces of HCN on the monitoxes. An outside operator stated that she had her monitox fail high after getting it wet or dropping it. The manufacturer's operating manual specifies operating limit of 20%–95% relative humidity. The Drager detector tubes also have a relative humidity range of less than 80%. Employee interviews indicate that no one smelled or tasted cyanide or anything else, except that the union representative stated that some of the Rohm and Haas employees could smell 81R from the semi-works next to N-5.

The accident of October 1999, the 14th: A leak occurred in a pipe between a condenser and a liquid separator process with a normal temperature and atmospheric pressure. About 200 kg HCN was emitted due to a ruptured stump, which was plugged with the polymerized product. However, due to a high level in the crude product storage tank and the omission of opening a proper release valve, an overpressure took place. Immediate preventive actions were taken including shutting off the plant.

The accident of March 29th, 2000 explosion: The accident occurred on the HCN production installations, at the level of the separator tank on the residual gas transfer line between the head of the HCN absorption tower and the boiler. Several incidents were noted on previous days and finally, the workshop was evacuated on March 29th at 4.15 p.m. before the stripping operation was launched. The order to evacuate was canceled 15 min later and 9 employees came back to the workshop. The explosion occurred at 4.40 p.m. at the level of the separator tank (H1191) at 5 m high on the gas line. Two analyzers (Toxguard) nearby (10 to 15 m) indicated 30 ppm of HCN; the concentration dropped after 5 min. Other equipment located 100 m from the unit detected values in the order of 4 to 5 ppm of HCN from time to time. The shop employees were evacuated to a withdrawal zone and the stripping operation was shut down immediately. The unit's Toxguard detector which triggered the alarm at 4.45 p.m. for an HCN concentration greater than 5 ppm, activated a gas alert and the 300 people present in the establishment confined themselves. The POI was put into operation at the same time and the site firemen intervened with self-contained breathing apparatus (SCBA). Water curtains were set up around the installations. Concentration measurements conducted on the periphery of the workshop recorded values less than 2 ppm of HCN. The header line of the HCN absorption tower was isolated from the HCN desorption tower (H2241) at 5.30 p.m. Only the detector located near the separator unit indicated, still 30 min later, a concentration of 15 ppm of HCN. As no consequences for the personnel or the environment exposure was signaled at 7.00 p.m., the operator felt there was no need to call the authorities to initiate the PPI (the special intervention plan) and lifted its POI (internal contingency plan), as well as the employee confinement order. The classified installations inspectorate and the mayors of the three surrounding communities were informed, although the "préfecture" was not informed by the operator.

15.15 Safety measures

- *Skin Contact*: If skin or hair contact occurs, immediately remove any contaminated clothing and place it in a sealed bag for decontamination or disposal. Wash skin and hair thoroughly with running water.
- *Eye Contact*: Immediately wash in and around the eye area with large amounts of water for at least 15 min. Eyelids to be held apart. Remove clothing if contaminated and wash skin.
- *Breathing*: If a person breathes an overdose of HCN, immediately move that person into the fresh air. Perform artificial respiration if breathing ceases.
- *Swallowing*: When HCN is swallowed and the person is conscious, immediately give the person a large amount of water. Try to induce vomiting in the person after drinking water.
- *Rescue*: Remove the person affected from the hazardous location. If the exposed person is away, notify someone else, and implement established emergency rescue procedures. Do not be a casualty. Understand the procedures leading to rescue and facility in case of emergency and take the rescue equipment to the appropriate places before the need arises.

15.16 Storage and handling

15.16.1 Storage

- Store cylinders in a cool, dry, well-ventilated location away from sources of heat, ignition, and direct sunlight.
- This gas should be discarded by the manufacturer if the shelf life exceeds 90 days.
- The boiling point, toxicity, and instability in the presence of moisture should not be ignored while storing HCN.
- No attempts to repair or remove cylinders containing gas mixture should be made. In case of malfunctioning or any operational issue, the nearest distributor must be called up immediately.
- Cylinders should be positioned so as they do not fall or knock over.
- These cylinders are nonrefillable. Therefore, no such attempts must be made.

15.16.2 Handling

Handling must be carried in the presence of trained personnel. The handler should be well equipped. If in any case HCN is released, the place must be evacuated until the severity of its release is determined by trained personnel. Emergency equipment should be on standby always. Liquid HCN should be used in a fume hood. Protective equipment like chemical splash goggles and impermeable gloves made from Teflon or butyl rubber must be worn during handling this chemical. Do not use polyvinyl chloride (PVC) or polyethylene.

Liquid HCN is highly flammable so it should be kept away from ignition sources. Welding operations, flames, or hot surfaces must be avoided where HCN is being handled. A stabilizer (usually phosphoric acid) is added which might decompose over the period of time leading to an explosion in old kept samples. Hence, the sample shelf life must not be more than 90 days. For this, the date should be properly mentioned on the cylinders. All valves must be closed and properly checked after withdrawing HCN from the cylinder. HCN should never be trapped between the valves.

15.17 Precautions

- Personnel employed should use impermeable clothing, face covering, gloves, and other appropriate protective clothing to prevent skin exposure.
- Eating and smoking should not be permitted in areas where cyanide or liquids are processed or stored.

- Employees must use dust and safety goggles so that HCN does not come in contact with their eyes.
- Clothes that are likely to contain HCN should be replaced and kept in containers till disposed of.
- Employees handling HCN should wash their hands carefully with soap or detergent and water before eating, smoking, and using toilet facilities.

References

Al-Azmi, A., Elassar, A. Z. A., & Booth, B. L. (2003). The chemistry of diaminomaleonitrile and its utility in heterocyclic synthesis. *Tetrahedron*, *59*(16), 2749–2763.

Asiah, M. N., Adon, M. Y., Normi, M., Amir, K. M., Tahir, A., Haniza, M. S. , et al. (2014). Cyanide level in the environment and occupational settings: A systematic review. *British Journal of Applied Science & Technology*, *4*(20), 2851–2863.

Bieging, J. H., Shaked, S., & Gensheimer, P. D. (2000). Submillimeter and millimeter-wavelength observations of SiO and HCN in circumstellar envelopes of AGB stars. *Astrophysical Journal*, *543*(2), 897–921.

Borowitz, J. L., Gunasekar, P. G., & Isom, G. E. (1997). Hydrogen cyanide generation by mu-opiate receptor activation: Possible neuromodulatory role of endogenous cyanide. *Brain Research*, *768*(1–2), 294–300.

Cipollone, R., & Visca, P. (2007). Is there evidence that cyanide can act as a neuromodulator? *IUBMB Life*, *59*(3), 187–189.

El Ghawabi, S. H., Gaafar, M. A., El-Saharti, A. A., Ahmed, S. H., Malash, K. K., & Fares, R. (1975). Chronic cyanide exposure: A clinical, radioisotope, and laboratory study. *Occupational and Environmental Medicine*, *32*(3), 215–219.

Gail, E., Gos, S., Kulzer, R., Lorösch, J., Rubo, A., & Sauer, M. (2011). Cyano compounds, inorganic. In *Ullmann's encyclopedia of industrial chemistry*. Weinheim: Wiley-VCH.

Gunasekar, P. G., Borowitz, J. L., Turek, J. J., Van Horn, D. A., & Isom, G. E. (2000). Endogenous generation of cyanide in neuronal tissue: Involvement of a peroxidise system. *Journal of Neuroscience Research*, *61*, 570–575.

Gunasekar, P. G., Prabhakaran, K., Li, L., Zhang, L., Isom, G. E., & Borowitz, J. L. (2004). Receptor mechanisms mediating cyanide generation in PC 12 cells and rat brain. *Neuroscience Research*, *49*, 13–18.

Houeto, P., Hoffman, J. R., & Imbert, M. (1995). Relation of blood cyanide to plasma cyanocobalamin concentration after a fixed dose of hydroxocobalamin in cyanide poisoning. *Lancet*, *346*, 605–608.

Iino, T., Sagawa, H., Tsukagoshi, T., & Nozawa, S. (2020). A belt-like distribution of gaseous hydrogen cyanide on Neptune's equatorial stratosphere detected by ALMA. *The Astrophysical Journal Letters*, *903*(1), L1.

Ko, S. Y., & Wang, T. L. (2005). Chemical poisoning: Cyanide poisoning. *Annals of Disaster Medicine*, *4*(1), S19.

Matthews, C. N. (2000). The HCN world: Establishing protein-nucleic acid life. *Origins of Life and Evolution of the Biosphere*, *30*(2/4), 292–293.

National Center for Biotechnology Information (2020). *PubChem Compound Summary for CID 768, Hydrogen cyanide.* Available from: (2020). https://pubchem.ncbi.nlm.nih.gov/compound/Hydrogen-cyanide.

Newhouse, K., Chiu, N., Haber, L., Strawson, J., & Stern, B. R. (2010). *Toxicological review of hydrogen cyanide and cyanide salts.* Washington, DC: U.S. Environmental Protection Agency.

Schilke, P., & Menten, K. M. (2003). Detection of a second, strong sub-millimeter HCN laser line toward carbon stars. *Astrophysical Journal*, *583*(1), 446–450.

Scholes, C. A. (2020). Hydrogen cyanide recovery by membrane gas separation. *Chemical Engineering Journal*, *386*, 124049.

Singh, J., & Chawla, M. (2018). Inorganic toxicity: Cyanide. In *Inorganic toxicity: Environment and human health* (pp. 180–192). Lambert Academic Publishing.

Snyder, L. E., & Buhl, D. (1971). Observations of radio emission from interstellar hydrogen cyanide. *Astrophysical Journal*, *163*, L47–L52.

Sykes, A. H. (1981). Early studies on the toxicology of cyanide. In *Cyanide in biology*. New York: Academic Press.

Todd, Z. R., & Öberg, K. I. (2020). Cometary delivery of hydrogen cyanide to the early earth. *Astrobiology*, *20*(9), 1109–1120.

Traber, D. L., Herndon, D. N., Enkhbaatar, P., Maybauer, M. O., & Maybauer, D. M. (2012). The pathophysiology of inhalation injury. In *Total burn care* (pp. 219–228)(4th ed.). .

Vetter, J. (2000). Plant cyanogenic glycosides. *Toxicon*, *38*(1), 11–36.

CHAPTER

16

Hydrogen selenide: Risk assessment, environmental, and health hazard

C. S. Mathela[a] *and Payal Devi*[b]

[a]Former Professor and Head, Department of Chemistry, Kumaun University, Nainital, Uttarakhand, India

[b]Department of Chemistry, Gurukula Kangri (Deemed to be University), Haridwar, Uttarakhand, India

Abbreviations

ACGIH	American Conference of Governmental Industrial Hygienists
AIGA	Asia Industrial Gases Association
EDTA	ethylenediaminetetraacetic acid
ERPG	Emergency Response Planning Guideline
IDLH	Immediately Dangerous to Life or Health
NIOSH	National Institute for Occupational Safety and Health
OSHA	Occupational Safety and Health Administration
PPE	personal protective equipment
US EPA	United States Environmental Protection Agency
WHO	World Health Organisation

16.1 Introduction

Hydrogen selenide (H_2Se) is an inorganic hydrogen chalcogenide existing in nature in the gaseous form (Malczewska-Toth, 2012). This hydride of selenium is highly flammable and toxic at room temperature and atmospheric pressure (AIRGAS, 2016). It is chiefly transported as a liquefied compressed gas under vapor pressure of 0.950 MPa (138 psi) at 20°C (68°F). Moreover, it possesses a disagreeable penetrating odor with a threshold limit of less than 0.1 ppm and with higher concentrations causes olfactory fatigue (Asia Industrial Gases Association (AIGA), 2017). H_2Se

Hazardous Gases
https://doi.org/10.1016/B978-0-323-89857-7.00002-5

rapidly breaks down into elemental selenium and water, in order to minimize its toxic effects. It is only toxic when exposed to a concentration of 0.05 ppm for more than 8 h (Dudley & Miller, 1941; Friberg, Nordberg, & Vouk, 1979; Wilber, 1980).

16.2 Occurrence

Hydrogen selenide may be synthesized directly from elements by heating at elevated temperatures. It is prepared by heating selenium metal in a sealed tube with hydrogen at 440°C (Eq. 16.1) (Langner, 2000; Pace et al., 2017) or with selenium vapor over pumice stone (Hoffmann & King, 2010):

$$H_2 + Se \rightarrow H_2Se \quad (16.1)$$

Alternatively, the compound may be prepared by treating an alkali metal selenide with hydrochloric acid during mild heating, Eq. (16.2):

$$NaSe + 2HCl \rightarrow H_2Se + NaCl \quad (16.2)$$

Several suitable synthetic routes to H_2Se for both large- and small-scale preparations have been reported. In the laboratory, H_2Se is usually prepared by the action of water on aluminum selenide (Al_2Se_3), associated with the formation of hydrated alumina (Brauer, 1963; O'Neil, 2006), as shown in Eq. (16.3).

$$Al_2Se_3 + 6\,H_2O \rightleftharpoons 2\,Al(OH)_3 + 3\,H_2Se \quad (16.3)$$

16.3 Physicochemical properties

Hydrogen selenide is a strong reducing agent and has been found to be an extremely toxic gas with a reported LC50 of 0.3 ppm per 8 h breathing. It has a bent structure with a 91 degree angle between all their atoms that is also demonstrated by the three IR vibrational frequencies at 1034, 2345, and 2358 cm^{-1}, respectively. H_2Se has a pK_a value of 3.89. On the contrary, it is more acidic and less stable than hydrogen sulfide. It reacts violently with nitric acid and ignites to give off selenium dioxide and water (Yaws, 2001).

Additionally, it is also found to be incompatible with oxidizing materials, members of the halogen family, acids, and other combustible materials. However, it reacts vigorously with oxidizers such as potassium permanganate, sodium hypochlorite, oxygen, ozone, chlorine, fluorine, and nitric oxide. It also tends to react with the alkali metal family. However, H_2Se, like other hydrides, is not known to polymerize.

The physical properties exhibited by the hydrogen selenide are given in Table 16.1.

TABLE 16.1 Physical properties of hydrogen selenide.

Serial no.	Property name	Property
1	Chemical formula	H_2Se
2	Molar mass	80.99 g/mol
3	Physical state	Colorless gas
4	Odor	Extremely disagreeable
5	Boiling point	−41.25°C (at 207 K)
6	Melting point	−65.73°C (at 231 K)
7	Solubility	Less soluble in water (about 0.70 g/100 mL of H_2O); Strongly soluble in carbon disulfide, carbonyl chloride, and phosgene
8	Density	3.553 g/dm^3
9	Vapor pressure	9.5 atm (at 294 K)

16.4 Physiological role

Se compounds have been found to be more reactive in comparison with counterpart selenium compounds due to a kinetically more labile and easy oxidizable nature. Selenium (Se) has been recognized to likely degrade to an array of organoselenium compounds in microorganisms, plants, and animals. H_2Se is a highly reactive intermediate amongst all, where several pathways converge for the biosynthesis of Se-compounds.

Some studies indicate glutathione (GSH) as a vital cofactor. The enzyme GSH reductase decomposes GSSeSG to H_2Se by means of GSH selenopersulfide (GSSeH) (Alexander, 2015). It is followed by methyltransferases-catalyzed reaction to form exhalable dimethyl selenide. Alternatively, TrxR and thioredoxin may also catalyze the conversion of selenite to H_2Se (Björnstedt, Kumar, & Holmgren, 1992). H_2Se and methylselenol, enhanced consumption of NADPH, and generation of reactive oxygen species (ROS) are chiefly resultant of reduction of selenite in the presence of oxygen (Anundi, Ståhl, & Högberg, 1984; Björnstedt et al., 1992; Fernandes et al., 2012; Garberg & Högberg, 1987; Lu, Berndt, & Holmgren, 2009; Ståhl, Anundi, & Högberg, 1984). Furthermore, the reduction to selenite takes place by sulfate reduction pathway (Boehler, Raines, & Sunde, 2013); however, the pathway is still a matter of research.

Selenium is taken up from dietary supplements either replaces methionine into proteins (Schomburg, Schweizer, & Köhrle, 2004) or coverts to selenocysteine by cystathione β-synthase and γ-lyase via selenocystathione. Consequently, Selenocysteine converts to H_2Se by selenocysteine

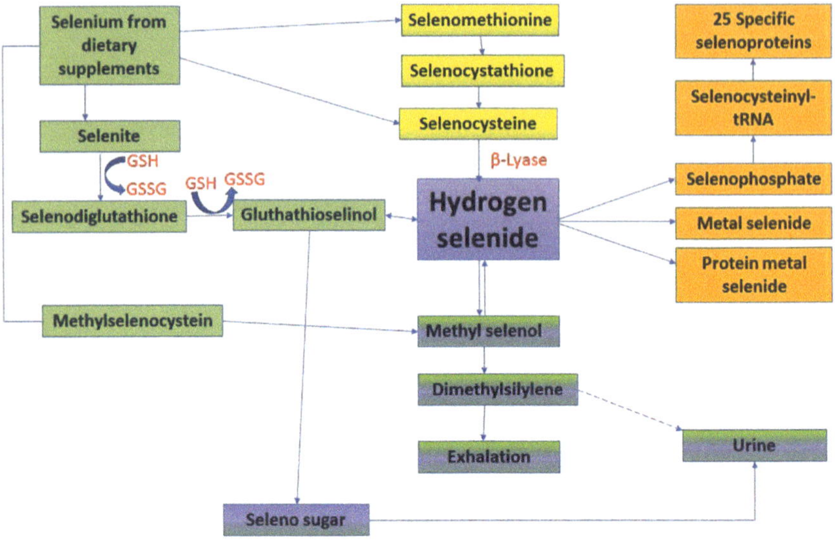

FIG. 16.1 Bioconversion pathways of selenium compounds to excretable metabolite. *Credit: Payal Devi.*

β-lyase (Esaki et al., 1981). On the other hand, Suzuki and the team suggested further degradation of hydrogen selenide to selenosugars via GSSeH, with the major species being SeGalNAc (Kobayashi et al., 2002), as an alternative pathway.

The essential role of H_2Se in the biochemical phase and various bioconversion pathways of selenium compounds to excretable metabolites is shown in Fig. 16.1 (Kuganesan, Samra, Evans, Singer, & Dyson, 2019).

16.5 Uses

Hydrogen selenide is used to prepare metallic selenides and organoselenium compounds. It is also used in doping gas mixtures for the preparation of semiconductor materials containing a controlled amount of a significant impurity. It is used in the manufacturing of photovoltaic cells such as copper-indium selenide/sulfide (CIS) or copper-indium-gallium-selenide (CIGS) as in thin-film deposition of multiple junction photovoltaic cells. The Eq. (16.4) shows the synthesis of selenoureas from cyanimides (Cohen, 1980).

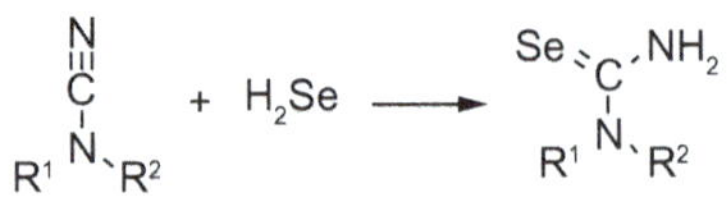

(16.4)

16.6 Risk assessment

H_2Se is a highly flammable gas, hence presents a severe fire hazard. Any probable cause of fire or leakage may consequently cause explosive reactions. Also, the by-products of the H_2Se combustion can be easily absorbed by the skin. The major concern is that fire can further damage adjacent piping and equipment that can create devastating conditions for not only the employees working in the facility but also for the nearby human population. For the evaluation of the risk, the citation of chemical hazards with respect to employees should be essentially defined (Hazardous Substance Fact Sheet, 1999). Broadly the risk assessment comprises of (1) toxicity of the substance; (2) extent of toxic effect; and (3) hazardousness to human and the environment on exposure.

H_2Se is considered as immediately dangerous to life or health (IDLH) on high concentration exposure that one can discharge within 30 min or without symptoms of any irreversible health effects (National Research Council (NRC), 2014). ACGIH threshold limit value certification illustrates that the assessment is determined to reduce the onset of burns, chronic H_2Se-related syndrome, and digestive effects (National Research Council (NRC), 2014). In order to assess, eliminate, and mitigate the related risks, emergency response programs are established to protect the public and the environment from any possible hazardous substance released. With consideration of the toxic effects of hydrogen selenide and selenium, a number of competent bodies have set various exposure limits (Table 16.2).

TABLE 16.2 Workplace exposure limits for hydrogen selenide and selenium.

Statutory bodies	Values for hydrogen selenide	Values for selenium
Occupational Safety and Health Administration (OSHA) permissible exposure limit (PEL) Time Weighted Average (TWA)	0.05 ppm	0.20 mg/m^3
National Institute for Occupational Safety and Health (NIOSH) recommended exposure limit (REL) TWA	0.05 ppm	0.20 mg/m^3
NIOSH Immediately Dangerous to Life or Health (IDLH)	1.00 ppm	1.00 ppm
American Conference of Governmental Industrial Hygienists (ACGIH) Threshold Limit Value (TLV)	0.05 ppm	0.05 ppm

16.7 Environmental effects

If H_2Se discharged into the atmosphere via an accidental release, it can contaminate water, air, and land with potentially devastating consequences. H_2Se and other compounds of Se metal are considered dangerous contaminants in the air. These contaminants may thereupon cause respirational disorder, liver disorder, other severe health effects, and antagonistic ecological effects.

After recognition of H_2Se as a gas of health concern at certain high levels of exposure, US EPA constituted drinking water standards. In nature, selenium is chiefly present in some dietary products and soils. Keeping this in view, the drinking water limit for selenium has been restricted to 0.05 mg/L. Whereas, in Japan and Germany, the drinking water limit for hydrogen selenide has been set at 0.01 mg/L and 0.008 mg/L, respectively (Langner, 2005).

16.8 Hazardousness

H_2Se is widely contemplated as a systemic poison. The major route of exposure is by either inhalation or dermal. Primarily, inhaled H_2Se may cause dizziness, fatigue, nausea, vomiting, and diarrhea, rashes, irritation in the eyes, nose, lungs, and throat (Dudley & Miller, 1937; Dudley & Miller, 1941), frostbite, coughing, and difficulty in breathing. Repeated exposure to this gas can cause garlic breath, metallic taste, deformities in nails, mood changes, loss of sense of smell, and tremor. It can also be the cause of anemia. Acute exposures can result in chemical bronchitis and delayed pulmonary edema. H_2Se acute toxicity appears at a high exposure level, which was inspected at various ranges of concentrations that is, 0.002–0.57 mg/mL at different time intervals as 10, 30, and 60 min in guinea pig. It was observed that animals exposed for 60 days died, whereas 93% died within 10 min of exposure (World Health Organisation (WHO), 1987).

H_2Se is the most toxic Se compound and its related compounds are far more toxic than hydrogen sulfide. The threshold limit value is 0.05 ppm. At concentrations greater than 0.3 ppm, the gas acts as a barrier, which is the main warning signal of risk; below 1 ppm, gas is "insufficient to prevent exposure", while burning at 1.5 ppm is "unbearable." Exposure in high concentrations, even for less than a minute can affect the eyes and mucosa. It is suspected that this is due to the tendency of the oxidizing gas to form red selenium in the mucous membrane. Elemental selenium is less toxic than selenide.

16.9 Treatment

There is no specific treatment or antidote available for the treatment of selenide toxicity. The major concern is the antagonizing the exposure and providing supportive care for symptoms (Agency for Toxic Substances and Disease Registry (ATSDR), 2003). It is important to know that one should avoid D penicillamine, 2,3-Dimercaptopropanol, and EDTA in case of selenide poisoning, as they form nephrotoxic complexes. However, studies have shown effective use of bromobenzene for the treatment of dermatitis caused due to selenosis. Though thiosulphate does not systematically affect, however its use may aid pain produced by selenium dioxide under the nails. There are also some contradictory reports suggesting the role of vitamin C in altering the selenium levels. Glutathione may act as a detoxifying agent, but this has not been demonstrated conclusively. The use of ammonia fumes to relieve the burning associated with acute inhalation and methionine in conjunction with vitamin E to prevent selenium-induced liver damage has also been put forward for consideration (Lillian, Alderman, James, & Bergin, 1986).

16.10 Risk management

Risk management is an essential step in order to mitigate any loss of life as well as property damage due to possible accidental hazards. Also, conduct surveillance in order to take all possible approaches to avoid any after effects. The primary concern is the engineering controls that can effectively avoid any related risks. The facilities must install a diversity of equipment to avoid the direct discharge of a pollutant into water bodies (Asia Industrial Gases Association (AIGA), 2017). Moreover, the storage building must have proper ventilation. The operations performed in isolated areas can also avoid risk. The mandatory utilization of respirators and personal protective equipment can help reduce exposure. The area with H_2Se should be equipped with release concentration monitoring equipment.

16.11 Current research

H_2Se has been recognized as a hazardous pollutant due to its toxic behavior. This mandates its determination in order to prevent human and the environment. In consideration of the similar information, Salih and Ayesh used graphene and graphene oxide to detect H_2Se and some other molecules based on Atomistic ToolKit Virtual NanoLab package. They critically investigated the adsorption energy, adsorption distance, charge transfer, density of states, and band structure. The studies were based on the adsorption phenomenon (Salih & Ayesh, 2020).

In addition, Li et al. have prepared novel molecules HMSeH, H_2MSe, $H_2M(SeH)_2$, and HMSe (M = Ti, Zr, Hf). The products were obtained after reacting laser-ablated Ti, Zr, and Hf atoms with H_2Se. Particularly, the obtained products are rare examples of multiple bonding between a heavier main group element and a group IV transition metal (Li et al., 2020).

16.12 Selenide toxicity: Case study

Twenty-five laborers were exposed to H_2Se engaged in various types of metal etching, buffering, and polishing work (Buchan, 1947) when ink reacted with selenius acid in the presence of a metal. Maximum workers had a distinct garlic smell in their breath. Generally, vomiting, laziness, nausea, fatigue, dizziness, and metallic taste are symptoms of H_2Se exposure. There was no relation between selenium symptoms and urine concentrations. It was also reported that (Glover, Levander, Parizek, & Vouk, 1979) the exposure of less than 0.07 mg/m^3 of H_2Se happened. Initial reports summarizing the symptoms observed in workers from exposure included tearing, runny nose, coughing, and sneezing, and chest tightness followed by pulmonary edema (Glover, 1970; International Programme on Chemical Safety (IPCS), 1987). But no risk concentration was measured to the recovery of affected workers (Rosenfeld & Orville, 1964).

16.13 Safety measures

- Direct skin interaction of H_2Se should be avoided.
- Protecting gloves of fabric materials must be recommended for the operation by the manufacturers and suppliers.
- The manufacturers and suppliers should provide specially designed apparel to avoid the freezing of tissue that protects the employee's body with the liquid or vapor of hydrogen selenide.
- The recognized safety guidelines must be followed to work safely.

16.14 Handling and storage

- Only trained personnel should be allowed to work with hydrogen selenide.
- H_2Se containers must be stored at the ground level to diminish the collapsing and under dried environment.

- The stored amount of hydrogen selenide must not be exceeded with the facility design and should follow the guidelines.
- H_2Se interaction with strong oxidizing agents such as perchlorates, chlorates, peroxides, bromine, fluorine, and chlorine should be avoided to prevent intense reactions.
- Throughout the handling and storing of H_2Se source of flame, ignition, and smoking should be avoided.
- The metal containers that are used for the hydrogen selenide transference must be confined and grounded.
- In the case of gaseous H_2Se, the handler should always wear PPE, safety goggles with a face shield, shoes, and leather gloves.

16.15 Precautions

- The valves of gas containers filled with H_2Se should be opened carefully and very slowly.
- H_2Se gas containers should always be protected and closed properly and there should be sufficient ventilation throughout the transportation of containers.
- The H_2Se gas containers must always be checked for leaks before its usage.
- Eating, drinking, and smoking during the handling of H_2Se should be avoided and hands be washed carefully after its handling.

References

Agency for Toxic Substances and Disease Registry (ATSDR). (2003). *Toxicologic profile for selenium*. Atlanta, GA: US Department of Health and Human Services, Public Health Service.

AIRGAS. (2016). *Hydrogen selenide*. https://www.airgas.com/msds/001152.pdf. (Accessed 7 February 2020).

Alexander, J. (2015). Selenium. In G. F. Nordberg, B. A. Fowler, & M. Nordberg (Eds.), *Handbook on the toxicology of metals* (4th ed., pp. 1175–1208). Academic Press.

Anundi, I., Ståhl, A., & Högberg, J. (1984). Effects of selenite on O_2 consumption, glutathione oxidation and NADPH levels in isolated hepatocytes and the role of redox changes in selenite toxicity. *Chemico-Biological Interactions*, *50*(3), 277–288.

Asia Industrial Gases Association (AIGA). (2017). *Code of practice hydrogen selenide (H_2Se)* (pp. 1–37). http://asiaiga.org/uploaded_docs/aiga%20096_16%20hydrogen%20selenide%2019_10_2016.pdf. (Accessed 7 February 2020).

Björnstedt, M., Kumar, S., & Holmgren, A. (1992). Selenodiglutathione is a highly efficient oxidant of reduced thioredoxin and a substrate for mammalian thioredoxin reductase. *Journal of Biological Chemistry*, *267*(12), 8030–8034.

Boehler, C. J., Raines, A. M., & Sunde, R. A. (2013). Deletion of thioredoxin reductase and effects of selenite and selenate toxicity in caenorhabditis elegans. *PLoS One*, *8*(8), e71525.

Brauer, G. (1963). *Handbook of preparative inorganic chemistry* (2nd ed., p. 418). New York, London: Academic Press.

Buchan, R. F. (1947). Industrial selenosis. A review of the literature, report of five cages and a general bibliography. *Occupational Medicine, 3*(5), 439–456.

Cohen, V. I. (1980). A convenient synthesis of mono-, N, N′-Di-, and trisubstituted selenoureas from methyl carbamimidothioates (s-methylpseudothioureas). *Synthesis, 1980*(01), 60–63.

Dudley, H. C., & Miller, J. W. (1937). *Toxicology of selenium: IV. Effects of exposure to hydrogen selenide* (pp. 1217–1231). Public Health Reports, (1896-1970).

Dudley, H. C., & Miller, J. W. (1941). Toxicology of selenium. VI. Effects of subacute exposure to hydrogen selenide. *The Journal of Industrial Hygiene and Toxicology, 23*(10), 470–477.

Esaki, N., Nakamura, T., Tanaka, H., Suzuki, T., Morino, Y., & Soda, K. (1981). Enzymic synthesis of selenocysteine in rat liver. *Biochemistry, 20*(15), 4492–4496.

Fernandes, A. P., Wallenberg, M., Gandin, V., Misra, S., Tisato, F., et al. (2012). Methylselenol formed by spontaneous methylation of selenide is a superior selenium substrate to the thioredoxin and glutaredoxin systems. *PLoS One, 7*(11), e50727.

Friberg, L., Nordberg, G. F., & Vouk, V. B. (1979). *Handbook on the toxicology of metals*. Amsterdam, The Netherlands: Elsevier/North-Holland Biomedical Press.

Garberg, P., & Högberg, J. (1987). The role of hypoxia in selenium metabolism. *Biochemical Pharmacology, 36*(8), 1377–1379.

Glover, J. R. (1970). Selenium and its industrial toxicology. *Industrial Medicine & Surgery, 39*(1), 50.

Glover, J., Levander, O., Parizek, J., & Vouk, V. (1979). Selenium. In *Handbook on the toxicology of metals* (pp. 555–557). Amsterdam: Elsevier/North Holland Biomedical Press.

Hazardous Substance Fact Sheet. (1999). *Hydrogen selenide.* https://nj.gov/health/eoh/rtkweb/documents/fs/1016.pdf. (Accessed 7 January 2020).

Hoffmann, J. E., & King, M. G. (2010). Selenium and selenium compounds. *Kirk-Othmer Encyclopedia of Chemical Technology*, 1–36. https://doi.org/10.1002/0471238961.1905120508150606.a01.pub3.

International Programme on Chemical Safety (IPCS). (1987). *Environmental health criteria 58*. Geneva, Switzerland: World Health Organization.

Kobayashi, Y., Ogra, Y., Ishiwata, K., Takayama, H., Aimi, N., & Suzuki, K. T. (2002). Selenosugars are key and urinary metabolites for selenium excretion within the required to low-toxic range. *Proceedings of the National Academy of Sciences of the United States of America, 99*(25), 15932–15936. https://doi.org/10.1073/pnas.252610699.

Kuganesan, M., Samra, K., Evans, E., Singer, M., & Dyson, A. (2019). Selenium and hydrogen selenide: Essential micronutrient and the fourth gasotransmitter? *Intensive Care Medicine Experimental, 7*(71). https://doi.org/10.1186/s40635-019-0281-y.

Langner, B. E. (2000). Selenium compounds. In *Ullmann's Encyclopedia of Industrial Chemistry* (7th ed., pp. 1999–2011). New York, NY: John Wiley & Sons.

Langner, B. E. (2005). Selenium and selenium compounds. *Ullmann's Encyclopedia of Industrial Chemistry, 32*, 343–355. https://doi.org/10.1002/14356007.a23_525.

Li, H., Beckers, H., Li, L., Thiele, G., Stüker, T., & Riedel, S. (2020). Reactions of titanium, zirconium and hafnium atoms with hydrogen selenide: A matrix-isolation study. *Chemical Physics Letters, 740*, 137063.

Lillian, C., Alderman, M. D., James, J., & Bergin, M. D. (1986). Hydrogen selenide poisoning: An illustrative case with review of the literature. *Archives of Environmental Health: An International Journal, 41*(6), 354–358. https://doi.org/10.1080/00039896.1986.9935778.

Lu, J., Berndt, C., & Holmgren, A. (2009). Metabolism of selenium compounds catalyzed by the mammalian selenoprotein thioredoxin reductase. *Biochimica et Biophysica Acta, 1790*, 1513–1519.

Malczewska-Toth, B. (2012). Phosphorus, selenium, tellurium, and sulfur. In *Patty Toxicology* (pp. 841–884). New York: Wiley. https://doi.org/10.1002/0471435139.tox044.pub2.

National Research Council (NRC). (2014). *Hydrogen selenide. Acute exposure guideline levels for selected airborne chemicals. Vol. 16*. Washington, DC: The National Academies Press (US). Available from: https://www.ncbi.nlm.nih.gov/books/NBK224929/.

O'Neil, M. J. (2006). *The Merck Index—An encyclopedia of chemicals, drugs, and biologicals* (p. 831). Whitehouse Station, NJ: Merck and Co., Inc.

Pace, E. J., Binns, J., Peña Alvarez, M., Dalladay-Simpson, P., Gregoryanz, E., & Howie, R. T. (2017). Synthesis and stability of hydrogen selenide compounds at high pressure. *The Journal of Chemical Physics, 147*(18), 184303. https://doi.org/10.1063/1.5004242.

Rosenfeld, I., & Orville, A. (1964). *Selenium: Geobotany, biochemistry, toxicity, and nutrition*. New York: Academic Press.

Salih, E., & Ayesh, A. I. (2020). First principle investigation of H_2Se, H_2Te and PH_3 sensing based on graphene oxide. *Physics Letters A, 384*(29), 126775.

Schomburg, L., Schweizer, U., & Köhrle, J. (2004). Selenium and selenoproteins in mammals: Extraordinary, essential, enigmatic. *Cellular and Molecular Life Sciences: CMLS, 61*(16), 1988–1995. https://doi.org/10.1007/s00018-004-4114-z.

Ståhl, A., Anundi, I., & Högberg, J. (1984). Selenite biotransformation to volatile metabolites in an isolated hepatocyte model system. *Biochemical Pharmacology, 33*(7), 1111–1117.

Wilber, C. G. (1980). Toxicology of selenium: A review. *Clinical Toxicology, 17*(2), 171–230.

World Health Organisation (WHO). (1987). *Environmental health criteria 58: Selenium*. Available from: http://www.inchem.org/documents/ehc/ehc/ehc58.htm#SectionNumber:7.1.

Yaws, C. L. (2001). *Hydrogen selenide. Matheson gas data book* (7th ed.). Basking Ridge, NJ: Matheson.

CHAPTER

17

Hydrogen sulfide: Risk assessment, environmental, and health hazard

Ali Mohammed[a] *and Payal Devi*[b]

[a]Department of Applied Chemistry, Faculty of Engineering and Technology, Aligarh Muslim University, Aligarh, Uttar Pradesh, India

[b]Department of Chemistry, Gurukula Kangri (Deemed to be University), Haridwar, Uttarakhand, India

Abbreviations

3MST	3-mercaptopyruvate sulfurtransferase
CAT	cysteine aminotransferase
CBS	cystathionine B-synthase
CFD	computational fluid dynamic
CSE	cystathionine-Γ-lyase
DAO	D-amino acid oxidase
GIT	gastrointestinal tract
OSHA	Occupational Safety and Health Administration
PET	permissible exposure limits
PVC	poly vinyl chloride
SCBA	self-contained breathing apparatus

17.1 Introduction

Hydrogen sulfide (H_2S) is a colorless, toxic, corrosive, and flammable chalcogen hydride gas with a rotten egg-like odor (Greenwood & Earnshaw, 1997). It was discovered in 1777 by a Swedish chemist, Carl Wilhelm Scheele. This hydride gas is produced through the anaerobic metabolism of organic matter by sulfate-reducing microorganisms. It

https://doi.org/10.1016/B978-0-323-89857-7.00008-6

occurs in natural gases, volcanic gases, and in some sources of well water. Some amount of H_2S is even produced in the human body and used as a signaling molecule (Bos, van Goor, Joles, Whiteman, & Leuvenink, 2015).

The explosive range of H_2S in air is 4.5–45.5% that is comparatively higher than the Permissible Exposure Limits (PEL). The half-life of H_2S in air varies from 12 to 37h. However, the presence of photoactive pollutants in ambient air and the variation in temperature influence the half-life of H_2S. At low temperatures, the half-life of H_2S prolonged over 37h in dry air. The water solubility of H_2S at 20°C is 1g in 242mL, whereas further elevation in temperature increases its solubility.

Because hydrogen sulfide tends to escape rapidly from water to cause an odor; hence it can be eliminated from the water by aeration. The process involves bubbling air through a water tank, then separating or "stripping" hydrogen sulfide into the air on the outside. Well-designed aeration tanks or bladderstyle pressure tankwith an old-style pressure tanks maintain a pocket of air in the upper third or upper half of the tank that allows the water to be distributed throughout the house before hydrogen sulfide gas is distributed outside the house. The process is more efficient when hydrogen sulfide concentrations are less than 2.0ppm. This process by itself cannot reduce hydrogen sulfide. Adding a GAC filter following the aeration system may remove the remaining trace amounts.

Aeration systems designed specifically for hydrogen sulfide removal have the added benefit of removing high levels of iron and manganese, provided a sediment filtration system is added after the aeration process to filter the sediment.

17.2 Occurrence

Hydrogen sulfide occurs both naturally and through artificial processes. In nature, it is found in abundant amounts due to anaerobic decay of sulfur-containing organic materials, and desulfurization of large amounts of petroleum fractions in oil refineries. It is generally present in the gaseous form in volcanoes, sulfur springs, undersea vents, swamps, stagnant bodies of water, crude petroleum, natural gas, municipal sewers, sewage treatment plants, swine containment, manure-handling operations, pulp, and paper. Some of the major origins of hydrogen sulfide are industries such as oil refinery, natural gas plants, petrochemical plants, coke oven plants, and food processing plants (World Health Organization, 2003).

The presence of bacteria in the mouth and GIT is also responsible for the formation of this gas by the dissolution process of protein-containing food.

The hydrolysis of sulfide minerals is responsible for the emission of H_2S in hot spring that is,

$$MS + H_2O \rightarrow MO + H_2S \text{ (where } M = \text{metal ion)}$$

Due to the presence of sulfate-reducing bacteria in water, hydrogen sulfide may be created in the water. The breakdown of bacteria with sulfur-containing proteins in the intestinal tract and mouth leads to the production of hydrogen sulfide in small amounts by the human body. Human activity contributes to the major global portion of H_2S. In the process known as hydrodesulfurization, the wide-ranged source of H_2S that is petroleum refineries play an important role in releasing sulfur from fuel by the action of hydrogen. Further, the H_2S is converted into elemental sulfur by combustion in the Claus process. Depending upon the eco-friendly conditions, it is accountable for the deterioration of the material through the action of certain sulfur-oxidizing microorganisms. This process is called biogenic sulfide corrosion.

17.3 Historical background

Interstellar medium has been beneficial in the detection of hydrogen sulfide (Despois, 1997). The clouds of planets in our solar system have also marked the presence of hydrogen sulfide (Irwin et al., 2018). It has also been confirmed that 252 million years ago, this gas was one of the reasons behind the extinction of Permian–Tricic as oxygen-free oceans, home to species of shallow plankton metabolized H_2S. Generally, H_2S is a by-product present in abundance and hence there is little need to produce it under laboratory conditions. H_2S reduces to nontoxic elemental sulfur via the Claus process, invented by a German chemist, Carl Friedrich Claus in 1883.

Hydrogen sulfide was first described as a natural compound by alchemists in the middle ages. The successful development of the sour gas field at Lacq (France) by the Societe Nationale des Petroles d'Aquitaine in 1957 modified the economy of hydrogen sulfide and allowed worldwide exploitation of natural gas with a high hydrogen sulfide level. Before this, hydrogen sulfide was regarded as a toxic, corrosive component of natural gas that needs to be eliminated.

17.4 Physicochemical properties

Hydrogen sulfide is slightly denser than air and it becomes extremely toxic and explosive when present in higher concentrations. It reacts with oxygen giving blue flame to form sulfur dioxide (SO_2) and water.

Hydrogen sulfide forms -SH bond in the presence of a base by virtue of acting as a reducing agent.

In an aqueous solution, the dissociation constant (K_a) for the first proton is $\approx 1 \times 10^{-7}$. The K_a of the second proton is a negative 1×10^{-12}. The property to form a precipitate with inorganic cations such as Cu(II), Pb(II), and Hg(II) helps in qualitative identification of these ions (American Chemical Society, 2015). Some other properties of hydrogen sulfide are presented in Table 17.1.

Hydrogen sulfide forms a colorless solution with water and acts as a weak acid, forming hydrosulfide ion SH^-. When it comes in contact with the atmosphere, it slowly converts into elemental sulfur, which is insoluble in water. Sulfide anion does not exist in aqueous solution (May, Batka, Hefter, Könignberger, & Rowland, 2018). However, they are formed on dissociation of H_2S or by considering hydrogen with molten primary sulfur at approximately 450°C, where hydrocarbons act as a source of hydrogen (Pouliquen et al., 2000). The oxidation of organic compounds or hydrogen by sulfates leads to the formation of hydrogen sulfide in presence of deficient oxygen.

Air, however, is the medium where H_2S is most commonly found and poses a good degree of risk for individuals. These interrelated chemical processes can be considered as simplified forms of the global sulfur cycle in Fig. 17.1 (Malone Rubright, Pearce, & Peterson, 2017).

TABLE 17.1 Physical properties of hydrogen sulfide.

Formula	H_2S
Molecular weight (lb/mol)	34.08
Critical temp. (°F)	212.8
Critical pressure (psia)	1306.5
Boiling point (°F)	−74.9
Melting point (°F)	−122.0
Psat @ 70 °F (psia)	263.6
Liquid density @ 70 °F (lb/ft^3)	48.52
Gas density @ 70 °F 1 atm (lb/ft^3)	0.0888
Specific volume @ 70 °F 1 atm (ft^3/lb)	11.26
Specific gravity	1.207
Specific heat @ 70 °F	8.29

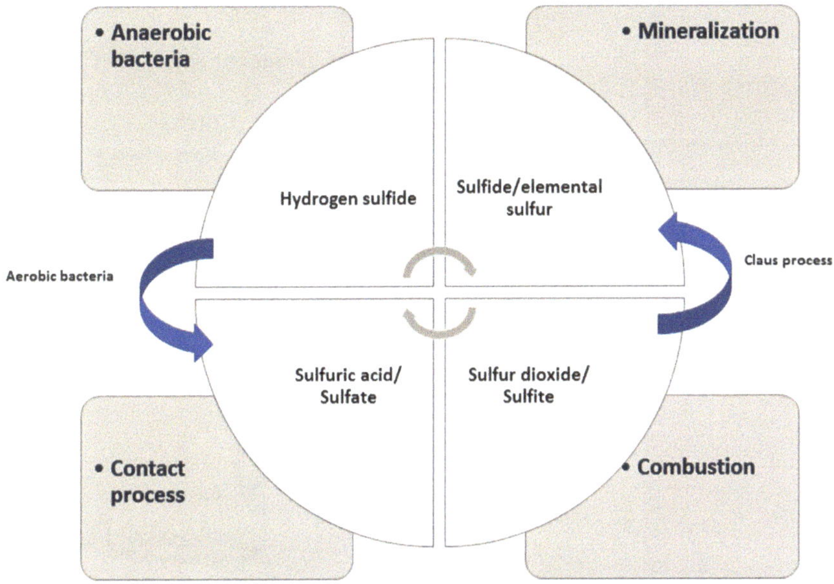

FIG. 17.1 Role of hydrogen sulfide in the global sulfur cycle. *Credit: Payal Devi*

17.5 Physiological role

H_2S acts both as a signaling and cytoprotective agent, where the former role augments the physiological function in the immune system, heart, and neuronal (Wang, 2012). H_2S modulates neurotransmission, regulates vascular tone, protects various tissues and organs, regulates inflammation, induces angiogenesis, and detects cellular oxygen levels. H_2S is produced by enzymes from L-cysteine by cystathionine β-synthase (CBS) and 3-mercaptopyruvate sulfurtransferase (3MST) together with cysteine aminotransferase (CAT) (Shibuya & Kimura, 2013). Recently, the production of H_2S has been achieved from D-cysteine by following a novel pathway including D-amino oxidase (DAO) along with 3MST (Fig. 17.2) (Kolluru, Shen, Bir, & Kevil, 2013). Extensively, the H_2S action has been examined by the process of sulfhydration, which adds excess of sulfur atoms to the cysteine of the targeted proteins in order to convert the activity of the protein.

In addition to this freely diffusible form, H_2S oxidation can produce reductant-labile sulfane sulfur pools containing hydrosulfide/sulfides (RSSH), hydropolysulfides (RS (S_n) SH), and polysulfides (RS (S_n) SR) that are generally present in micromolar concentrations. In accumulation to their potential roles in the storage of H_2S, S-sulfhydration of protein-Cys residues has been postulated as an important oxidative posttranslational modification that contributes to H_2S signaling and action.

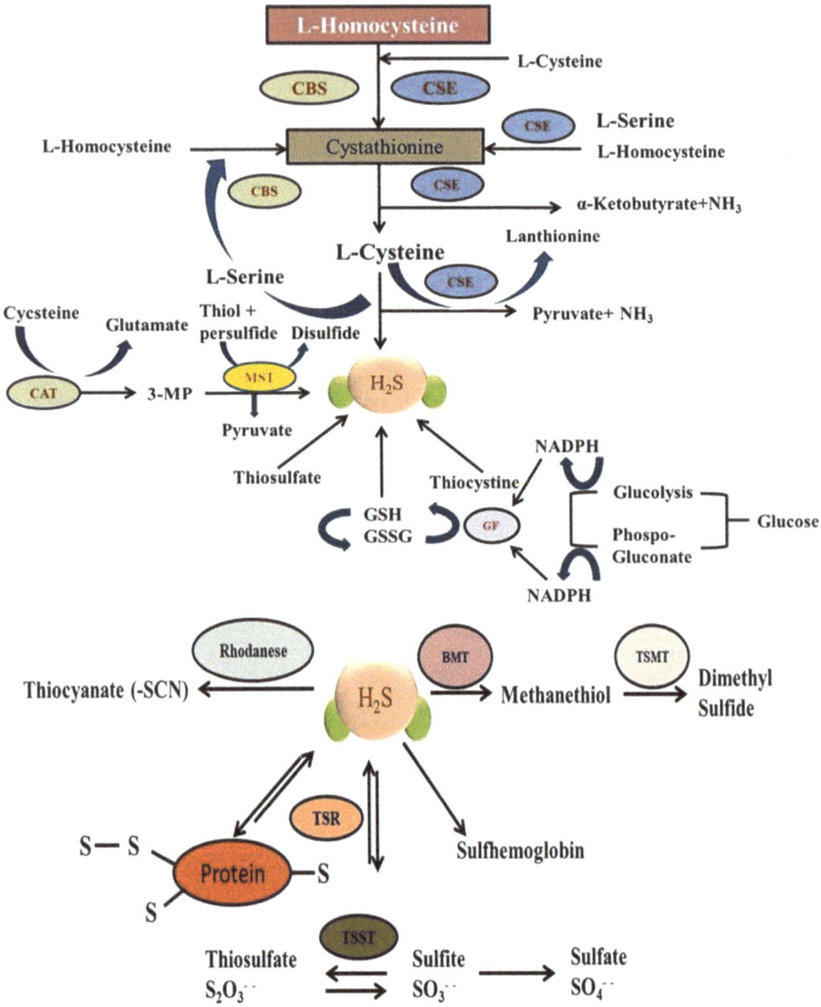

FIG. 17.2 Pathophysiological role of hydrogen sulfide. *Reprinted from Nitric Oxide, 35, Gopi K. Kolluru, Xinggui Shen, Shyamal C. Bir, Christopher G. Kevil, Hydrogen sulfide chemical biology: Pathophysiological roles and detection, 5–20, Copyright (2013), with permission from Elsevier.*

H_2S gas is a toxicant to most organs in the human body. It works as a gaseous gesture molecule and a chemical substance, which implicates different physiological routes, comprising the pathogeneses for several diseases such as heart failure, diabetes, and neurodegenerative syndrome (Holwerda, Karumanchi, & Lely, 2015; Kashfi & Olson, 2013). In the last few years, the favorable characteristics of H_2S gas were ignored due to its noxious nature. Conversely, in recent years, the importance of gasotransmitters such as a nitric oxide (NO) and carbon monoxide (CO) had served as a filler to test the compatibility of H_2S. Moreover, a re-evaluation of the endogenous levels of H_2S confirmed its existence and advantages in mammalian tissues (Furne, Saeed, & Levitt, 2008).

As an alternative to reductant-labile sulfane sulfur, sulfides can be deposited in acid-labile sulfur pools, such as iron-sulfur cluster-containing proteins. Although there are some controversial reports around the basic reaction of sulfan sulfur, specifically persulfides, despite available separation methods for the free, acid-labile, and reductant-labile sulfide pools. For example, although some reports suggest that reductant-labile sulfane sulfur is also acid-labile, it is more likely that acid-labile and reductant-labile sulfurs create two chemically distinct pools.

17.6 Uses

Hydrogen sulfide was earlier used for the manufacturing of sulfur, sulfuric acid, sodium sulfide, and sodium hydrosulfide, which were further used to form a diversity of products, including dyes, pesticides, analgesics, and antibiotics. It can also be used for the purification of nickel and manganese in addition to its use in designing, nuclear industry, laboratory experiments, agricultural disinfectants as well as petroleum production and refining. A number of organosulfur composites have been synthesized using hydrogen sulfides such as methanithol, ethnithole, and thioglycolic acid. In combination with alkali metal bases, hydrogen sulfide can be converted into alkali hydrosulfides such as sodium hydrosulfide and sodium sulfide as evident from reactions 17.1 and 17.2, respectively, which are further used in papermaking.

$$H_2S + NaOH \rightarrow NaSH + H_2O \tag{17.1}$$

$$NaSH + NaOH \rightarrow Na_2S + H_2O \tag{17.2}$$

Specifically, the salts of SH^- are responsible for breaking the bond between the lignin and cellulose constituents of the plugin through the Kraft process. It is also employed to separate deuterium oxide or heavy water from normal tap water through the girdler sulfide process.

Hydrogen sulfide can be used by various types of bacteria as fuel, oxidizing it to elemental sulfur or sulfating using dissolved oxygen, metal oxides such as Iron oxyhydroxide and manganese oxide, or electron acceptor nitrate (Nelson, Amend, Edwards, & Lyons, 2004).

Sulfur bacteria such as violet and green utilizes H_2S as an electron transfer agent in photosynthesis. In the iron-sulfur world, the biochemistry of hydrogen sulfide is an important component of the chemical sciences. In this model of the origin of life on Earth, geologically produced hydrogen sulfide is postulated as an electron donor driving the reduction of carbon dioxide (Wächtershäuser, 1988).

A study at the University of Exeter revealed that an optimum volume of hydrogen sulfide gas can avoid mitochondrial damage from cell exposure. The enzymes present in cells initiate the release of hydrogen sulfide once the cells are attacked by antigen, which further prevents the occurrence of any heart-related diseases or arthritis.

17.7 Sources of contaminants

The occurrence of hydrogen sulfide is often associated with sulfur springs, swamps, salt marshes, decomposition of organic material, sewage treatment plants, tanneries, piggeries, or manure-handling operations. Some coastal communities of Western Australia have been exposed to hydrogen sulfide generated by the breakdown of seaweed. The individuals are prone to exposure to hydrogen sulfide by nearby wastewater treatment facilities, landfills, and manure storage, except for proximity to gas and oil drilling operations (Agency for Toxic Substances and Disease Registry (ATSDR), 2017). Various sources of contamination of H_2S along with their outcomes are mentioned in Table 17.2.

17.8 Risk assessment

Hazard identification: For safe use of this product, a user must (1) notify employees, agents, and contractors regarding known product hazards and safety information, (2) present this information to each purchaser of the product, and (3) request each buyer to inform their employees

TABLE 17.2 Some possible sources of contamination along with their outcomes.

Source	Outcome
The major contributors to hydrogen sulfide in the atmosphere are from swamps, volcanoes, and industrial sources like petroleum/natural gas plants, paper mills, manure treatment facilities, polluted water treatment systems, and tanneries	It stays in the environment for about 1 to 42 days, converting into sulfur dioxide and sulfates in the presence of air
Liquid wastes from the industrial facility or well water	The surface water has a low concentration of hydrogen sulfide as it usually evaporates
By atmospheric growth or diffusion processes, hydrogen sulfide can easily enter the soil	Hydrogen sulfide is destroyed by bacteria present in the soil, which converts it into sulfur

and customers about product hazards and safety information (Safety Data Sheet P-4611, 2016).

Risk analysis: The explosive possibility with air and oxidizing agents may be considered and the danger zones should be recognized accordingly (Safety Data Sheet P-4611, 2016).

Risk evaluation: The monitoring of air for hydrogen sulfide gas should be periodically done by an experienced person. For the same, detector tubes, direct analysis of gas monitors, alarm gas monitors, and explosion meters should be used. Regular air observing should be completed during some work activity where exposure to hydrogen sulfide is probable. Air monitoring must comply with applicable OSHA standards when working in confined spaces (Occupational Safety and Health Administration (OSHA), 2001). It is desirable to estimate the efficiency of an alternative response to avoid the hydrogen sulfide reaction with wells of high-pressure natural gas. Toxic gas in huge quantities could diffuse into the environment in both horizontal and vertical directions, in case of any accidental leakage from wells of high-pressure natural gas. Prediction of spills of toxic gas based on different gas spill prototypes could deliver the basis for estimation of health outcomes and determination of emergency response zones. For simulation studies, the CFD model is one such accurate technique for modeling air transport of contaminants (Gai, Jia, Xi, & Deng, 2020).

17.9 Environmental effects

Hydrogen sulfide gas is a major contributor to the biogeochemical cycle of sulfur on Earth (Barton, Fardeau, & Fauque, 2014). As organic matter falls into low oxygen situations such as swamps, eutrophic lakes, or oceans dead zones, sulfate-reducing bacteria use sulfates existing in water to oxidize organic materials, producing waste hydrogen oxide. Some of the hydrogen sulfides react with the metal ions in the water to generate insoluble metal sulfides, often black or brown, leaving a darker color of sediment.

Though hydrogen sulfide is toxic to most animals (Tobler, Riesch, Garcíade León, Schlupp, & Plath, 2008), its higher levels in the deep sea are a medium of survival for a number of species ranging from bacteria to fish (Bernardino, Levin, Thurber, & Smith, 2012).

Hydrogen sulfide-rich freshwater springs are primarily home to invertebrates but contain very small amounts of fish such as Cyprinodonbobelmari (a pupil from Mexico), Limiasulfophila (a shell of the Dominican Republic), Gambusiauristoma (a venomous from Mexico) Substance) and some Poecilia (poeciliids from Mexico) (Palacios et al., 2013; Tobler et al., 2008). In some caves, such as Move cave, invertebrates, and microorganisms are adapted to high levels of hydrogen sulfide (Kumaresan, Wischer, Stephenson, Hillebrand-Voiculescu, & Murrell, 2014).

The major contribution to hydrogen sulfide gas is from erupting volcanoes that result in the death of oxygen-producing plants and affect the ozone layer, leading to further stress.

17.10 Hazardousness

Hydrogen sulfide enters the body through breathing or by penetrating into the skin from where it is absorbed into the bloodstream and distributed throughout the body.

The health effects of hydrogen sulfide depend upon the amount and duration of its exposure. Studies have shown that the respiratory tract and nervous system are the most vulnerable and easily accessible to hydrogen sulfide toxicity. No serious health problems have been recognized in humans exposed to typical environmental concentrations of hydrogen sulfide.

Respiratory effects: Hydrogen sulfide may cause irritation to the eyes, nose, or throat, causing difficulty in the breathing of individuals suffering from asthma or respiratory distress.

Nervous system effects: The long-term exposure to H_2S gas may mainspring weak memory, and motor function directing headaches, memory loss, tiredness, balance diseases, and loss of consciousness.

17.11 Risk management

Some of the risk management techniques are as follow:

1. Usage of corrosion-resistant equipment, lighting, and explosion-proof local exhaust systems.
2. Adequate availability of local exhaust and general ventilation to meet exposure standards (Safety Data Sheet P-4611, 2016).
3. Use of air monitoring equipment, like a hydrogen sulfide detector tube or a gas detection multi-gas meter in the areas susceptible to hydrogen sulfide. OSHA's limited space standard with specific requirements for identification, monitoring, and entry of confined spaces should be followed (Occupational safety and health administration (OSHA), 2005).
4. Exhaust and external respiration systems to decrease hydrogen sulfide levels. Ensure that the system is nonsparking.

17.12 Treatment

If H_2S enters in body, then the person should be moved immediately to fresh air/environment. While working in the H_2S environment, care must be taken to avoid exposure to the gas. Chemical Hazards Emergency Medical Management recommendations for recognizing and responding to chemical perversities should be followed. Precautions for using respirator devices (self-contained breathing apparatus [SCBA]) should be included. For airway and ventilatory protection, intubation could be used in severe cases (Hydrogen sulfide toxicity treatment and management, 2019).

17.13 Current research

In addition to sulfide salts, several sulfide-release compounds have been synthesized, including organosulfur compounds, Lawesson's reagents, analogs, and plant-derived natural products. Based on garlic extracts, S-propargyl-L-cysteine and its analogs have been synthesized contributing to the drug development of sulfide-containing compounds (Wen, Wang, & Zhu, 2018).

17.14 Sulfide toxicity: Case study

Case 01: At 4:30 p.m. on January 31, 2018, an employee was working for a building finishing contractor. He was repairing metal hog pens within hog confinement. This rig was manufactured by Hobart and consisted of a generator/welder with a 305 cc (18.6 cubic inches) 19.7 Newton-meter (14.50 pound-foot) gross torque Brigg and Stratton Series 1450 gasoline combustion engine. Earlier that day, at approximately 3:30 p.m., one of his co-workers working at the same facility completed his shift for the day and went to use the restroom located within the job site's breeding room. He didn't notice anything unusual upon entering the restroom. Shortly after resting on the stool, his heart began racing with stressed breathing; he quickly exited the restroom and went back into the breeding room, where he collapsed onto the ground. The exposure to carbon monoxide during this activity, along with the presence of high levels of hydrogen sulfide, was thought to be the probable cause of this fatality.

Case 02: On July 7, 2017, at 11:00 am, an employee was attempting to remove a 24-in. rubber plug from an existing 22-ft diameter sewer pipe located inside a 24-ft deep wet well. Employees outside the well were pulling a 1/4-in. nylon cord attached to a 24-in. diameter plug lodged inside a T-shaped PVC fitting. Without testing the atmosphere of the workspace, one of the employees climbed down along the ladder and removed the 24-in. diameter rubber plug, which was about 8 ft below the top of the well. On his way up the ladder, he lost consciousness and fell into the 24 ft deep well. Another employee descended the ladder to provide emergency rescue, but lost consciousness and went underwater. The wastewater level was about 3 ft deep at this point. One more employee climbed down the ladder to provide emergency rescue but lost consciousness as well. All three workers were asphyxiated by hydrogen sulfide (H_2S) gas (Occupational Safety and Health Administration (OSHA), 2017).

17.15 Safety measures

- In the presence of a strong smell of hydrogen sulfide, outdoor activities should be avoided, escaping from the particular areas is a preferred action and opening of windows should be done.
- If the hydrogen sulfide is being produced by human activities, then the reduction in those activities may help to solve the problem at the source.
- The industrial emissions must be controlled by filter ventilation.
- The bore water contamination should be treated with agents responsible for the removal of transition metals such as iron interfering bacterial activity, which happens to produce the gas.
- Water treatment companies should avoid the usage of agents responsible for the liberation of gas.
- The untreated water for irrigation purposes has the potential to release H_2S gas. This can be reduced by decreasing the quantity of used water or desynchronizing with other similar bore water use.
- Consultation with the government bodies should be done in case of possible conflict with watering rosters.
- Usually, hydrogen sulfide is the result of seagrass deposition and decay. During winters, deposition occurs on some beaches which are removed after high storms naturally. Prolonged accumulation can generate an odor problem for nearby residences or passers-by especially if the deposit is disturbed or when there are light onshore winds blowing.

17.16 Handling and storage

Handling:

- All requirements of protecting equipment, engineering controls, and individual sanitation procedures must be followed during working with hydrogen sulfide due to its flammable, compressible, toxicity, and eye-irritating nature.

- For working with hydrogen sulfide, every person should be well trained because of its hazardous effects.
- Always use a suitable respirator and in the case of leaks, areas should be left till the severity of the hydrogen sulfide release disappeared.
- The discharge of gas into the air of working place should be precluded. Explosion-resistant permitted equipment, corrosion-resistant, and nonsparking ventilation systems should be used at the workplace of hydrogen sulfide.
- The installation of fire and leak detection equipment having an automatic fire suppression system should be considered during working with hydrogen sulfide at a large scale.
- Incompatible tackles such as oxidants and different metal oxides must not be used during handling this gas.
- Hydrogen sulfide compressed gas cylinders or/and any part of its cylinder should not be exposed to heat.
- Cylinders filled with H_2S gas should be checked regularly to avoid leakage.

Storage:

- Hydrogen sulfide gas should be stored in a well-ventilated, dry, and cool area and avoid the direct interaction of sunlight.
- Avoid the contact of flame, static discharge, heat, and smoking around the indoor and outdoor storage place.
- This gas should be stored away from the workplace.
- Hydrogen sulfide is colder and heavier than air, it should not be stored below the ground level to avoid the accumulation of leaked gas around low areas.
- The corrosion and fire-resistant materials must be used for making storage cylinders or facilities. Stainless steel cylinders are recommended for the storage of H_2S gas.
- The protection device systems must be used with the liquid H_2S storage tanks.

17.17 Precautions

- Wear impervious clothing, face shields, metatarsal shoes, gloves, and other appropriate protective clothing to prevent any possibility of skin contact with hydrogen sulfide or liquids containing hydrogen sulfide.
- Avoid eating and smoking in areas where sulfide is handled, processed, or stored.
- Use dust and splash-proof safety goggles where there is any possibility of sulfide or liquids containing hydrogen sulfide contacting the eyes.
- Thoroughly wash hands with soap or mild detergent and water before eating, smoking, and using toilet facilities.

References

Agency for Toxic Substances and Disease Registry (ATSDR). (2017). *Toxic substances portal—Hydrogen sulfide carbonyl sulfide.* (2017). https://www.atsdr.cdc.gov/toxfaqs/tf.asp?id=388&tid=67 (Accessed 08 November 2020).

American Chemical Society. (2015). https://www.acs.org/content/acs/en/molecule-of-the-week/archive/h/hydrogen-sulfide.html (Accessed 06 September 2020).

Barton, L. L., Fardeau, M. L., & Fauque, G. D. (2014). Hydrogen sulfide: A toxic gas produced by dissimilatory sulfate and sulfur reduction and consumed by microbial oxidation. In *The metal-driven biogeochemistry of gaseous compounds in the environment* (pp. 237–277). Dordrecht: Springer.

Bernardino, A. F., Levin, L. A., Thurber, A. R., & Smith, C. R. (2012). Comparative composition, diversity and trophic ecology of sediment macrofauna at vents, seeps and organic falls. *PLoS ONE*, *7*(4), 33515.

Bos, E. M., van Goor, H., Joles, J. A., Whiteman, M., & Leuvenink, H. G. (2015). Hydrogen sulfide: Physiological properties and therapeutic potential in ischaemia. *British Journal of Pharmacology*, *172*(6), 1479–1493.

Despois, D. (1997). Radio line observations of molecular and isotopic species in Comet C/1995 O1 (Hale-Bopp). *Earth, Moon, and Planets*, *79*(1–3), 103–124.

Furne, J., Saeed, A., & Levitt, M. D. (2008). Whole tissue hydrogen sulfide concentrations are orders of magnitude lower than presently accepted values. *American Journal of Physiology-Regulatory, Integrative and Comparative Physiology*, *295*(5), R1479–R1485.

Gai, W., Jia, H., Xi, X., & Deng, Y. (2020). Shelter-in-place risk assessment for high-pressure natural gas wells with hydrogen sulphide and its application in emergency management. *Journal of Loss Prevention in the Process Industries*, *63*, 103993.

Greenwood, N. N., & Earnshaw, A. (1997). *Chemistry of the elements* (2nd ed.). Butterworth-Heinemann.

Holwerda, K. M., Karumanchi, S. A., & Lely, A. T. (2015). Hydrogen sulfide: Role in vascular physiology and pathology. *Current Opinion in Nephrology and Hypertension*, *24*(2), 170–176.

Hydrogen sulfide toxicity treatment and management. (2019). https://emedicine.medscape.com/article/815139-treatment (Accessed 28 July 2020).

Irwin, P. G., Toledo, D., Garland, R., Teanby, N. A., Fletcher, L. N., Orton, G. A., et al. (2018). Detection of hydrogen sulfide above the clouds in Uranus's atmosphere. *Nature Astronomy*, *2*(5), 420.

Kashfi, K., & Olson, K. R. (2013). Biology and therapeutic potential of hydrogen sulfide and hydrogen sulfide-releasing chimeras. *Biochemical Pharmacology*, *85*(5), 689–703.

Kolluru, G. K., Shen, X., Bir, S. C., & Kevil, C. G. (2013). Hydrogen sulfide chemical biology: Pathophysiological roles and detection. *Nitric Oxide*, *35*, 5–20.

Kumaresan, D., Wischer, D., Stephenson, J., Hillebrand-Voiculescu, A., & Murrell, J. C. (2014). Microbiology of Movile cave-a chemolithoautotrophic ecosystem. *Geomicrobiology Journal*, *31*(3), 186–193.

Malone Rubright, S. L., Pearce, L. L., & Peterson, J. (2017). Environmental toxicology of hydrogen sulfide. *Nitric Oxide*, *71*, 1–13.

May, P. M., Batka, D., Hefter, G., Könignberger, E., & Rowland, D. (2018). Goodbye to S^{2-} in aqueous solution. *Chemical Communications*, *54*(16), 1980–1983.

Nelson, D. C., Amend, J. P., Edwards, K. J., & Lyons, T. W. (2004). *Sulfide oxidation in marine sediments: Geochemistry meets microbiology:* Special Paper 379: Sulfur Biogeochemistry-Past and Present. (pp. 63–81).

Occupational Safety and Health Administration (OSHA). (2001). https://www.osha.gov/SLTC/hydrogensulfide/exposure.html#:~:text=Use%20respiratory%20and%20other%20personal%20protective%20equipment.&text=For%20exposures%20below%20100%20ppm,respirator%20will%20provide%20eye%20protection. (Accessed 29 July 2020).

Occupational safety and health administration (OSHA) (2005). *OSHA fact sheet.* (2005). https://www.osha.gov/OshDoc/data_Hurricane_Facts/hydrogen_sulfide_fact.pdf. (Accessed 29 July 2020).
Occupational Safety and Health Administration (OSHA) (2017). (2017). https://www.osha.gov/pls/imis/accidentsearch.accident_detail?id=97526.015. (Accessed 07 November 2019).
Palacios, M., Arias-Rodriguez, L., Plath, M., Eifert, C., Lerp, H., Lamboj, A. , et al. (2013). The rediscovery of a long described species reveals additional complexity in speciation patterns of poeciliid fishes in sulfide springs. *PLoS ONE, 8*(8), 71069.
Pouliquen, F., Blanc, C., Arretz, E., Labat, I., Tournier-Lasserve, J., Ladousse, A. , et al. (2000). Hydrogen sulfide. *Ullmann's Encyclopedia of Industrial Chemistry* (pp. 429–449)Vol. 18, (pp. 429–449). https://doi.org/10.1002/14356007.a13_467.
Safety Data Sheet P-4611 (2016). *Hydrogen Sulfide.* USA: Praxair, Inc.. Available from(2016). http://www.praxair.co.in/-/media/corporate/praxairus/documents/sds/hydrogen/hydrogen-sulfide-h2s-safety-data-sheet-sds-p4611.pdf?la=en&rev=1322a7d72ebc46808bbafb550de12060.
Shibuya, N., & Kimura, H. (2013). Production of hydrogen sulfide from d-cysteine and its therapeutic potential. *Frontiers in Endocrinology (Lausanne), 4,* 87. https://doi.org/10.3389/fendo.2013.00087.
Tobler, M., Riesch, R., Garcíade León, F. J., Schlupp, I., & Plath, M. (2008). Two endemic and endangered fishes, Poeciliasulphuraria (Alvarez, 1948) and Gambusiaeurystoma miller, 1975 (Poeciliidae, Teleostei) as only survivors in a small sulphidic habitat. *Journal of Fish Biology, 72*(3), 523–533.
Wächtershäuser, G. (1988). Before enzymes and templates: Theory of surface metabolism. *Microbiological Reviews, 52*(4), 452.
Wang, R. (2012). Physiological implications of hydrogen sulfide: A whiff exploration that blossomed. *Physiological Reviews, 92*(2), 791–896.
Wen, Y. D., Wang, H., & Zhu, Y. Z. (2018). The drug developments of hydrogen sulfide on cardiovascular disease. *Oxidative Medicine and Cellular Longevity, 2018,* 1–21.
World Health Organization (WHO) (2003). *Hydrogen sulfide: Human health aspects.* Geneva.

CHAPTER

18

Methane: Risk assessment, environmental, and health hazard

Nitish Kumar[a] *and Himanshu Gupta*[b]

[a]Department of Chemistry, Gurukula Kangri (Deemed to be University), Haridwar, Uttarakhand, India

[b]Department of Physics, Gurukula Kangri (Deemed to be University), Haridwar, Uttarakhand, India

Abbreviations

AHP	analytical hierarchy process
CLD	chemical liquid deposition
CT	computed tomography
FTA	fault tree analysis
LNG	liquefied natural gas
NFPA	National Fire Protection Association
NIOSH	National Institute for Occupational Safety and Health
NOAA	National Oceanographic and Atmospheric Administration
PPE	personal protective equipment
SCBA	self-contained breathing apparatus

18.1 Introduction

Methane is one of the major gases that came into existence with the Earth and contributes to the origin of life. It is a common colorless, odorless fuel that is the chief component of natural gas. CH_4 is conventionally used as a source of heat and electricity production by many countries. Also, it is used as an alternative to carbon dioxide, a short-lived greenhouse gas, in order to reduce global warming. Some of the primary sources of its development are generally "firedamp" coal mines, anaerobic bacterial destruction of plants and animals, and production by enteric

Hazardous Gases
https://doi.org/10.1016/B978-0-323-89857-7.00009-8

bacteria in humans (Pimentel et al., 2006). Methane gas is also called marsh gas because it is present in high content (50%–80%) in marsh gas (Kocasoy, 1999). The concentration of methane present in the atmosphere is about 2.2 ppm by volume of methane. The impact of methane on climate change over 20 years is 84 times greater than CO_2; over a 100-year period, it is 28 times greater. Recent research suggests that the contribution of methane emissions to global warming is 25% higher than previous estimates. Generally, methane is structurally tetrahedral and a less toxic, inflammable, and explosive gas, Fig. 18.1.

18.2 Occurrence

During the study of major components of natural gas, i.e., marsh gas from Lake Maggiore, Alessandro Volta discovered and isolated CH_4. The gas originates from geological deposits called natural gas fields, which are extracted from shallow levels created by anaerobic decay of organic matter. In general, natural gas sediments are buried deeper and at higher temperatures than those containing petroleum. Methane is produced in bulk by natural gas pipelines or by liquefied LNG carriers.

The synthesis of methane can occur by both biological and industrial pathways. The industrial route involves the hydrogenation of carbon dioxide (CO_2) through the Sabatier process (Liu, Wang, Tao, & Sun, 2012). Physiologically, methanogens in the colon use CO_2 and H_2 to generate methane (Roccarina et al., 2010). However, due to their abundance in natural gas, methane is relatively rarely intentionally produced. The naturally occurring methane is formed by both biological and geological processes and found below ground as well as under the seafloor. A large amount of methane is found under the seafloor in the form of methane clathrates. The systematic representation of methane produced through the biological process is shown in Fig. 18.2.

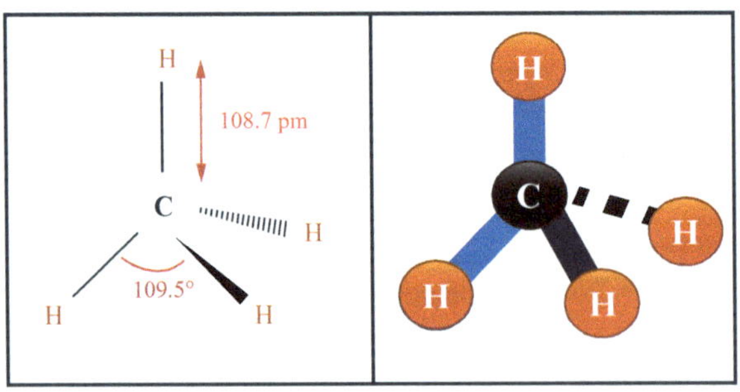

FIG. 18.1 The tetrahedral structure of methane (CH_4) as explained in the VSEPR theory. *Credit: Nitish Kumar.*

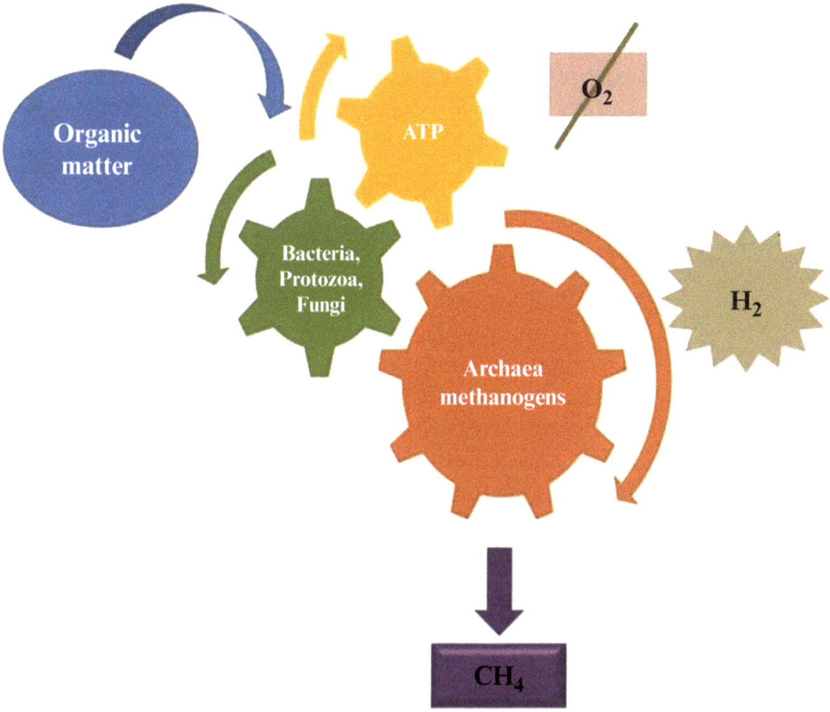

FIG. 18.2 Generation of methane during ATP production by anaerobic microbes. *Credit: Nitish Kumar.*

18.3 Historical background

CH_4 was successfully identified scientifically from the marshes of Lake Maggiore in 1776 by Italian physicist Alessandro Volta. He was inspired by Benjamin Franklin's search for "flammable air." Volta collected the lake's marsh gas and extracted pure gas in 1778, which was demonstrated by the combustion of gas with an electric spark. August Wilhelm von Hofmann, a German pharmacist, coined the name "methane" in 1866 due to its synthesis from methanol.

Etymologically, the word "methane" is coined from the chemical suffix "-ane," which denotes substances belonging to the alkane family; and the word "methyl," which is derived from the German "methyl" (A.D. 1840) or directly from the French "methyle," which is a back-formation from the French "methylene" (corresponding to English "methylene"), the root of which is coined from the Greek "methy" (related to English "mead") and "hyle" (meaning "wood"). The radical is named after this because it was first detected in wood alcohol. The chemical suffix "-ane" is from the coordinating chemical suffix "-ine," which is from the Latin feminine suffix "-ina" which is applied to represent abstracts.

18.4 Physicochemical properties

Methane is typically a stable gas but becomes explosive if present 5%–14% by volume. Some of the physicochemical properties of methane are mentioned in Table 18.1.

18.5 Physiological role

Methane's physiological function has long been ignored in the human body. However, the function of CH_4 in the treatment of diseases is little known. Some studies reported that CH_4 had positive effects on diabetic retinopathy (Ye et al., 2015), animal intestinal ischemia-reperfusion injury (IRI) (Song et al., 2015), and liver IRI via its antioxidative, antiinflammatory (Chen et al., 2016), and antiapoptotic activities (Wu et al., 2015).

Different data suggest that the breath of mammals represent bacterial fermentation by excretion of CH_4; furthermore, the mitochondrial dysfunction may be correlated with variable quantities. The information available indicates that CH_4 generation with hypoxia may be necessary for aerobic life and maybe a surviving developmental feature of plants and animals (Boros, Tuboly, Mészáros, & Amann, 2015).

18.6 Uses

- For domestic and industrial heating, it is used as an oil, because it does not smell and doesn't leave residue (i.e., it is compatible with the environment).

TABLE 18.1 Physicochemical properties of methane.

Physicochemical properties, methane	
Formula	**CH_4**
Molecular weight	16.04 lb./mol
Critical temperature	−116.2 °F
Critical pressure	673.0 psia
Boiling point	−258.7 °F
Melting point	−296.5 °F
Gas density	0.0416 @ 70 °F 1 atm (lb/ft^3)
Specific volume	24.06 @ 70 °F 1 atm (ft^3/lb)
Specific gravity	0.565
Specific heat	8.53 @ 70 °F (Btu/lbmol °F)

- Used to produce other compounds. Examples include black carbon, hydrogen, alkynes, carbohydrate, HCl acid, $CHCl_3$ (also referred to as chloroform), and CCl_4.
- Utilized in saline as a medicinal gas.
- In industry, methane is used as an ingredient in the preparation of common fabric, plastic, antifreeze, and fertilizer.
- Methane is also used to generate electricity and make energy. It can produce enough power for a home or building.
- Methane is a renewable source of energy; the conversion of methane into bio-oil and natural product gives green energy (Fig. 18.3).

Uses of halogenated products of methane:

- CH_3Cl gas is used primarily in the manufacture of silicone resin and rubber. Much of it consists of hydrogen and methyl chloride.

$$CH_3OH_{(l)} + HCl_{(aq)} \rightarrow H_2O_{(l)} + CH_3Cl_{(g)}$$

- Methylene chloride (dichloromethane), CH_2Cl_2, is a liquid, used in paint and varnish removers and as a solvent for aerosol paints.
- Chloroform (trichloromethane), $CHCl_3$—The chief use of chloroform is the production of chlorodifluoromethane (Freon-22). Also used as an anesthetic in surgical operations.
- Tetrachloromethane or carbon tetrachloride, CCl_4—This is the most important chlorinated methane. The industrial usefulness of this compound is based largely on its conversion to dichlorodifluoromethane (Freon-12), trichlorofluoromethane (Freon-11), and tetrachloroethylene. It is also used as a solvent and for the removal of grease in the industry.

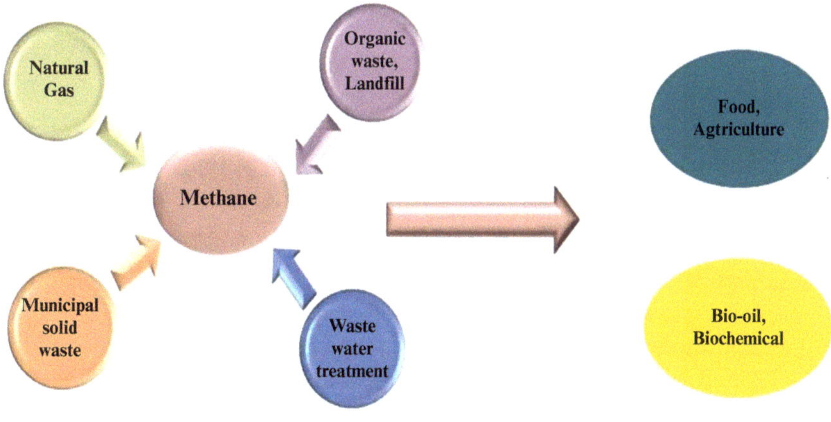

FIG. 18.3 Methane as a renewable source of energy. *Credit: Nitish Kumar.*

18.7 Sources of contaminants

Anaerobic microbial decomposition of vegetable matter underwater (sometimes referred to as marsh or swamp gas), wetays, termites (as a result of digestive processes), volcanoes, ocean floor vents, and methane deposits, found along the continent and under Antarctic ice and arctic permafrost, contribute to the natural source of methane. This comprises 50%–90% (depending on the source) of natural gas and exists along coal seams as a firedamp component.

Certain human activities associated with the production of methane include burning biomass, livestock production, and the management of waste (in which bacteria produce methane while sludge decomposes in waste treatment plants and rotting matter in waste disposal plants).

The average level of atmospheric methane slowed down until 1999 and remained largely stable between 2000 and 2007. The growth rate slightly increased from 1866 parts per billion (ppb) in 2018 to 1875 ppb in 2019 according to a preliminary estimate released by the United States National Oceanographic and Atmospheric Administration (NOAA) (THE HILL, 2020). It should be noted that CH_4 does not persist in the atmosphere the way CO_2 persists. In 10 years, methane undergoes chemical atmospheric reactions and degrades to CO_2 and water. While evaluating methane's overall impact, the inclusion of this resulting CO_2 is known as feedback and atmospheric methane depletion is known as CH_4 sinks. Where the quantity of methane consumed by the sink effect exceeds the number of methane emissions per year, their total methane levels decrease.

There are three main types of CH_4 emissions:

- Biogenic—wetlands, rice paddies, cows, landfills, etc.
- Thermogenic—deriving from either natural seep from fossil fuel or as a result of exploration and production (E&P) and coal mining activities.
- Pyrogenic—the result of incomplete combustion of biomass (for example forest fires), biofuels, and fossil fuel.

18.8 Risk assessment

Ignition or explosion of methane is associated with one of the most natural or sometimes accidental hazards. The preliminary risk assessment is related to the ratio of methane concentration between surface and atmosphere and identification of methane source. The occupational risk assessment of methane is very complicated because it requires minimum energy for ignition and/or explosion. Nevertheless, the risk assessment is based upon the workplace like a coal mine, oil and natural gas refineries, and other domestic and organic waste plants or landfills. The fault tree analysis especially the 3-T matrix tool (Table 18.2) and analytical hierarchy process are the two excellent methods for the quantitative analysis of methane-related risk in a coal mine (Shia, Wang, Zhanga, Cheng, & Zhao, 2017). Moreover, a new approach is developed by Krausel and Krzemien (2014) for the assessment of methane risk in each area of coal mine. They have studied different factors like the activity of initiators (electric, friction spark, and blasting), shaping factor of methane hazard, human casualties and material losses during the explosion, and detection and prevention of methane risk.

18.9 Environmental effects

CH_4 has the most noticeable environmental impacts in production by hydraulic fracturing of natural gas. The well-headed sites require several acres to manage and disrupt the land with supplies and equipment. The fracking fluid takes an enormous amount of water and is typically transported to the drilling plant in tanker transport convoys. Fracking can use large amounts of water for food and other human uses in areas where water is scarce. For this purpose, the waste fluid from the source gets contaminated and become unusable.

Oklahoma saw a seismically silent condition when disruptions came in the form of tornadoes and droughts most frequently from above the

TABLE 18.2 3T risk assessment matrix tool.

The current level of prevention and control	The potential severity of injuries and diseases		
	Slight	Serious	Major
Control is sufficient	1: Risk is insignificant	1: Slight risk. Keep observing the situation	2: Low risk. Check that problem remains under control
Some control needs improvement	2: Low risk. Keep observing the situation and carry out easy to implement measures	3: Medium risk. Plan and carry out suitable measurements	4: High risk. Plan and carry out corrective measurements quickly
Considerable need for improvement	3: Medium risk. Plan and carry out suitable measurements	4: High risk. Plan and carry out corrective measurements quickly	5: Major risk. Plan and carry out corrective measurements immediately

surface, but tremors are now common, given that they have a large fracking concentration. Methane position in global warming and climate change is a less obvious consequence. As already indicated, the combustion of gas produces CO_2 that is supplemented by the combustion of other fossil fuels. However, CH_4 is in itself a greenhouse gas that is, possibly, even more, strong than carbon dioxide. In contrast to carbon dioxide, methane lasts in the atmosphere for many centuries with a half-life of about 10 years. Some studies have predicted that over a 20-year period, 1 kg of methane warms the planet as much as 80 times more than the same amount of carbon dioxide. The atmospheric methane is slowly oxidized by the formation of $OH^{\bullet}$ radical in the troposphere (Wuebbles & Hayhoe, 2002).

The concentration of methane in most of the natural atmosphere has been about 700 ppb (parts per billion) for nearly every 10,000 years. Yet, as human activities were introduced to natural sources since the Industrial Revolution, methane has increased to about 1870 ppb (in 2016). At present, the greenhouse effect due to atmospheric carbon dioxide is around one-fifth of the atmospheric methane.

Approximately 33% of the emissions of CH_4 are from wetlands, one-third from human-flowing rice fields, with a remainder from natural swamps and moors. About 20% come from the output of fossil fuel (coal mines and not captured from oil and gas sources) and from leakage of the huge system of distribution of natural gas. An additional 20% is the consequence of massive cattle rising (for bovine beef and dairy products). There is much cellulose that cattle (and other ruminant animals such as bovine, goat, and camel) consume, including grass and rice. Animals cannot digest cellulose, but ruminants (and termites that also eat cellulose-wood) are equipped with digestive systems of cellulose breakdown microbes. Unfortunately, one of the by-products of this method is a large amount of methane, which is to be extracted from the livestock, largely caused by burping. Solid and liquid waste is often deposited in waste ponds where decay bacteria create methane.

18.10 Hazardousness

Until about 100 years ago, the levels of CH_4 in the atmosphere began to rise stably. In 1992, methane was predicted to warm globally by 25 times of CO_2 and accounted for 15%–17% of global warming during the next century (Johnson & Johnson, 1995; Shindell et al., 2009). Nevertheless, methane inhalation, due to its flammability and difficulty of transport, is ideal for the protection against acute oxidative stress, but it cannot be applied daily for the prevention of disease.

Where the waste disposal system works, the methane becomes toxic on combining with certain chemicals that can be smelled. The fragrance of methane itself indicates the presence of a higher concentration in ambient air, which can be fatally inflammable. Underground natural methane has caused a devastating explosion in the mining and oil factory.

Asphyxiation: The oxygen needed for breathing can be replaced with CH_4 as levels below 10%–16% can be dangerous and cause immediate loss of consciousness and inevitably death. There are no regulations for approved methane in the atmosphere either at home or at work, whereas the global minimum oxygen content is 18%.

Carbon monoxide poisoning: If combined with other materials, methane has the ability to become harmful. Natural gas contains carbon monoxide (CO), which is a deadly gas that is difficult to detect (refer Chapter 8). Relatively low carbon monoxide levels can cause dizziness and nausea within 20 min and death in 2 h of exposure. The Centers for Disease Control and Prevention estimates that 500 people are killed each year by carbon monoxide intoxication.

18.11 Risk management

The implementation of risk-based management has the capability to mitigate the methane explosion. It goes through by observation and controlling of the process following regulatory guidelines and good practices. Some practices like methane concentration (10–15 vol%), paying attention to rock having 30% quartz and 70 μm particle size, restrictive speed of coal pick (1.5 m/s), and use of high-quality equipment (high melting point or noninflammable) have to be considered for avoiding methane explosion in a coal mine (Shia et al., 2017). The number of hazards is generally caused by human errors. National Institute for Occupational Safety and Health (NIOSH) recommends giving specific education and training to employees (NIOSH Pocket Guide to Chemical Hazards, 2005). NIOSH also recommends 1 ppm methane concentration during 8 h exposure for workers. In the domestic area, the management of risk involves the use of a gas detector in the vicinity of the methane source.

18.12 Current research

Massie, Stewart, Gregor, and Gilchrist (2006) have developed the portable optical gas sensor for methane detection. It is a near IR-LED-based low cost sensor with high sensitivity ~1% (below lower exposure limit (LEL)) for methane gas (500 ppm) and also able to operate in harsh environment (−20°C to 50°C). Based on the performance of this sensor, a prototype instrument was also prepared to have vital sensitivity of 0.2% LEL methane gas (100 ppm).

Furthermore, a cyclic voltammetry-based gas sensor has been prepared by Gross, Jaramillo, and Pruitt (2018) for the detection of methane. It is a microfabricated electrochemical gas sensor that is applicable to a sense of methane signal (10%–50% in N_2/air).

Now, real-time monitoring of methane leaks is possible; Gibson et al. (2017) have demonstrated real-time imaging of methane gas leak by using a single-pixel camera. The image can show the methane gas leak of 0.2 L/m from a distance of 1 m.

For the adsorption of methane, a chemical liquid deposition-modified 4A zeolite has been prepared by Wang and Yang (2019). The 4A zeolite is a size-selective adsorbent having 0.01 nm resolution for gas separation. Commercially, this adsorbent may be very useful for the separation of CO_2/CH_4, which is very crucial in biogas upgrading and natural gas processing.

Jackson, Solomon, Canadell, Cargnello, and Field (2019) have proposed a new technique for the reduction of methane emission by converting it into CO_2. Such a method was not previously suggested: all extraction methods were commonly used only on carbon dioxide. Zeolites are one suitable catalyst family. These are crystalline materials composed of an extremely porous, molecular structure, of aluminum, silicon, and oxygen that can act as a sponge to absorb methane.

The scope of electric ventilation systems powered by renewable energy pushing large volumes of air into the chamber where the catalyst is exposed to air. The catalyst (Cu-ZSM-5 and Fe-ZSM-5) is then oxygenally heated and CO_2 released. These fans can be installed wherever there are renewable energy and ample storage. One can even calculate in-line with mitigation scenarios that hold global warming below 2°C and also can be economically practicable and even cost-effective if the cost of eliminating CO_2 increases rapidly from US$50 to US$500 or more in this century. The occupational conversion of methane into CO_2 using catalysts is shown in Fig. 18.4.

18.13 Methane toxicity: Case study

A 21-year-old nonsmoker exposed to CH_4 gas was admitted to the hospital in a drowsy state. He had been working in the medical gas company without using any breathing protective equipment for 3 months. As he opened a CH_4 gas tank, mistakenly he was exposed to the gas supply room, which had no window and doors were closed, thinking that it was a tank of nitrogen. A bilateral decrease in breathing sounds without wheezing, tightening, and crackling on lung auscultation was observed in

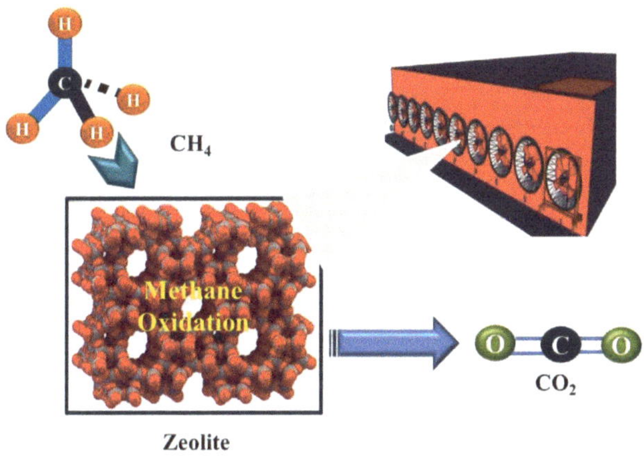

FIG. 18.4 Occupational array to oxidize methane to carbon dioxide. *Credit: Nitish Kumar.*

preliminary diagnosis. The original chest X-rays revealed an unexplained bilateral airspace accumulation in both perihilar areas that imitate pulmonary edema and that is rare for the heart and large vessels. A CT chest scan showed a bilateral symmetrical convergence of the air-space and ground-glass opacity in the lung's dependent section. He was treated with an albuterol nebulizer. Upon 10 days of discharge, his symptoms were resolved (Jo et al., 2013).

Terazawa, Tomii, Takatori, and Nakano (1985) have reported a case of methane asphyxia, a gas outburst that happened in a coal mine of Yubari, Hokkaido on October 16, 1981, where 93 people were killed in this accident. The methane concentration in blood and tissue was measured by gas chromatography of 22 victims. The cardiac blood showed a methane concentration in the range of 6.8–26.8 μL/g. The autopsy of two victims was performed and results revealed that the accumulation of methane in the lungs causes the high concentration of methane (163.8–281.3 μL/g).

Giorgio et al. (2012) have studied the case of a 53-year-old woman who was killed by her husband with methane gas. The histological examination of the lungs has shown microcirculation congestion, emphysema, and pulmonary edema. The concentration of methane in blood was found to be 0.094 mg/mL.

18.14 Safety measures

- Provide adequate and sufficient ventilation of all pits and manure storage areas.
- Smoking around manure pits should not be permitted.

- Use an explosion meter to test methane levels in the barn.
- Do not turn on fans to the manure pit because that might lead to an explosion of CH_4.
- Prohibit open sparks or fires from being used in areas near pits or warehouses.
- To avoid the combustion of methane from a spark, electrical motors, equipment, and wiring, it should be kept close to manure storage structures.
- Entry into a confined space without adequate respiratory equipment should not be carried out.
- Post signs to prevent people from confined dangerous spaces.
- Have someone on call for assistance outside the manure pit, if necessary.
- Make sure the instruments like gas detector, alarm, and methane destruction system are in proper working condition.
- Proper implementation of safety guidelines and operational procedures to minimize methane emissions to the atmosphere should be done.

18.15 Handling and storage

- First, remove all the ignition sources like spark generating equipment, open flame, and static discharge. Make a sign of "No Smoking Zone." Place methane cylinder in an upright position, and use hand truck to move. Do not drag, roll, slide, or drop the cylinder.
- Do not overfill the cylinder.
- Store in a cool, dark, temperature-controlled, and properly ventilated area, and away from other combustible materials like old rags and cardboard.

18.16 Precautions

Methane hazard is a subject of genetic and potential factors related to the inflammable and explosive properties of methane and always creates a dangerous risk to employees and the surroundings. The National Fire Protection Association (NFPA) has released a safety label for methane to avoid hazards (Fig. 18.5) (National Fire Protection Association (NFPA), 2010).

- Health (Blue): 2
- Flammability (Red): 4
- Reactivity (Yellow): 0
- Special (White): None

FIG. 18.5 NFPA 704 label for methane. *Reproduced with the permission of NFPA from NFPA 704,* Standard System for the Identification of the Hazards of Materials for Emergency Response, *2017 edition. Copyright© 2016, National Fire Protection Association.*

Some important precaution and suggestion are listed below:

- If you are working with methane, then always wear personal protective equipment (PPE).
- At the site of oxygen deficiency or failure of equipment, there may be sudden high in concentration of methane; in that situation respiratory protection system must be available.
- If you hear the hissing sound of a gas leak or smell a natural gas odor, remain calm and do not smoke or light a match/candle or other flames. Do not turn on any electrical equipment or other devices that could generate a spark.
- Immediately evacuate the area where gas leaks, go to a safe zone and inhale the fresh air.
- If a person is exposed to methane, give first aid as needed (eye: remove the contact lenses or eyeglasses if worn by the exposed person, then wash the eyes for 20–30 min; skin: remove clothes and wash gently with cold water and soap; inhalation: first leave the site and take a deep breath in fresh air).
- If the victim has symptoms like wheezing, shortness of breath, coughing, or burning in the throat, immediately call the doctor. Possibly, self-contained breathing apparatus should be used; it is very effective in that type of injury.

References

Boros, M., Tuboly, E., Mészáros, A., & Amann, A. (2015). The role of methane in mammalian physiology—Is it a gasotransmitter? *Journal of Breath Research, 9*(1), 014001.

Chen, O., Ye, Z., Cao, Z., Manaenko, A., Ning, K., Zhai, X., et al. (2016). Methane attenuates myocardial ischemia injury in rats through anti-oxidative, anti-apoptotic, and anti-inflammatory actions. *Free Radical Biology and Medicine, 90*, 1–11.

Gibson, G. M., Sun, B., Edgar, M. P., Phillips, D. B., Hempler, N., Maker, G. T., et al. (2017). Real-time imaging of methane gas leaks using a single-pixel camera. *Optics Express, 25*(4), 2998–3005.

Giorgio, F. D., Grassi, V. M., Vetrugno, G., Rossi, R., Fucci, N., Aloja, E. D., et al. (2012). Homicide by methane gas. *Forensic Science International, 221*, e1–e3.

Gross, P. A., Jaramillo, T., & Pruitt, B. (2018). Cyclic-voltammetry-based solid-state gas sensor for methane and other VOC detection. *Analytical Chemistry, 90*, 6102–6108.

Jackson, R. B., Solomon, E. I., Canadell, J. G., Cargnello, M., & Field, C. B. (2019). Methane removal and atmospheric restoration. *Nature Sustainability*, *2*, 436–438.

Jo, J. Y., Kwon, Y. S., Lee, J. W., Park, J. S., Rho, B. K., & Choi, W. I. (2013). Acute respiratory distress due to methane inhalation. *Tuberculosis and Respiratory Diseases*, *74*(3), 120–123.

Johnson, K. A., & Johnson, D. E. (1995). Methane emissions from cattle. *Journal of Animal Science*, *73*(8), 2483–2492.

Kocasoy, G. (1999). Marsh gas (methane). *Environmental Geology*, 395–417.

Krausel, E., & Krzemien, K. (2014). Methane risk assessment in underground mines by means of a survey by the panel of experts (SOPE). *Journal of Sustainable Mining*, *13*(2), 6–13.

Liu, W., Wang, D., Tao, H., & Sun, X. (2012). Is methane a new therapeutic gas? *Medical Gas Research*, *2*(1), 25.

Massie, C., Stewart, G., Gregor, G. M., & Gilchrist, J. R. (2006). Design of a portable optical sensor for methane gas detection. *Sensors and Actuators B*, *113*, 830–836.

National Fire Protection Association (NFPA). (2010). Fire protection guide to hazardous materials. In *NFPA 704 identification of the hazards of materials for emergency response (1996 edition)* (p. 704)(14th ed.). .

NIOSH Pocket Guide to Chemical Hazards. (2005). *National institute for occupational safety and health*: (pp. 2005–2149). NIOSH Publications.

Pimentel, M., Lin, H. C., Enayati, P., van den Burg, B., Lee, H. R., Chen, J. H., et al. (2006). Methane, a gas produced by enteric bacteria, slows intestinal transit and augments small intestinal contractile activity. *American Journal of Physiology-Gastrointestinal and Liver Physiology*, *290*(6), G1089–G1095.

Roccarina, D., Lauritano, E. C., Gabrielli, M., Franceschi, F., Ojetti, V., & Gasbarrini, A. (2010). The role of methane in intestinal diseases. *American Journal of Gastroenterol*, *105*, 1250–1256.

Shia, L., Wang, J., Zhanga, G., Cheng, X., & Zhao, X. (2017). A risk assessment method to quantitatively investigate the methane explosion in underground coal mine. *Process Safety and Environmental Protection*, *107*, 317–333.

Shindell, D. T., Faluvegi, G., Koch, D. M., Schmidt, G. A., Unger, N., & Bauer, S. E. (2009). Improved attribution of climate forcing to emissions. *Science*, *326*(5953), 716–718.

Song, K., Zhang, M., Hu, J., Liu, Y., Liu, Y., Wang, Y., et al. (2015). Methane-rich saline attenuates ischemia/reperfusion injury of abdominal skin flaps in rats via regulating apoptosis level. *BMC Surgery*, *15*(1), 92.

Terazawa, K., Tomii, S., Takatori, T., & Nakano, K. (1985). Methane asphyxia. *The American Journal of Forensic Medicine and Pathology*, *6*(3), 211–214.

THE HILL (2020). *2019 atmospheric methane increase greatest in five years: preliminary data.* Retrieved from https://thehill.com/policy/energy-environment/491424-2019-atmospheric-methane-increase-greatest-in-five-years.

Wang, Y., & Yang, R. T. (2019). Chemical liquid deposition modified 4a zeolite as a size-selective adsorbent for methane upgrading, co2 capture and air separation. *ACS Sustainable Chemistry Engineering*, *7*, 3301–3308.

Wu, J., Wang, R., Ye, Z., Sun, X., Chen, Z., Xia, F., et al. (2015). Protective effects of methane-rich saline on diabetic retinopathy via anti-inflammation in a streptozotocin-induced diabetic rat model. *Biochemical and Biophysical Research Communications*, *466*(2), 155–161.

Wuebbles, D. J., & Hayhoe, K. (2002). Atmospheric methane and global change. *Earth-Science Reviews*, *57*, 177–210.

Ye, Z., Chen, O., Zhang, R., Nakao, A., Fan, D., Zhang, T., et al. (2015). Methane attenuates hepatic ischemia/reperfusion injury in rats through antiapoptotic, anti-inflammatory, and antioxidative actions. *Shock*, *44*(2), 181–187.

CHAPTER

19

Methyl bromide: Risk assessment, environmental, and health hazard

R. D. Kaushik

Department of Chemistry, Gurukula Kangri (Deemed to be University), Haridwar, Uttarakhand, India

Abbreviations

DOE	Department of Energy
DWEL	Drinking Water Equivalent Level
IDLH	Immediately Dangerous to Life or Health
IRIS	Integrated Risk Information System
MeBr	Methyl Bromide
NIOSH	National Institute for Occupational Safety and Health
OSHA	Occupational Safety and Health Administration
PAC	Protective Action Criteria
PEL	Permissible Exposure Limit
RfC	Inhalation Reference Concentration
RfD	Oral Reference Dose

19.1 Introduction

Methyl bromide is a colorless gaseous pesticide custom-built to fumigate soil, crops, commodity warehouses, and commodity-shipping facilities. The innate origin of methyl bromide is the ocean, presumably produced by algae or kelp, and is likely propelled into the air. However, most of the methyl bromide is generally human-made and utilized for fumigation of homes, foods, or soil. For fumigation, it is customarily mixed with 2% chloropicrin or 0.3% amyl acetate as warning agents or up to 70% chloropicrin or other fumigants or hydrocarbons as inert

Hazardous Gases
https://doi.org/10.1016/B978-0-323-89857-7.00003-7

diluents. Hundred percent of methyl bromide is used in the case of commodity fumigation.

Some of its nonpesticidal handlings include degreasing wool, extracting oils from nuts, seeds, and flowers, methylating agents in the chemical industry, refrigerant, and fire-extinguishing agent in aircraft.

The efficiency of bromomethane to deplete the ozone layer phased out its production (EPA, 2014, 2015) and use in several countries such as the United States in 2005 under the *Montreal Protocol* (UNEP, 2018). The protocol was established on a scientific concurrence worrying the damage caused to the ozone layer on ceaseless usage of methyl bromide, which in turn would concede greater amounts of UV radiations to reach the earth's atmosphere and bring about vandalism to human health and the environment (Budnik, Kloth, Velasco-Garrido, & Baur, 2012; FAO, 2003; IPCS, 2009).

As per TRI reports, methyl bromide accounted for about 99.98% of the predicted total environmental releases from estimated releases of approximately 133.78 metric tons of bromomethane to the atmosphere (TRI17, 2019). The propelled methyl bromide in ambient air is substantially culpable for inhalation exposure to agricultural workers and the general population, primarily developing neurotoxic conditions in them. Some of the pertinent international and national regulations, advisories, and guidelines regarding bromomethane in air, water, and other media are summarized in Table 19.1.

19.2 Physicochemical properties

Methyl bromide (IUPAC name: bromomethane) classified as an alkyl bromide is a heavier, colorless, and volatile gas with slight water solubility and is handled as a liquid (14.4 lb./gal) under moderate pressure (Table 19.2). Practically odorless but has a sweetish chloroform-like odor at high concentrations with an odor threshold of 80 mg/m^3 (IPCS, 2009).

Methyl bromide is readily soluble in lower alcohols, ethers, esters, ketones, halogenated hydrocarbons, aromatic hydrocarbons, and carbon disulfide. It can easily pervade via concrete, leather, rubber, and certain plastics. Depending on the pH in the aqueous solution, methyl bromide hydrolyses to methanol and hydrobromic acid (Ioffe & Kampf, 2002; O'Neil, Smith, & Heckelman, 2001).

19.3 Physiological role

After subsequent exposure, methyl bromide is rapidly absorbed in the lower respiratory tract and expeditiously dispersed among major organs such as the lungs, liver, and kidneys (ATSDR, 2020). Initially, methyl

TABLE 19.1 Regulations and guidelines applicable to methyl bromide.

Agency	Description	Information	Reference
Air			
EPA	RfC (Inhalation Reference Concentration)	5×10^{-3} mg/m[3a]	IRIS (2002)
Water and Food			
EPA	Drinking Water Equivalent Level (DWEL)	0.05 mg/L	EPA (2018)
	RfD (Oral Reference Dose)	1.4×10^{-3} mg/kg/day[b]	IRIS (2002)
Occupational			
OSHA	Ceiling[c] PEL for general industry, shipyards, and construction	20 ppm[d]	OSHA (2018a), OSHA (2018b), and OSHA (2018c)
NIOSH	Immediately Dangerous to Life or Health (IDLH)	250 ppm	ATSDR (2020)
Emergency criteria			
DOE (Department of Energy)	PAC (Protective Action Criteria)'s air		DOE (2018)
	PAC-1	19 ppm	
	PAC-2	210 ppm	
	PAC-3	740 ppm	

[a] *The RfC is derived from the identification of LOAEL.*
[b] *The RfD is derived from the identification of NOAEL.*
[c] *The ceiling was assessed as a 15-min TWA exposure.*
[d] *Skin designation.*

bromide encounters metabolism by nucleophilic displacement of the bromide ion (Eq. 19.1).

$$HOH + CH_3Br \rightarrow CH_3OH + H^+ + Br^- \qquad (19.1)$$

Further, the formed methanol oxidizes to give formaldehyde and formate. Moving ahead this may get oxidized to carbon dioxide (40%–50% of the administered dose) and water, or other unidentified nonvolatile metabolites (20%–25% of the dose) or form amino acids such as cysteine or homocysteine via reactions in the oxalate or tricarboxylic acid cycles

TABLE 19.2 Physical and chemical properties.

Molecular weight	94.95
Formula	CH_3Br
Common synonyms	Bromomethane, monobromomethane, embafume
Melting point	−93.6°C
Boiling point	3.56°C
Density	$1.73\,g/cm^3$ at 0°C
Vapor pressure	1893 kPa at 20°C
Partition coefficient	1.94 ± 0.31 at 25°C
Water solubility	16.1 g/L at 25°C
Flashpoint	194°C
Specific gravity	3.355 at 25°C
Odor	Odorless, but at high concentrations sweetish, chloroform-like odor
Reactivity	Reacts with dimethyl sulfoxide, ethylene oxide, and strong oxidizers. Attacks various metals in the presence of water

(Bulathsinghala & Shaw, 2014) or form *S*-methyl derivatives via reaction with organic thiols, as shown in the following equation:

$$R-SH+CH_3Br \rightarrow R-SCH_3+H^{+}+Br^{-} \quad (19.2)$$

The methyl bromide exposure to *S*-methyl cysteine derivatives or by analogy with methyl chloride leads to the formation of *S*-methyl glutathione as the final product. Moreover, the production of methanethiol via intermediates *S*-methyl cysteine and methylthioacetic acid takes place on the metabolism of *S*-methyl derivatives (Bulathsinghala & Shaw, 2014).

Excretion occurs mainly by the expiration of CO_2 and urinary excretion, as shown in Fig. 19.1. Only small amounts are excreted in the feces. The metabolized remains in the tissue significantly represent adducts and incorporation into metabolic pools (Committee on Acute Exposure Guideline Levels & National Research Council, 2012).

19.4 Uses

In the earlier decades, principally methyl bromide was used as a soil or space fumigant for the control of insects, fungi, and rodents (ATSDR, 2020). During the 1920s, it was efficiently utilized as fire extinguishers

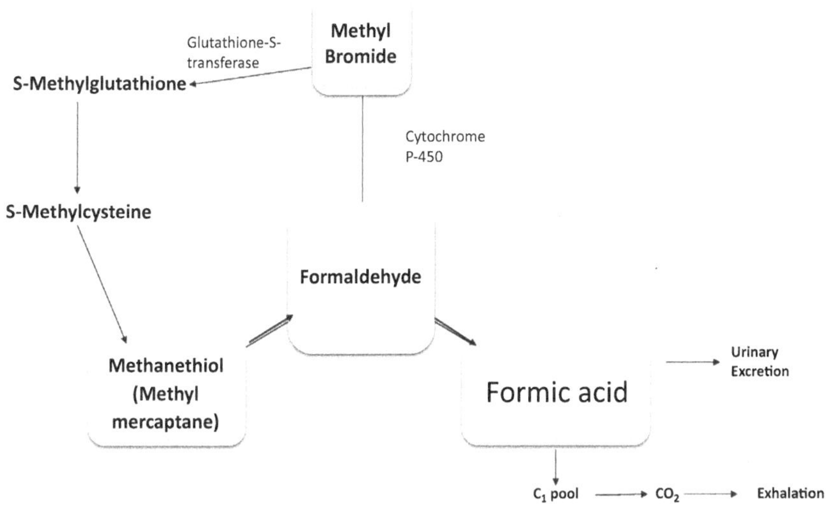

FIG. 19.1 Flow diagram illustrating the physiological role of methyl bromide. *Credit: Rajneesh Dutt Kaushik*

in Europe (O'Neil, Heckelman, Dobbelaer, et al., 2014) but never gained widespread consideration. It has also been used in ionization chambers; wool degreasing; oil extraction of nuts, seeds, and flowers; as a methylating agent in the chemical industry (Larrañaga, Lewis, & Lewis, 2016; O'Neil et al., 2014); as a disinfectant of potatoes, tomatoes, and other crops (Larrañaga et al., 2016).

19.5 Sources of contaminants

The considerable amount of methyl bromide accounts from the fumigation process of homes/fields/warehouses/silos/mills/vaults/ships/freight cars/drinking water, degreasing, and extraction operations, as well as automobile and turbine exhaust, whereas some of it is naturally produced in the ocean by algae or kelp (ATSDR, 2020). Often it has been detected in air and water including surface water, drinking water, and groundwater. The atmospheric air accommodates <0.025 ppb of methyl bromide, whereas industrial areas have slightly higher levels ranging up to 1.2 ppb. Customarily the annual methyl bromide contribution in the atmosphere is due to biomass burning and polar macroalgae like brown, red, and green. It is generally released from automobiles by catalytic decomposition of ethylene dibromide in leaded gasoline or from turbines.

19.6 Risk assessment

Despite the availability of several known treatments, globally, tons of methyl bromide is used for the fumigation of freight containers destined for export, particularly in Asia (Baur, Poschadel, & Budnik, 2010; Budnik, Fahrenholtz, Kloth, & Baur, 2010; Fahrenholtz, Hühnerfuss, Baur, & Budnik, 2010). The fumigated containers are transported within continents and countries, initiating widespread risk dispersal of methyl bromide in air. Therefore, this necessitates conducting risk assessment protocol prior to and after shipments. The risk assessment in terms of adverse health effects, as well as intensity, frequency, and duration of exposure, should be determined, which can conclusively govern the risk analysis by the use of methyl bromide.

However, conducting risk assessment at the port is a challenge in itself due to limited communications and higher costs. The risk assessment in this situation is limited to containers and staff in the logistics companies, customs officers, and workers unloading containers. However, some countries have adopted diverse methods correlating risk assessment such as the provision of instructions/information sheets from organizations and employers on the safe handling of containers, measurement of pollutants with the container door closed, and screening with multifunction devices.

Some of the risk assessment followed by different countries include "The Technical Rules for Hazardous Substances" on fumigations (de Souza, Narvencar, & Sindhoora, 2013); safe handling of gasses in shipping containers (Lemoine, Schoolman, Jackman, & Vernon, 2011); WorkSafe New Zealand; Quick guide: Keeping Safe from Harmful Substances while Inspecting or Unpacking Containers. The Department of Pesticide Regulation (DPR), California, follows the margin-of-exposure (MOE) approach for estimating risks. The risk analysis infers detailed insights into exposure levels and associated risks.

19.7 Environmental effects

The availability of methyl bromide in the environment is mainly due to its release through various waste streams and also its usage as an insecticide, acaricide, nematicide, soil sterilant, and rodenticide. In the air, it exists in the gas phase degraded by reaction with photochemically produced hydroxyl radicals with a half-life of about 1.5 years.

The fumigation of soil results in the availability of methyl bromide to depths of about 0.8 m, which further gets hydrolyzed to methanol and bromide ion with a half-life of 20–26 days. With K_{oc} values ranging between 9 and 22, methyl bromide expects high mobility in soil (Yates, Gan, & Papiernik, 2003), whereas its vapor pressure suggests its volatility from the soil, predominating nucleophilic substitutions on soil organic matter for degradation, potentially initiating this chemical to leach. In environmental conditions with pH 5.0–9.0, the fate of methyl bromide in water could be due to the volatilization by hydrolysis.

19.8 Hazardousness

Acute effects: Studies stipulate that the profound inhalation of methyl bromide severely injures the lung, may have neurological effects, and damages the kidney and liver. Respiration in high concentrations may cause pulmonary edema and impaire respiratory functions (ATSDR, 2020). A study of bromomethane applicators (69% did not use protective equipment) reported a 36% incidence of dyspnea, cough, and phlegm (Akca et al., 2009). Inhalation studies in laboratory animals suggested that the nasal cavity, particularly the olfactory epithelium, is the most sensitive target in the respiratory tract (Holbrook, Iwema, Peluso, & Schwob, 2014; Youngentob & Schwob, 2006). The signs of acute vulnerability involve headache, vertigo, fainting, infirmity, weakness, confusion, speech impairment, optical effects, numbness, twitching, tremors, paralysis, and convulsions. However, methyl bromide may irritate eyes, cause itching, redness, and blisters on the skin, and aggravate the mucous membrane of the upper respiratory tract (ACGIH® Threshold Limit Values (TLVs®) and Biological Exposure Indices (BEIs®), 2012). Acute massive or prolonged exposure may also lead to permanent debilitation or death (Langård, Rognum, Fløtterød, & Skaug, 1996).

Bromomethane exposure may show signs of skin irritation. The case of conjunctivitis has also been detailed by exposed individuals. Animal studies have revealed a decrease in body weight gain and weight loss.

Chronic effects: An occupational study suggests that chronic inhalation exposure of methyl bromide may result in neurological impairment (ATSDR, 2020) including lethargy, forelimb twitching, tremors, and paralysis.

Carcinogenic effects: Methyl bromide has been categorized as carcinogenic by the US Department of Health and Human Services (NTP, 2016). Additionally, IRAC also classified it as a Group 3 carcinogen (not classifiable as to its carcinogenicity to humans) (IRAC, 2020), whereas EPA (IRIS, 2002) determined it as a Group D carcinogen (not classifiable as potent to human carcinogenicity). On the basis of animal studies, NIOSH has categorized methyl bromide as a "potential occupational carcinogen." Moreover, it has also been found that workers occupationally exposed to methyl bromide are comparatively more vulnerable to testicular cancer (IARC, 2018).

Cardiovascular effects: Detailed studies are available concluding that exposure to bromomethane causes high blood pressure and minute myocardial hemorrhages, although more detailed literature is still unavailable. Some other reported effects include fibrosis, myocardial degeneration, cardiomyopathy, and cardiac thrombi at exposure levels of 89.1–150 ppm.

Gastrointestinal effects: Some of the GIT effects noted in exposed individuals are nausea, vomiting, and/or diarrhea (Kulkarni et al., 2015; Yamano & Nakadate, 2006).

Hematological effects: Inconsistent reports have been observed showing decreased erythrocyte, hematocrit, and hemoglobin levels, increased leukocyte levels, and splenic hemosiderosis, although there are studies with not much significant hematological effects.

Hepatic effects: There are cases of humans exposed to bromomethane with tenderness and swelling in the liver, congestion, fatty degeneration, or atrophy of the liver. In other cases, either no significant liver injury or mild signs of liver injury or no hepatocellular lesions were reported.

Renal effects: The individuals exposed to bromomethane have often reported congestion, anuria or oliguria, proteinuria, and histological alterations in the kidney (ATSDR, 2020). However, there are many cases where renal effects were minimal or absent, kidney effects—enlarged and paled kidney and nephrosis.

19.9 Risk management

Exposure to methyl bromide can be eliminated or minimized, by the following methods:

Measurement of fumigant in containers: To determine the amount of each chemical compound in the ambient air portable instrumentations such as Fourier Transform Infrared Spectroscopy (FTIR) and Photoionization Detector (PID) can be used for online monitoring. Some other methods include the use of different kinds of adsorbent tubes; a standardized screening/monitoring procedure; the use of a gas-free/safe container certificate; etc.

Degasification/Ventilation of containers: The ventilation or off-gassing by forced ventilation of containers should be conducted. The time of degasification is dependent on factors, for instance, how the goods are stacked inside the container, the container's filling grade, the nature of the goods, the climatic conditions, and fumigant concentration. The process of ventilation must be an ongoing process and prioritized during stripping or at the time of the reopening of the container.

Personal protective equipment: The ability of fumigant to enter via respiration process or skin makes methyl bromide more vulnerable. Therefore, use of PPE i.e., respirators, gloves, suits, boots, and safety goggles should be recommended. The use of PPE is also recommended if personnel enter the container without prior follow-up of risk analysis procedures or ventilation protocols, for instance, inspection by the food inspector or customs.

19.10 Methyl bromide toxicity: Case study

There are details of cases of severe malignity with some fatalities post-injection of 30–3000 ppm methyl bromide into the soil. However, the study shows that 200 ppm of the injected methyl bromide remains intact for only a few seconds and deteriorates to 4 ppm post 5 days of injection. The tilling process can although maintain the level up to 15 ppm for 9 days after injection. The case of two South Australian carpenters engaged by a stevedoring company in March 2002 has also been addressed with headache, nausea, numbness, and profuse sweating and they were hospitalized for 4 days.

19.11 Safety measures

Some of the specific measures to reduce the risk of MeBr exposure are enlisted below:

- The instructions provided on the label or permit for storage, use, and disposal should be judiciously followed.
- The Material Safety Data Sheets (MSDS) should be kept at an easy-to-reach place.
- The approved labels on containers should be used while handling pesticides.
- Only the required quantity of dosage should be utilized.
- While fumigating, the atmospheric conditions should be preliminary looked over, in order to avoid drifting of MeBr outside the target area.
- It should be ensured that the personnel involved in the handling of MeBr are wearing PPE and are well acquainted with emergency procedures.
- It should also be ensured that people are not present downwind of spray.
- Adequate buffer areas should be provided between the application of MeBr and dwellings or sensitive areas.
- Eyewash fountains should be provided in the immediate work area for emergency use.

19.12 Handling and storage

- The cool, well-ventilated, and free from moisture and combustibles area should be used for the storage of tightly closed methyl bromide containers.
- Methyl bromide should be stored away from aluminum, as it tends to form spontaneously flammable aluminum trimethyl.
- It should not be stored in the vicinity of oxidizing agents (such as perchlorates, peroxides, permanganates, chlorates, nitrates, chlorine, bromine, and fluorine) since this may result in fire and explosion.
- The storage of MeBr should be avoided in the proximity of metals (such as tin, magnesium, and zinc), dimethyl sulfoxide, and ethylene oxide.

19.13 Precautions

- While handling, processing, or storage of methyl bromide, activities like smoking, drinking, or eating should be prohibited.
- Avoid wearing contact lens while handling the gas.
- Wear nonvented, impact-resistant goggles when working with fumes, gases, or vapors.
- Wear a face shield along with goggles when working with this gas.

References

ACGIH® Threshold Limit Values (TLVs®) and Biological Exposure Indices (BEIs®) (2012). (2012). https://www.nsc.org/Portals/0/Documents/facultyportal/Documents/fih-6e-appendix-b.pdf. (Accessed 12 May 2020).

Agency for Toxic Substances and Disease Registry (ATSDR) (2020). *Toxicological profile for bromomethane.* Atlanta, GA: U.S. Department of Health and Human Services, Public Health Service.

Akca, E. T., Serpil, S., Sezer, U., Ozlem, E., Ayse, G., Canan, C., et al. (2009). Health profiles of methyl bromide applicators in greenhouses in Turkey. *Annals of the Academy of Medicine, Singapore, 38*, 707–713.

Baur, X., Poschadel, B., & Budnik, L. T. (2010). High frequency of fumigants and other toxic gases in imported freight containers—An underestimated occupational and community health risk. *Occupational and Environmental Medicine, 67*(3), 207–212. https://doi.org/10.1136/oem.2008.043893.

Budnik, L. T., Fahrenholtz, S., Kloth, S., & Baur, X. (2010). Halogenated hydrocarbon pesticides and other volatile organic contaminants provide analytical challenges in global trading. *Journal of Environmental Monitoring, 12*(4), 936–942. https://doi.org/10.1039/b918428g.

Budnik, L. T., Kloth, S., Velasco-Garrido, M., & Baur, X. (2012). Prostate cancer and toxicity from critical use exemptions of methyl bromide: Environmental protection helps protect against human health risks. *Environmental Health: A Global Access Science Source, 11*, 5. https://doi.org/10.1186/1476-069X-11-5.

Bulathsinghala, A. T., & Shaw, I. C. (2014). The toxic chemistry of methyl bromide. *Human & Experimental Toxicology*, *33*(1), 81–91. https://doi.org/10.1177/0960327113493299.

Committee on Acute Exposure Guideline Levels & National Research Council (2012). Methyl bromide: Acute exposure guideline levels. *Acute exposure guideline levels for selected airborne chemicals*. Vol. 12. National Academies Press (US).

de Souza, A., Narvencar, K. P., & Sindhoora, K. V. (2013). The neurological effects of methyl bromide intoxication. *Journal of the Neurological Sciences*, *335*(1–2), 36–41.

DOE (2018). *Table 3: Protective Action Criteria (PAC) Rev. 29a based on applicable 60-minute AEGLs, ERPGs, or TEELs. The chemicals are listed by CASRN*. Oak Ridge, TN: U.S. Department of Energy. (2018). https://edms.energy.gov/pac/docs/Revision_29A_Table3.pdf. (Accessed 10 June 2020).

EPA (2014). *The phaseout of methyl bromide. Ozone layer protection-regulatory programs*. U.S. Environmental Protection Agency. (Accessed 10 June 2020).

EPA (2015). Protection of stratospheric ozone: The 2016 critical use exemption from the phaseout of methyl bromide. *Federal Register*, *80*(199), 61985–61993.

EPA (2018). *2018 Edition of the drinking water standards and health advisories*. Washington, DC: U.S. Environmental Protection Agency. EPA822S12001 (2018). https://www.epa.gov/sites/production/files/2018-03/documents/dwtable2018.pdf. (Accessed 15 July 2020).

Fahrenholtz, S., Hühnerfuss, H., Baur, X., & Budnik, L. T. (2010). Determination of phosphine and other fumigants in air samples by thermal desorption and 2D heart-cutting gas chromatography with synchronous SIM/scan mass spectrometry and flame photometric detection. *Journal of Chromatography A*, *1217*(52), 8298–8307. https://doi.org/10.1016/j.chroma.2010.10.085.

FAO: Food and Agriculture Organization of the United Nations, International standards for phytosanitary measures (2003). *Manual on pesticides. (2003)*. http://www.fao.org/fileadmin/templates/agphome/documents/Pests_Pesticides/Specs/qualitycontrol05.pdf. (Accessed 24 August 2020).

Holbrook, E. H., Iwema, C. L., Peluso, C. E., & Schwob, J. E. (2014). The regeneration of P2 olfactory sensory neurons is selectively impaired following methyl bromide lesion. *Chemical Senses*, *39*(7), 601–616. https://doi.org/10.1093/chemse/bju033.

IARC (2018). *Methyl bromide. IARC monographs on the evaluation of the carcinogenic risk of chemicals to humans. Some halogenated hydrocarbons and pesticide exposures*. Vol. 98. Lyon, France: International Agency for Research on Cancer Available from (2018). https://monographs.iarc.fr/wp-content/uploads/2018/06/mono98.pdf.

Ioffe, D., & Kampf, A. (2002). *Kirk-Othmer encyclopedia of chemical technology*. Indianapolis, IN: John Wiley & Sons.

IPCS (2009). *International chemical safety card entry for methyl bromide, ISCS 0109*. Geneva: World Health Organization.

IRAC (2020). *IRAC mode of action classification scheme. Version 9.4. Insecticide resistance action committee. (2020)*. MoA-Classification_v9.4_3March20.pdf. Accessed 04 August 2020).

IRIS (2002). *Bromomethane (CASRN 74-83-9). Integrated Risk Information System*. Washington, DC: U.S. Environmental Protection Agency. (2002). https://cfpub.epa.gov/ncea/iris/iris_documents/documents/subst/0015_summary.pdf. (Accessed 05 July 2020).

Kulkarni, P. A., Duncan, M. A., Watters, M. T., Graziano, L. T., Vaouli, E., Cseh, L. F. , et al. (2015). Severe illness from methyl bromide exposure at a condominium resort—US Virgin Islands, March 2015. *MMWR Morbidity and Mortality Weekly Report*, *64*(28), 763.

Langård, S., Rognum, T., Fløtterød, Ø., & Skaug, V. (1996). Fatal accident resulting from methyl bromide poisoning after fumigation of a neighbouring house; leakage through sewage pipes. *Journal of Applied Toxicology*, *16*(5), 445–448.

Larrañaga, M. D., Lewis, R. J., & Lewis, R. A. (2016). Methyl bromide. *Hawley's condensed chemical dictionary*. (16th ed.). Vol. 896. John Wiley & Sons, Inc. https://doi.org/10.1002/9781119312468.

Lemoine, T. J., Schoolman, K., Jackman, G., & Vernon, D. D. (2011). Unintentional fatal phosphine gas poisoning of a family. *Pediatric Emergency Care*, *27*(9), 869–871.

NTP (2016). *Substances listed in the fourteenth report on carcinogens. Report on carcinogens* (14th ed.). National Toxicology Program. (2016). https://ntp.niehs.nih.gov/ntp/roc/content/listed_substances_508.pdf. (Accessed 20 August 2020).

O'Neil, M. J., Smith, A., & Heckelman, P. E. (Eds.). (2001). *The Merck Index: An encyclopedia of chemicals, drugs, and biologicals*. In Vol. 13. Whitehouse Station, NJ: Merck.

O'Neil, M., Heckelman, P. E., Dobbelaer, R. , et al. (2014). 6103. Methyl bromide. *The Merck index*. (15th ed.). Vol. 119. London, England: Emerald Group Publishing Limited. https://doi.org/10.1108/rr-08-2014-0224.

OSHA (2018a). *Occupational safety and health standards. Subpart Z - Toxic and hazardous substances. Air contaminants. Table Z-1: Limits for air contaminants. Occupational safety and health administration. Code of Federal Regulations. 29 CFR 1910.1000. (2018a).* https://www.osha.gov/lawsregs/regulations/standardnumber/1910/1910.1000TABLEZ1. (Accessed 25 August 2020).

OSHA (2018b). *Occupational safety and health standards for shipyard employment. Subpart Z - Toxic and hazardous substances. Air contaminants. Occupational safety and health administration. Code of Federal Regulations. 29 CFR 1915.1000. (2018b).* https://www.osha.gov/lawsregs/regulations/standardnumber/1915/1915.1000. (Accessed 25 August 2020).

OSHA (2018c). *Safety and health regulations for construction. Subpart D - Occupational health and environment controls. Gases, vapors, fumes, dusts, and mists. Occupational safety and health administration. Code of Federal Regulations. 29 CFR 1926.55. Appendix A. (2018c).* https://www.osha.gov/laws-regs/regulations/standardnumber/1926/1926.55AppA. (Accessed 25 Auguat 2020).

TRI17 (2019). *TRI explorer: Providing access to EPA's toxics release inventory data*. Washington, DC: Toxics Release Inventory. U.S. Environmental Protection Agency. (2019). http://www.epa.gov/triexplorer/. (Accessed 05 September 2020).

UNEP. (2018). Montreal protocol on substances that deplete the ozone layer. https://ozone.unep.org/sites/default/files/2019-04/MBTOC-assessment-report-2018_1.pdf. (Accessed 25 December 2019).

Yamano, Y., & Nakadate, T. (2006). Three occupationally exposed cases of severe methyl bromide poisoning: Accident caused by a gas leak during the fumigation of a folklore museum. *Journal of Occupational Health*, *48*(2), 129–133.

Yates, S. R., Gan, J., & Papiernik, S. K. (2003). Environmental fate of methyl bromide as a soil fumigant. In *Reviews of environmental contamination and toxicology* (pp. 45–122). New York, NY: Springer.

Youngentob, S. L., & Schwob, J. E. (2006). Odorant identification and quality perception following methyl bromide-induced lesions of the olfactory epithelium. *Behavioral Neuroscience*, *120*(6), 1346. https://doi.org/10.1037/0735-7044.120.6.1346.

CHAPTER

20

Methyl isocyanate: Risk assessment, environmental, and health hazard

Manila

Department of Chemistry, Gurukula Kangri (Deemed to be University), Haridwar, Uttarakhand, India

Abbreviations

COPD	chronic obstructive pulmonary disease
ICMR	Indian Council of Medical Research
MIC	methyl isocyanate
NIOSH	National Institute for Occupational Safety and Health
OSHA	Occupational Safety and Health Administration
PIH	poison inhalation hazard
PEL	permissible exposure limit
UCIL	Union Carbide India Limited

20.1 Introduction

Methyl isocyanate (MIC) is a volatile, highly flammable colorless liquid that quickly evaporates into the air. It has a sharp, strong odor that causes tears. It is extremely toxic if inhaled and absorbed through the skin to cause burns. MIC undergoes an exothermic reaction with water violently. It attacks the respiratory system, eyes, and skin primarily and thus, has the potential to cause severe damage to eyes, lungs, and bronchial airways.

MIC is extensively used as a chemical intermediate for the production of carbamate pesticides (like carbofuran, carbaryl, methomyl, and aldicarb). It is also used for the manufacture of rubbers and adhesives and to a lesser extent in the production of polyurethane foams and plastics (Bingham, Cohrssen, & Powell, 2001).

https://doi.org/10.1016/B978-0-323-89857-7.00014-1

MIC is avowed as the most reactive and poisonous member of isocyanate family manifested by a dreadful industrial accident embroiling MIC at Union Carbide India Pesticide Plant, killing approximately 2500–5000 people and injuring around 200,000 (Dhara, Dhara, Acquilla, & Cullinan, 2002; Mehta, Mehta, Mehta, & Makhijani, 1990; Mishra et al., 2009; Sharma, 2005).

20.2 Occurrence

MIC has been found to occur in tobacco smoke, so people who smoke or are exposed to second-hand smoke are vulnerable to MIC (Agency for Toxic Substances and Disease Registry, Division of Toxicology ToxFAQs, 2002).

MIC is manufactured usually by the reaction of monomethyl amine and phosgene at high temperatures in the industrial units wherein people can be exposed by breathing or touching it (Agency for Toxic Substances and Disease Registry, Division of Toxicology ToxFAQs, 2002).

The evidence of the extra-terrestrial occurrence of MIC has been obtained by scientists on the surface of comet 67/P by Philae lander on July 30, 2015 (Bibring et al., 2015).

As it is artificially manufactured in industrial units, its exposure limit has been set to 0.02 ppm by the Occupational Safety and Health Administration (OSHA). The safe exposure limit does not pose any serious threat to human beings as MIC is degraded rapidly in the air owing to its reactive nature. If MIC comes in contact with water, it degrades into other compounds rapidly (within minutes to a few hours) (Agency for Toxic Substances and Disease Registry, Division of Toxicology ToxFAQs, 2002). The same degradation principle applies to soil also.

20.3 Historical background

A scientific report on MIC toxicity surfaced in 1964 in which a study conducted on human volunteers exposed to MIC vapors was reported. Chocking sensations, eye irritation, and difficulty in breathing were observed in these volunteers (Kimmerle & Eben, 1964).

MIC toxicity was almost unheard till the Bhopal disaster took place in 1984. This catastrophe stimulated the interest of scientists to investigate deep into this compound. A number of studies on this particular incident were done and are still underway in order to assess the impact of MIC in long term on generations of victims exposed to MIC in Bhopal gas tragedy (Dodd, 1987; Mishra, 2015; Varma & Mulay, 2006; Vijayan, 2010).

TABLE 20.1 Physicochemical properties of MIC.

Formula	C_2H_3NO or CH_3NCO
IUPAC name	Methylamino(*oxo*)methane
Molecular weight (g/mol)	57.05
Boiling point (°F)	102
Melting point (°F)	−112
Flashpoint (°F)	19
Density (at 68 °F)	0.9599
Vapor pressure (mm Hg at 68 °F)	348
Autoignition temperature (°F)	995
Ionization potential (eV)	10.67
Refractive index (at 64 °F)	1.3419

20.4 Physicochemical properties

MIC is a colorless low boiling liquid, highly flammable, with a sharp, unpleasant odor. It is a highly reactive gas that violently undergoes an exothermic reaction with water to produce carbon dioxide, dimethylurea, and trimethylbiuret. Below 68 °F, the reaction is relatively slow but it catches pace at elevated temperatures or in the presence of acid and base. Vapors of MIC may travel to the ignition source and flash back. Containers of MIC may explode violently under the effect of heat. It emits toxic fumes of nitrogen oxides if decomposes due to heating. MIC not only interacts with a large number of molecules but it undergoes self-polymerization also (Varadarajan, Doraiswamy, Ayyangar, et al., 1985). MIC is found to react with some plastics, rubbers, and coatings also. Some of the properties of MIC are discussed in Table 20.1.

20.5 Physiological role

MIC can affect human beings via two pathways:

- Breathing
- Absorption through the skin

MIC is a skin irritant and burns the skin and eyes. The exposure to MIC beyond the safe limit can lead to permanent eye damage. MIC irritates the nose, throat, and lungs causing dyspnea. It has been found to interact with different molecules as well as with itself (Varma & Mulay, 2015). MIC polymerizes as it enters the human body which has been evidenced by the autopsies conducted on Bhopal gas victims, wherein it trimer along with its other metabolites like dimethyl isocyanate and 2,4-dione of MIC (Varma & Mulay, 2015). As per the ICMR report, toxic adducts enter the bloodstream of the exposed people leading to pernicious influence on almost all organs (Dinham & Sarangi, 2002).

20.6 Uses

MIC is widely used in various fields enlisted as follow:

- Organic synthesis and production of carbamate pesticides
- Chemical intermediate for the manufacturing of insecticides and herbicides
- Manufacture of polyurethane foams and plastics
- It yields carbamate insecticide carbaryl on reaction with 1-napthol
- It is reacted with methylthioisobutyrladoxime for the large-scale production of Aldicarb (National Center for Biotechnology Information, 2020)

20.7 Sources of contaminants

Human beings can be exposed to MIC by any of the following ways:

- MIC has been found to be present in tobacco smoke. Thus, smokers and those who inhale second-hand smoke are prone to MIC exposure.
- People living nearby to manufacturing facilities, storage areas, or the places where MIC is used may breathe it at low levels.
- People may be exposed to MIC by either inhaling or touching it at workplaces where this compound is being used or manufactured.

20.8 Risk assessment

Risk assessment is a vital exercise required to be done in facilities that are going to manufacture, store, or use MIC. It is done to ensure the complete safety of the people associated with the facility along with the risk cover of the population residing in nearby areas and ecosystems. As MIC is a highly toxic chemical, so following risk assessment procedure is needed to be in place.

Identification of hazard:

- DOT Poison Inhalation Hazard (PIH)
- DOT Number: UN 2480

- Flammable liquid
- Corrosive
- Acute toxic
- Irritant
- Health hazard

Who is at risk?
- Workers working in industrial units utilizing or producing MIC
- If anyone touches it at the workplace or breathe in where this compound is produced, used, or analyzed
- People living in the vicinity of facilities involved in the manufacture, store, or use of MIC

How can it affect health?

MIC is very harmful if breathed in. The effects will depend on how much concentration was one exposed and up to how long. Exposure to low limits may cause eye and throat irritation but if exposure reaches beyond the limit, it can cause swelling in the lungs making it difficult to breathe. This may happen instantly or take a day or two. Exposure to high concentration can be fatal. Long-term exposure could result in permanent lung damage. Chemical burns may be experienced if MIC comes in contact with skin. Permanent eye damage might be induced at high concentrations. If swallowed accidentally, the mouth, throat, esophagus, and stomach could be damaged.

Based on the identification of the vulnerable class of people and health implications that stand before them on exposure, all safety measures should be adopted to avoid any mishappening. Risk assessment procedure should be followed from time to time in order to ensure the complete safety of employees, the population residing nearby industrial units, and the ecosystem as a whole.

20.9 Environmental effects

MIC exists solely as a gas when released into the air. It degrades very rapidly by reacting with commonly found substances in the air. MIC breaks down by moisture present in clouds or rainfall. Its persistence in the atmosphere ranges from few hours to few days.

MIC degrades very swiftly, if present in the water, into other compounds as it undergoes an exothermic reaction with water. The production of MIC and its use in the preparation of carbamate-based and urea-based insecticides and fungicides eventually release it into the environment via water streams. Its hydrolysis products may include *N*-carboxymethyl amine, carbon dioxide, methylamine, and *N,N′*-dimethyl urea, which have toxic effects on human beings and aquatic life as well. Although MIC degrades rapidly as it enters the environment at low concentrations but its toxic levels, could lead to devastating outcomes. For instance, MIC leakage in toxic levels at the Bhopal gas tragedy not only created havoc for humans but also flora and fauna existing nearby to Union Carbide Plant. The instant effects of MIC on the plants in its proximity comprised defoliation, burnt patches, and drying of most of the aerial parts of the plant. Succulent and xerophytic plants were seen to undergo maximum damage as their stomata open during the night. Chlorosis and necrosis were observed in newly formed leaves (Kumar, Tripathi, & Roy, 1989). Another study revealed that plant species within 0.5 km of the site of this incident were heavily burned and those plant species at a distance of 3–4 km showed signs of scorching, necrosis, and chlorosis (Saify & Bhat, 1985). A large number of animals also died in this incident.

20.10 Hazardousness

MIC is extremely flammable and its vapors may rove to the source of ignition and flash back. It is extremely dangerous if exposed to heat or flame oxidizers. It may prove fatal if inhaled or absorbed through the skin. Beyond its safe exposure limit of 0.02 ppm, it is lachrymatory and could lead to permanent eye damage. It can cause burns to the skin. It can lead to shortness of breath as it affects the lungs also. Exposure to higher concentrations can lead to pulmonary edema (fluid build-up in the lungs). MIC may also cause asthma-like allergy. The exposure after effects of MIC can last for months or even years. MIC has the potential to cause mutations also. There have been reports related to miscarriages and infant mortality in the victims exposed to MIC gas leakage (Varma, 1987).

MIC exerts a broad range of toxic effects, which may lead to severe organ-system damage as it enters the bloodstream once it finds a way into the human body as reported by a study of ICMR (Dinham & Sarangi, 2002).

20.11 Risk management

The magnitude of risk is quite high in the case of MIC but certain preventive measures can be taken to avoid any disaster. Workplace exposure limits should be as follows:

OSHA: The legal airborne permissible exposure limit (PEL) is 0.02 ppm, averaged over 8-h work shift.

NIOSH: The recommended airborne permissible exposure limit is 0.02 ppm, averaged over a 10-h work shift.

MIC is a potential mutagen, so minimal contact with this chemical is recommended. Containers of MIC should be checked regularly for any leakage. Transportation of MIC should be done through a registered person/transportation company that is licensed to accept hazardous material and MIC should be properly closed, described, packaged, labeled, and should be in condition for shipment as per regulations.

Industries processing MIC should be regulated properly as per government guidelines in order to ensure the complete safety of the environment.

20.12 Current research

A number of studies to unfold the long-lasting pattern of various disorders and deformities on the victims exposed to MIC in the Bhopal gas tragedy have been done and are still undergoing. A study of a population-based cancer register study of ICMR reported a higher incidence of gall bladder carcinoma (GBC) in the surviving population of the Bhopal gas tragedy (Mishra et al., 2009; Sriramachari, 2004). Another hospital registration-based study on 1261 cancer sufferers and their offsprings exposed to MIC was conducted which revealed a gradual rise of cancer in different areas nearby to Union Carbide Plant post-Bhopal gas tragedy (Senthilkumar, Malla, Sah, & Ganesh, 2011). Molecular studies in reference to MIC have been conducted to manifest persistent DNA damage in ovarian epithelial cells upon treatment with MIC (Raghuram et al., 2010).

Furthermore, a study to investigate the effect of MIC among a transection of current survivors encountering COPD by using ISSR-PCR was conducted which revealed a weak association between microsatellite instability and exposure distance from the gas-release site and smoking status along with age, although regression studies did provide supporting evidence (Bose & Bathri, 2012). A combination of immuno-FISH and other molecular techniques in diagnosing the offspring of victims of MIC in Bhopal gas tragedy was conducted to get a deep insight into the deadly accident and its aftermath.

20.13 Methyl Isocyanate: Case study

Case study—The Bhopal gas tragedy: The smallest, highly reactive, and noxious member of the isocyanate family was virtually unknown until the intervening night of December 2–3, 1984, when approximately 40 metric tons of this deadly chemical leaked out of Union Carbide India Limited (UCIL) pesticide plant at Bhopal within an hour. This accident accounted for about 2500–5000 deaths and affected as many as 200,000 people along with animals and plants in Bhopal. Most of the fatalities happened between 24 h and 48 h proving that HCN was not the primary suspect in the case as thought initially. This chemical disaster not only affected human beings but also sent shock waves to the ecosystem near the accident spot. MIC was the prime culprit of this disaster. MIC encompassed a wide range of toxic effects initiating from lacrimation, chocking, dyspnea, skin burns, and progressing toward pulmonary edema, convulsions, and ultimately cardiorespiratory arrest (Kamat, Mahashur, Tiwari, et al., 1985; Misra, Pathak, Gaur, et al., 1987). The pattern of deaths observed in this case were seen to be delayed for about 24 h manifesting gradual access of MIC into different body organs through the bloodstream as evidenced by the ICMR report (Dinham & Sarangi, 2002). The effects of this exposure did not end after a few days or months but were observed in future generations of the exposed population. High pregnancy loss was observed in the pregnant women exposed to MIC with an augmentation in 5-year mortality of the offspring born to exposed parents and delayed development of male progeny (Sarangi et al., 2010). A study conducted on lymphocytes of 31 gas-exposed individuals after two and half months postdisaster revealed a significant degree of chromosomal; aberrations (Saxena, Singh, & Nagle, 1988). MIC released not only affected humans but also the ecosystem. It was reported that the extent of total chlorophyll loss was found to be proportional to the distance from the incident spot. Although plants near to the site of the accident were heavily burned, 17 out of 26 plant species showed a loss of more than 50% of total chlorophyll. Necrosis, scorching, and chlorosis were observed in some plants like *Phaseolus mungo*, *Glycine max*, *Cajanus cajan*, *Mirabilis jalapa*, *Helianthus annus*, *Luffa cylindrica*, *Lycospescium esculentum*, and *Solanum melongeana* in the range of 3–4 km from the Union Carbide Plant.

20.14 Safety measures

- The exposure limit of MIC should not cross the permissible limit.
- Containers of MIC should be kept away from water, alkalis, acids, and heat.
- MIC should be stored in a cool and dry place.
- Smoking should be strictly prohibited in areas near the MIC storage, usage, production, and analysis.
- Workers should be guided properly regarding the precautions to be taken while working with the chemical.
- Proper awareness program regarding MIC hazards should be held for workers.
- Eyewash fountains must be provided in the work area for emergency use.

- Emergency shower facilities should be provided.
- Industrial plants using or manufacturing MIC should be constructed away from the residential area.
- Transportation of MIC should be done keeping public safety at priority.
- Proper safety equipment should be installed and made available to avoid any accident.
- PPE kits and eye-protecting goggles should be made available in industrial facilities handling MIC to control minor leakages from turning into big tragedies.

20.15 Handling and storage

- The person handling MIC should be properly trained.
- All equipment must be grounded and bonded while handling MIC.
- Nonsparking tools and equipment should be used especially during opening and closing containers of MIC.
- Damaged containers or spilled material should not be touched unless equipped with appropriate protective clothing.
- MIC must be stored in such a way to avoid contact with water, strong acids, alkalis, amines, copper, iron, and oxidizing agents.
- MIC should be stored in firmly closed containers in a cool, dry, properly ventilated area avoiding heat as well as sunlight.
- Prohibition of smoking, open flames, and other sources of ignition should be done around the area of storage of MIC.

20.16 Precautions

- Workers working in industrial units related to MIC use or production should change into clean clothing promptly if clothes have been contaminated.
- If skin contact with MIC has taken place, immediately washing and showering should be exercised to remove the chemical.
- At the end of the work shift, all body parts exposed to MIC should be washed properly.
- DO NOT smoke, eat, or drink where MIC is handled, stored, or processed.
- For eye protection, indirect-vent, impact, and splash-resistant goggles should be worn while handling MIC.
- Contact lenses should be strictly avoided while working with MIC.

References

Agency for Toxic Substances and Disease Registry, Division of Toxicology ToxFAQs (2002, April). *Methyl isocyanate CAS #624-83-9*. www.atsdr.cdc.gov/toxfaq.html. (Accessed 20 August 2020).

Bibring, J. P., Taylor, M. G., Alexander, C., Auster, U., Biele, J., & Finzi, A. E. (2015). Philae's first look. Philae's first days on the comet. *Science, 349*(6247), 493.

Bingham, E., Cohrssen, B., & Powell, C. H. (2001). *Patty's Toxicology*: (p. 4) (5th ed.). (Vols. 1–9). New York: John Wiley & Sons. 1434.

Bose, P., & Bathri, R. (2012). Association of microsatellite instability and chronic obstructive pulmonary disorder in isocyanate-exposed population of Bhopal. *Indian Journal of Human Genetics, 18*, 172–176.

Dhara, V. R., Dhara, R., Acquilla, S. D., & Cullinan, P. (2002). Personal exposure and long-term health effects in survivors of the union carbide disaster at Bhopal. *Environmental Health Perspectives, 110*, 487–500.

Dinham, B., & Sarangi, S. (2002). The Bhopal gas tragedy 1984 to? The evasion of corporate responsibility. *Environment and Urbanization, 14*(1), 89–99.

Dodd, D. E. (1987). Toxicology of methyl Isocyanate. G. W. Ware (Ed.), *Reviews of environmental contamination and toxicology*Vol. 105, .

Kamat, S. R., Mahashur, A. A., Tiwari, A. K. , et al. (1985). Early observations on pulmonary changes and clinical morbidity due to the isocyanate gas leak at Bhopal. *Journal of Postgraduate Medicine, 31*(2), 63–72.

Kimmerle, G., & Eben, A. (1964). Zur Toxicitaet Von Methylisocyanat Und Dessen Quantitative Bestimmung In Der Luft [On the toxicity of methyl isocyanate and its quantitative determination in the air]. *Archives of Toxicology, 20*, 235–241.

Kumar, G., Tripathi, D. S., & Roy, S. K. (1989). Cytological effects on plants by an accidental leakage of methylisocyanate (MIC) gas. *Environmental and Experimental Botany, 29*(2), 261–271.

Mehta, P. S., Mehta, A. S., Mehta, S. J., & Makhijani, A. B. (1990). Bhopal tragedy's health effects: A review of methyl isocyanate toxicity. *JAMA, 264*(21), 2781–2787.

Mishra, B. (2015). Health issues among MIC exposed population 30 years post MIC disaster at Bhopal. *International Journal of Health System and Disaster Management, 3*, 115–128.

Mishra, P. K., Jatawa, S. K., Raghuram, G. V., Pathak, N., Jain, A., Tiwari, A. , et al. (2009). Correlation of aberrant expression of p53, Rad50, and cyclin-E proteins with microsatellite instability in gallbladder adenocarcinomas. *Genetics and Molecular Research, 8*, 1202–1210.

Mishra, P. K., Samarth, R. M., Pathak, N., Jain, S. K., Banerjee, S., & Maudar, K. K. (2009). Bhopal gas tragedy: Review of clinical and experimental findings after 25 years. *International Journal of Occupational Medicine and Environmental Health, 22*, 1–10.

Misra, N. P., Pathak, R., Gaur, K. J. , et al. (1987). Clinical profile of gas leak victims in acute phase after Bhopal episode. *Indian Journal of Medical Research, 86*, 11–19.

National Center for Biotechnology Information (2020). *PubChem Compound Summary for CID 7158, Fenoprop*. https://pubchem.ncbi.nlm.nih.gov/compound/Fenoprop. (Accessed 20 July 2020).

Raghuram, G. V., Pathak, N., Jain, D., Panwar, H., Pandey, H., Jain, S. K. , et al. (2010). Molecular mechanisms of isocyanate induced oncogenic transformation in ovarian epithelial cells. *Reproductive Toxicology, 30*, 377–386.

Saify, T., & Bhat, P. K. (1985). Effects of methyl isocyanate on the vegetation of Bhopal with special reference to total chlorophyll damage. *Journal of Scientific Research, 7*, 17–19.

Sarangi, S., Zaidi, T., Pal, R. K., Katgara, D., Gadag, V. G., Mulay, S. , et al. (2010). Effects of exposure of parents to toxic gases in Bhopal on the offspring. *American Journal of Industrial Medicine, 53*, 836–841.

Saxena, A. K., Singh, K. P., & Nagle, S. L. (1988). Effect of exposure to toxic gas on the population of Bhopal: Part IV—Immunological and chromosomal studies. *Indian Journal of Experimental Biology, 26*(3), 173–176.

Senthilkumar, C. S., Malla, T. M., Sah, N. K., & Ganesh, N. (2011). Cancer morbidity among methyl isocyanate exposed long-term survivors and their offspring: A hospital-based five year descriptive study (2006–2011) and future directions to predict cancer risk in the affected population. *Asian Pacific Journal of Cancer Prevention, 12*, 3443–3452.

Sharma, D. C. (2005). Bhopal: 20 years on. *Lancet, 365*, 111–112.

Sriramachari, S. (2004). The Bhopal gas tragedy. *An Environmental Disaster, 86*, 905–920.

Varadarajan, S., Doraiswamy, L. K., Ayyangar, N. R. , et al. (1985). *A scientific Enquiry into the Methyl Isocyanate Leak in Bhopal.* New Delhi: Council of Scientific and Industrial Research, Publication.

Varma, D. R. (1987). Epidemiological and experimental studies on the effects of methyl isocyanate on the course of pregnancy. *Environmental Health Perspective, 72*, 153–157.

Vijayan, V. K. (2010). Methyl isocyanate (MIC) exposure and its consequences on human health at Bhopal. *International Journal of Environmental Studies, 67*(5), 637–653.

Varma, D. R., & Mulay, S. (2015). Methyl isocyanate. In *Handbook of toxicology of chemical warfare agents* (pp. 287–299). Academic Press.

Varma, D. R., & Mulay, S. (2006). The Bhopal accident and methyl isocyanate toxicity. In *Toxicology of organophosphate & carbamate compounds* (pp. 79–88).

CHAPTER

21

Nitric oxide: Risk assessment, environmental, and health hazard

Krishan Chandra Singh[a] *and Kavita Rawat*[b]

[a]Department of Chemistry, Maharishi Dayanand University, Rohtak, Haryana, India

[b]Department of Chemistry, H.N.B. Garhwal University, Badshahithaul Campus, Tehri Garhwal, Uttarakhand, India

Abbreviations

cGMP	cyclic guanosine monophosphate
COVID-19	coronavirus
GC	guanylyl cyclase
GTP	guanosine triphosphate
MLC	myosin light chain
NO	nitric oxide
SARS	severe acute respiratory syndrome

21.1 Introduction

Nitric oxide is a vital signaling molecule and was identified as free radical gas in 1772 by Joseph Priestly. It is a major carcinogenic pollutant possessing high nitrosative and genotoxic potential. NO principally shares an essential part in human physiology by dilation of blood vessels, consequently affecting blood pressure, and hence acts as a vasodilator. A significant amount of nitric oxide synthase, an enzyme responsible for maintaining the body's blood pressure, is produced by amino acid L-arginine (Gardner, Gardner, Martin, & Salzman, 1998).

21.2 Occurrence

Most of the oxides of nitrogen are gases under normal conditions such as nitrous oxide, nitric oxide, NO_2, and some other forms of nitrogen exist as aerosols of nitric acid (Griffith & Stuehr, 1995). Among the oxides of

https://doi.org/10.1016/B978-0-323-89857-7.00020-7

nitrogen, nitrous oxide is the only oxide found in excess levels in the atmosphere, that is chiefly produced in soil by the bacterial action on nitrogenous compounds (Robinson & Robbins, 1974).

21.3 Historical background

Being a free radical colorless gas, NO has a very short life as discovered by Joseph Priestly. Bredt and Snyder isolated nitric oxide synthase for the first time in different cells as NOS_1, NOS_2, and NOS_3 (endothelial cell) (Griffith & Stuehr, 1995). Among three isoforms of nitric oxide, NOS_3 was separated from bovine endothelial cells. On the basis of isolation and characterization, human genes were officially categorized as bNOS (brain nitric oxide synthase), iNOS (induced nitric oxide synthase), and eNOS (endothelial nitric oxide synthase) which are termed as NOS_1, NOS_2, and NOS_3, respectively (Grurtter et al., 1979). Nitric oxide has a major impact on biology in the last 25 years (Hevel, White, & Marletta, 1990; Yetik-Anacak & Catravas, 2006).

21.4 Physicochemical properties

Nitric oxide is a colorless, water-soluble, and free radical uncharged species. It rapidly reacts with oxygen to form higher oxides of nitrogen, which are listed as the major environmental pollutants. Nitric oxide is generally produced in the various tissues and it plays an important role in vasodilation, inflammation, neurotransmission, and cell regulation in the biological systems (Harrison, 1996).

NO decomposes into the oxygen and nitrogen gases (Eq. 21.1).

$$2NO \rightarrow O_2 + N_2 \tag{21.1}$$

Nitric oxide reacts with oxygen to form NO_2 (Eq. 21.2).

$$2NO + O_2 \rightarrow 2NO_2 \tag{21.2}$$

Some of the physical and chemical properties of nitric oxide are detailed in Table 21.1 (National Center for Biotechnology Information, 2020).

21.5 Physiological role

Nitric oxide plays an important part in physiological systems, such as in the vasculature as nitric oxide induces vasodilation, restrains neutrophils, hinders platelet aggregation, inhibits flexible muscles cells generation and transformation, controls apoptosis, and maintains the NOS_3 barrier functions. In the recent two decades, nitric oxide has played an

TABLE 21.1 Physical and chemical properties of nitric oxide.

Molecular weight	**30.006 g/m**
Boiling point	−241.1 °F at 760 mmHg
Melting point	−262.5 °F
Solubility	7.38 mL/100 mL water at 0°C
Density	1.27 at −238.36 °F
Vapor density	1.04
Vapor pressure	26,000 mmHg at 68 °F
Viscosity	0.0188 cP at 25°C @ 101.325 kPa (gas)
Heat of vaporization	3.293 kcal/mol
Ionization potential	9.27 eV

unexpected role in the physiology of the reproductive system and controlled multiple functions in males as well as the female reproductive system. Nitric oxide is a well-known biological product in living organisms, which is present in microorganisms as well as plants (Rosselli, Keller, & Dubey, 1998).

Nitric oxide is a signaling molecule for transient paracrine and autocrine, transmitting the information in the neuro system (Stryer, 1995). In males, nitric oxide induces the erection of the penis and is also necessary for sore repair. Nitric oxide also plays an essential part in cell division and maturation as well as an intermediate for neovascularization, i.e., the formation of mature blood vessels to enrich the healing tissues. In blood vessels, NOS_3 (endothelial cells) moves to muscle cells and relaxes them by liberating nitric oxide. This enlarges the blood vessel cells and minimizes the blood pressure (Griffith & Stuehr, 1995).

Deficiency of nitric oxide synthase causes high blood pressure in the body and nitroglycerine drug is used to reduce the blood pressure by vasodilation in this condition. Nitric oxide synthase enzyme exists in three isoforms as NOS_1, NOS_2, and NOS_3. The isoform NOS_1 is referred to as the brain NO synthase enzyme, also known as neuronal isoform (nNOS). It regulates the blood pressure, induces smooth muscle cell relaxation, and vasodilation. NOS_2 is referred to as an inducible NO synthase enzyme (iNOS) and produces a high concentration of nitric oxide to kill microorganisms and is stimulated in various inflammatory diseases. However, NOS_3 is known as endothelial NO synthase (eNOS) and used in the dilation of blood cells and the regulation of blood pressure (Griffith & Stuehr, 1995).

Nitric oxide is a key factor of various biological phenomena (Roszer, 2012), also known as an endothelium-derived relaxing factor. Biosynthesis of NO is carried out by various nitric oxide synthase (NOS) enzyme.

Release of nitric oxide: Binding of acetylcholine to the vascular endothelial receptor (R) results in the excess release of Ca^{2+} from their indigenous cell. Calcium ion binds to calmodulin protein to form the calcium calmodulin complex, which activates the NO synthase enzyme. This active NO synthase enzyme catalyzes the conversion of arginine and oxygen into citrulline amino acid and NO molecule.

Intracellular mechanism: The newly produced nitric oxide diffuses into the nearby smooth muscle cells with the activation of the guanylyl cyclase (GC). Active GC enzyme in turn activates the conversion of guanosine triphosphate (GTP) to its active form cyclic guanosine monophosphate (cGMP), hence facilitating the formation of myosin light chain protein (MLC). This process induces the dissociation of myosin and results in smooth muscle relaxation (Fig. 21.1).

The pharmacological nitric oxide donors such as organic nitrates pass through the smooth muscle cells and release NO into their cytoplasm.

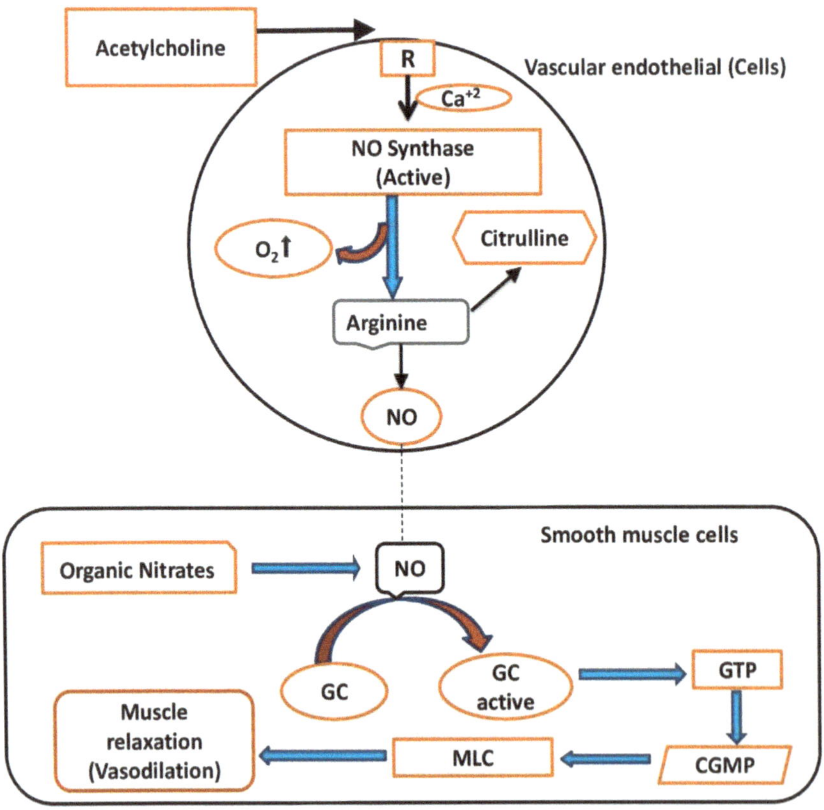

FIG. 21.1 Mechanism of action of nitric oxide. *Credit: Kavita Rawat*

The external NO mimics the physiological functions of the indigenous NO, which usually ends with smooth muscle relaxation.

21.6 Uses

Nitric oxide has several medicinal uses and can be used in the treatment of impotence and sexual diseases in men.

- Nitric oxide is an excellent blood pressure controller as it dilates the blood vessels, further allowing a smooth flow of blood into the system. Nitric oxide produced by neurons (nerve cells) is used by the central nervous system as a neurotransmitter to manage functions starting from digestion to blood flow to memory and vision.
- Nitric oxide released from macrophages kills various microorganisms like bacteria, parasites, and tumor cells by disturbing their metabolism.

21.7 Sources of contaminants

High-temperature fuel combustion process, as well as thunderstorm lightning, is responsible for the production of nitric oxide, due to extremely high and low temperatures (Schumann & Huntrieser, 2007). Oxidation of nitric oxide produces nitric acid due to lightning, which either induces acid rain or gets accumulated in the atmosphere. Weather, topographical location, and temperature variations are also responsible for the production of nitric oxide at a high level. Lightning is the main cause of the production of NO during summer months as reported through satellites (Hauglustaine et al., 2001). The main source of the carcinogenic nitric oxide gaseous radical production is the metabolic reduction of inorganic nitrates or nitrites in the microorganism (Gardner et al., 1998).

21.8 Risk assessment

Nitric oxide is a highly reactive species that oxidizes into nitrogen dioxide, which is highly toxic. In the case of inhalation or skin adsorption, nitric oxide can severely affect the nervous system, cardiovascular system, and respiratory system. This necessitates the conductance and documentation of NO-related risks. The risk can be evaluated on the basis of various factors such as activity involved in the release of NO, for example, oxides of nitrogen; hazardousness on human health like peeling skin from burns, etc.; and validation of employed safety measures.

Inhaled NO therapy is useful in the treatment of respiratory illness and lung diseases. However, it is difficult to separate the pharmacological properties of inhaled NO therapy from its toxicological effects. This therapy produces by-products as NO_2 and methemoglobin, which causes hypoxemia, and therefore monitoring is essential to ensure that the level of by-products lie within the safe limit.

The influence of exogenous nitric oxide on the development of tumors may be mediated by its action on the system of antitumor resistance and it is dose-dependent. The overall effect of inhaled NO therapy in potentiating inflammation and oxidative damage in diseased lungs is dependent on the dose administered. Therefore, efficacy and safety within a safe limit must be followed (Weinberger, Laskin, Heck, & Laskin, 2001).

21.9 Environmental effects

Nitric oxide adversely affects the wider ecosystem by inducing acid rain, due to the formation of nitrous acid on reacting with the reaction of NO and NO_2 (Eq. 21.3 and 21.4). Acid rain infiltrates effectively into the lungs causing respirational sicknesses like bronchitis, emphysema, and coronary illness, hence drastically affecting the health of a wide array of living organisms, including flora and fauna.

$$NO + HO_2 \rightarrow NO_2 + OH \tag{21.3}$$

$$NO_2 + OH \rightarrow HNO_3 \tag{21.4}$$

In addition to this, an increased level of nitrogen oxides in the environment causes global warming and ozone layer depletion (Badr & Probert, 1993) (Eq. 21.5).

$$NO + O_3 \rightarrow NO_2 + O_2 \tag{21.5}$$

21.10 Hazardousness

Exposure to nitric oxide can cause many short-term (acute) and long-term (chronic) health effects in the human body, such as:

- Nitric oxide present in the air readily reacts with moisture and atmospheric compounds, which results in the formation of nitric acid and other carcinogenic particles. Consequently, inhalation of which causes respiratory diseases, heart diseases, damage to lungs, or even premature death in some cases (Grurtter et al., 1979; Murray, 2016).
- Nitric oxide is highly toxic among the oxides of nitrogen as a high exposure level of NO can cause dimness of vision, nausea, headache, skin

irritation, and sometimes premature death and also lowers the blood pressure by reducing the ability of blood to transport oxygen to the lungs. A high level of nitric oxide can also harm teeth (Murray, 2016).

Various diseases can be encountered either by the deficiency or excess production of NO at various locations in the body and have different significant results. Dietary factors play a significant role in the regulation of NO production. A low level of NO production can promote vascular diseases and it can be prevented by controlling the decreased level of NO production. While inhibiting the excess level of nitric oxide production could also be a major therapeutic target. Various diseases such as sepsis and trauma, vascular diseases, etc. can affect the production level of nitric oxide (Melledo, 2019).

21.11 Treatment

NOx are the major sources of pollution in the environment and it can enter the human body either through inhalation or skin contact. Various pieces of literature have shown in-depth concern with the hazardous post-inhalation of nitric oxide. The sequence of main steps involved in the treatment include (1) oxygen therapy should be given to the patients in order to balance the capacity of blood oxygen, (2) reformation of methemoglobin by using methylene blue, and (3) pneumonia treatment and its inhibition (Becklake, Goldman, Bosman, & Freed, 1957; Lowry & Schuman, 1956). The use of corticosteroids and antibiotics is highly suggested with such treatments for better results.

Patients affected with sepsis disease have a very low level of nitric oxide and conversion of arginine to citrulline is an important way to estimate the NO production. Arginine balance or NO production can be maintained by the supplementation of arginine or citrulline. If a patient is suffering from arginine deficiency (trauma), then supplementation of citrulline is a suitable way to regulate the NO production. If nitric oxide is produced by macrophages in excess level, then it can be controlled by the glutamine supplementation (Gilani et al., 2007; Luiking, Engelen, & Deutz, 2010; Oshima & Bartsch, 1994).

Oxygen therapy: If an individual has been exposed to nitric oxide gas, oxygen therapy is widely used in order to maintain the decreased level of oxygen and arterial blood in the body. Due to more than 50% exposure of NOx, methemoglobin formation takes place resulting in intense cyanosis. It is therefore essential to supply maximum available oxygen to the patient in the inspired gas (Wamberg & Zeskov, 1966).

21.12 Risk management

In order to manage NO-induced risks, the primary reassurance providing safety measures is recommended. Fundamentally, Risk Management Programs for Chemical Accidental Release should be judiciously followed.

21.13 Current research

Nitric oxide promotes viral infection clearance as it can prevent the viral spread by inhibiting the viral replication process at the initial stage. Recent research on COVID-19 hypothesized that the high concentration of NO present in cigarette smoke could be effectively utilized against COVID-19 (Hedenstierna, Chen, Hedenstierna, Lieberman, & Fine, 2020). Although smoking is a major risk factor affecting human health, it can suppress COVID-19 and it has also been demonstrated that non-smokers are more susceptible to SARS coronavirus in comparison to smokers (Berlin, Thomas, Le Faou, & Cornuz, 2020).

21.14 Nitric oxide toxicity: Case study

Fifty-nine-year-old women were diagnosed with bronchiectasis in 1988 and *Mycobacterium bolletii* in 2013. Initially, she was treated with 160 ppm nitric oxide gas for only 50 min and later thrice a day for 15 weekdays. Chronic respiratory infection symptoms score improved by 19 after the treatment of 15 days. The patient was allowed to take nitric oxide gas for more than 100 days and it was observed that long-term treatment of NO illuminated potential side effects to the patient (Miller, Martins, McMullin, Road, & Miller, 2019).

21.15 Safety measures

Some of the safety measures are as follow:

- Personal protection care should be taken by wearing protective clothing that is impermeable and resistant to nitric oxide, such as Teflon gloves.
- The work area must be well ventilated and all the main exhaust lines must be checked regularly.
- Avoid the use of flammable sources near the workplace and nitric oxide must be used in the closed system to avoid gas exposure in the lab.
- All the connections and fitting of regulators must be checked for any leakage and cylinder pressure must be checked properly.

- All the personnel handling NO should be well trained or must work under the supervision of an experienced person.

21.16 Storage and handling

- Proper handling and storage of nitric oxide is essential before performing any action and should be done by trained persons.
- Due to the free radical gaseous nature of nitric oxide, it is highly reactive and reacts readily with atmospheric oxygen, air, and water or moisture to give toxic products as nitric acid and nitrogen dioxide; therefore NO should be carefully stored.
- Being a highly reactive molecule, nitric oxide also reacts with halogens to give nitrogen trichloride, ozone, and monoxides of chlorine, which are harmful to living organisms.
- Several oxidizing agents like periodates, peroxides, chlorates, etc. are not consistent with nitric oxide and it does not show any compatibility with olefins, metals, fuels, and metallic salts.
- Due to the high reactivity of nitric oxide, it must be stored in tight vessels in dry and cool places in order to avoid the effect of moisture, heat, and sunlight.

21.17 Precautions

- Wear splash-resistant safety eyeglasses.
- Emergency eyewash stations should be provided.
- Chemical-resistant clothes must be used.
- Wear chemical-resistant gloves.

References

Badr, O., & Probert, S. D. (1993). Environmental impacts of atmospheric nitrous oxide. *Applied Energy*, *44*, 197–231.

Becklake, M. R., Goldman, H. I., Bosman, A. R., & Freed, C. C. (1957). Long term effects of exposure to nitrous fumes. *American Review of Tuberculosis*, *76*, 398.

Berlin, I., Thomas, D., Le Faou, L. A., & Cornuz, J. J. (2020). COVID-19 and smoking. *Nicotine & Tobacco Research*, *9*(22), 1650–1652.

Gardner, P. R., Gardner, A. M., Martin, L. A. A., & Salzman, A. L. (1998). Nitric oxide dioxygenase: An enzymic function for flavo-hemoglobin. *Proceedings of the National Academy of Sciences*, *95*(18), 10378–10383.

Gilani, M., Kaiser, D. R., Bratteli, C. W., Alinder, C., Rajala, S., Bank, A. J. , et al. (2007). Role of nitric oxide deficiency and it's detection as a risk factor in pre-hypertension. *Journal of the American Society of Hypertension*, *1*(1), 45–55.

Griffith, O. W., & Stuehr, D. J. (1995). Nitric oxide synthases: Properties and catalytic mechanism. *Annual Review of Physiology*, *57*(1), 707–734.

Grurtter, C. A., Barry, B. K., McNamara, D. B., Gruetter, D. Y., Kadowitz, P. J., & Ignarro, L. (1979). Relaxion of bovine coronary artery and activation of coronary arterial guanylate cyclase by nitric oxide, nitro prusside and a carcinogenic nitroso amine. *Journal of Cyclic Nucleotide Research*, *5*(3), 211–224.
Harrison, D. G. (1996). Endothelial control of vasomotion and nitric oxide production: A potential target for risk factor management. *Cardiology Clinics*, *14*.
Hauglustaine, D., Emmons, L., Newchurch, M., Brasseur, G., Takao, T., Matsubara, K., et al. (2001). On the role of lightning NOx in the formation of tropospheric ozone plumes: A global model perspective. *Journal of Atmospheric Chemistry*, *38*(3), 277–294.
Hedenstierna, G., Chen, L., Hedenstierna, M., Lieberman, R., & Fine, D. H. (2020). Nitric oxide dosed in short bursts at high concentrations may protect against Covid-19. *Nitric Oxide*, *103*, 1–3.
Hevel, J. M., White, K. A., & Marletta, M. A. (1990). Purification of the inducible murine macrophage nitric oxide synthase. Identification as a flavoprotein. *Journal of Biological Chemistry*, *266*(34), 22789–22791.
Lowry, T., & Schuman, M. L. (1956). "Silo-Filer's disease": A syndrome caused by nitrogen dioxide. *Journal of the American Medical Association*, *162*, 153.
Luiking, Y. C., Engelen, M. P. K. J., & Deutz, N. E. P. (2010). Regulation of nitric oxide production in health and disease. *Current Opinion in Clinical Nutrition & Metabolic Care*, *13*(1), 97–104.
Melledo, J. M. (2019). Dysregulation of the nitric oxide pathway as a risk factor for increased cardiovascular risk in patients with major depression. *European Neuropsycho Pharmacology*, *29*, 501.
Miller, K., Martins, J., McMullin, B., Road, J. D., & Miller, C. (2019). Case study: Bronchiectasis patient with nontuberculous mycobacteria, improved outcomes with inhaled nitric oxide. *American Journal of Respiratory and Critical Care Medicine*, *199*, A2040.
Murray, L. T. (2016). Lightning NOx and impacts on air quality. *Current Pollution Reports*, *2*(2), 115–133.
National Center for Biotechnology Information (2020). *PubChem compound summary for CID 145068, nitric oxide*. https://pubchem.ncbi.nlm.nih.gov/compound/Nitric-oxide. Accessed 25 October 2020.
Oshima, H., & Bartsch, H. (1994). Chronic infections and inflammatory processes as cancer risk factors: Possible role of nitric oxide in carcinogenesis. *Mutation Research*, *305*, 253–264.
Robinson, E., & Robbins, R. C. (1974). Anthropogenic sources of NO and NO_2 are combustion processes in which temperatures are high enough to oxidize atmospheric nitrogen in conjunction with the rapid quenching of the hot gases, thereby minimizing decomposition of the nitrogen oxides. In W. Strauss (Ed.), *Air pollution control* (pp. 1–93). New York: Wiley.
Rosselli, M., Keller, R. J., & Dubey, R. K. (1998). Role of nitric oxide in the biology, physiology and pathophysiology of reproduction. *Human Reproduction Update*, *4*(1), 3–24.
Roszer, T. (2012). *The biology of subcellular nitric oxide*. Springer Science and Business Media.
Schumann, U., & Huntrieser, H. (2007). The global lightning-induced nitrogen oxide source. *Atmospheric Chemistry and Physics*, *7*, 3823–3907.
Stryer, L. (1995). *Biochemistry*: (p. 732) (4th ed.). W.H. Freeman and Company.
Wamberg, K., & Zeskov, B. (1966). Experimental studies on the course and treatment of aspiration pneumonia. *Anesthesia and Analgesia*, *45*(2), 230.
Weinberger, B., Laskin, D. L., Heck, D. E., & Laskin, J. D. (2001). The toxicology of inhaled nitric oxide. *Toxicological Sciences*, *59*, 5–16.
Yetik-Anacak, G., & Catravas, J. D. (2006). Nitric oxide and the endothelium: History and impact on cardiovascular disease. *Vascular Pharmacology*, *45*(5), 268–276.

CHAPTER

22

Nitrogen dioxide: Risk assessment, environmental, and health hazard

Sushil K. Pandey[a] *and Jaspal Singh*[b]

[a]Department of Chemistry, University of Jammu, Jammu, India

[b]Department of Chemistry, Gurukula Kangri (Deemed to be University), Haridwar, Uttarakhand, India

Abbreviations

CAA	Clean Air Act
CASAC	Clean Air Scientific Advisory Committee
CLMS	Copernicus Land Monitoring Service
CTM	chemical transport model
EPA	Environmental Protection Agency
GR	geostatistical regression
NASA	National Aeronautics and Space Administration
NIOSH	National Institute for Occupational Safety and Health
OMI	ozone monitoring instrument
VOC	volatile organic compounds
WHO	World Health Organization

22.1 Introduction

Nitrogen dioxide (NO_2) is a reddish-brown gas with a distinctive pungent, foul smell at higher temperature. It is toxic, a major pollutant, a strong oxidant, and one of the chemical species of oxides of nitrogen (NOx). The atmosphere consists of 78% nitrogen by volume. This means, there are more than 20 million tons of nitrogen over every square mile of the earth's surface. The aptness to absorb visible solar radiation, a major regulator of the oxidizing capacity of the free troposphere as well as estimation of ozone concentrations in the troposphere are some of the

Hazardous Gases
https://doi.org/10.1016/B978-0-323-89857-7.00001-3

distinguished properties that make nitrogen dioxide a vital trace gas. It is an intermediate in nitric acid industrial synthesis of which millions of tons are produced annually, and mainly used in fertilizer production.

Chiefly, the process of chemiluminescence, i.e., reaction emitting energy in the form of light, is used to estimate nitrogen dioxide. In this exothermic reaction, the nitric oxide oxidizes to nitrogen dioxide in the presence of ozone. Consequently, the end product of the reaction formed remains in the activated form that further releases energy in the form of light while returning to its normal state. Thus, chemiluminescence takes one molecule of NO to give one molecule of NO_2, and the process is positively related to the concentration of NO. The whole process can be further assessed to estimate the concentration of nitrogen dioxide by measurement of emitted light. However, it can be seen that there are slight chances of reaction between ozone and NO; therefore the measurement is generally performed by diverting the incoming air stream through a converter. This in a way reduces any NO_2 present to NO before entering the reaction cell. The difference between NO levels measured in the two streams is a measure of the amount of NO_2.

Also, it can be determined through catalytic ammonia combustion or oxidation of nitrosyl chloride. Furthermore, the reaction proceeds by establishing temperature-dependent equilibrium among NO_2 and its dimeric form, nitrogen tetroxide ($N_2O_4^-$) (Depayras et al., 2018). Overall, a photochemical reaction between NO_2 and hydroxyl radicals lead to the generation of ozone, as shown in Fig. 22.1.

Some of the other available techniques for the determination of nitrogen dioxide comprehend utilization of photolysis process (Kley & McFarland, 1980), LIF (Thornton, Wooldridge, & Cohen, 2000), fast gas

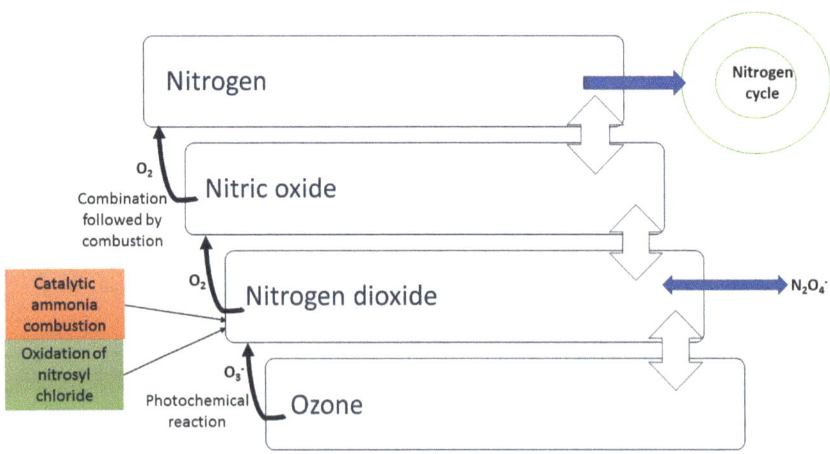

FIG. 22.1 Flowchart of a photochemical reaction. *Credit: Jaspal Singh.*

TABLE 22.1 The threshold limits of nitrogen dioxide by competent authorities.

Time (min)	10	20	30	60
Lethal effects	118ppm	98ppm	88ppm	73ppm
Irreversible effects	60ppm	55ppm	50ppm	40ppm
Reversible effects	5ppm	5ppm	5ppm	5ppm

chromatography (Marley et al., 2004), differential optical absorption spectroscopy (Platt & Perner, 1980), cavity ring down (Osthoff et al., 2006), infrared laser differential absorption spectroscopy (Li et al., 2004), cavity attenuated phase shift spectroscopy, and CL NOx monitors (Li et al., 2004; Dunlea et al., 2007).

The health effects after long-term exposure to NO_2 have also been reviewed by the Working Group of the World Health Organization (WHO). There have been recent studies on the long-term risk and mortality of NO_2. In the studies, particular consideration was given to the direct accountability of nitrogen dioxide or any other particulate matter present in the air for adverse health effects (Faustini, Rapp, & Forastiere, 2014). Table 22.1 shows some threshold limits of nitrogen dioxide (Depayras et al., 2018).

22.2 Occurrence

Generally, nitrogen dioxide originates from natural sources, motor vehicles, power plants, and other fuel-burning processes. It is collectively defined as a combination of nitric oxide (NO) and nitrogen dioxide (NO_2). The indoor household appliances such as gas stoves and gas or wood heaters are some of the major sources of origin of nitrogen gases in poorly ventilated conditions. The nitrate particles that result from NOx make the air hazy and difficult to see though. It is one of the crucial constituents of photochemical smog. It is also responsible for the yellow-brown color of the smog. Breathing air with a high concentration of NO_2 can irritate airways in the human respiratory system.

According to reports from the United States, nitrogen oxide emissions are highly correlated with population levels. Broadly, among the major contributors of nitrogen dioxide is cooking with indoor sources, such as gas or cigarette smoking. Maximum 30-min or 1-h average and maximum 24-h average outdoor nitrogen dioxide concentrations of up to 940 $\mu g/m^3$ (0.5 ppm) and 400 $\mu g/m^3$ (0.21 ppm), respectively, have been reported.

Nitrogen dioxide occurs when nitric oxide is oxidized in the air, Eq. (22.1) (Holleman & Wiberg, 2002).

$$2NO + O_2 \rightarrow 2NO_2 \quad (22.1)$$

Nitrogen dioxide is formed using air as an oxidant in most burning processes. Nitrogen mixes with oxygen to form nitric oxide at high temperature (Eq. 22.2).

$$O_2 + N_2 \rightarrow 2NO \quad (22.2)$$

A two-step process renders synthesis of NO_2 at laboratory scale, as shown in Eq. (22.3) and (22.4). Primarily, the dehydration of nitric acid produces dinitrogen pentoxide, which is followed by its thermal decomposition.

$$2HNO_3 \rightarrow N_2O_5 + H_2O \quad (22.3)$$

$$2N_2O_5 \rightarrow 4NO_2 + O_2 \quad (22.4)$$

22.3 Physicochemical properties

Nitrogen dioxide is an acidic, reddish-brown, and highly corrosive gas with an irritating odor. It has an odor threshold ranging between 0.1 and 0.4 ppm. It converts into nitric acid and nitric oxide (NO) on reacting with water (Budavari, O'Neil, Smith, & Heckelman, 1989). Other physical and chemical properties are enlisted in Table 22.2.

Nitrogen dioxide and nitric acid tend to react with combustible materials, chlorinated hydrocarbons, carbon disulfide, and ammonia and also may react violently with cyclohexane, fluorine, formaldehyde and alcohol, nitrobenzene, petroleum, and toluene.

22.4 Physiological role

According to the studies reported, little research has been done by scientists with respect to the kinetics and metabolism of nitrogen dioxide. However, limited or partial information is available on the deposition as well as the fate of NO_2 in the respiratory tract. Most of the studies suggest that about 70–90% of this gas gets absorbed in the respiratory tract. And, there is a positive correlation between absorption percentage and exercise activities (Wagner, 1970; Bauer, Utell, Morrow, Speers, & Gibb, 1986). Researchers have hypothesized the correlation by removing nitrogen dioxide from the nasopharynx of dogs and rabbits (EPA, 1993; Berglund et al., 1993; Advisory Group on The Medical Aspects of Air Pollution Episodes, 1993; Jarvis, Adamkiewicz, Heroux, Rapp, & Kelly, 2010).

TABLE 22.2 Physical and chemical properties of nitrogen dioxide.

Chemical formula	NO_2
IUPAC name	Nitrogen dioxide
Color	Red to brown gas above 21.1°C
Molecular weight	46.006 g/mol
Density	1.448 at 20°C
Melting point	−9.3°C
Boiling point	21.15°C
Odor	Pungent, acrid odor
Molecular shape	Bent
Vapor density	1.58
Solubility in water	Hydrolyses
Vapor pressure	900 mmHg at 25°C
Viscosity	0.0142 at 26.8°C
Heat of vaporization	9.110 kcal/mole
Ionization potential	9.75 eV

The computational approaches have also indicated the deposition of nitrogen dioxide in the respiratory tract, which is expected to be maximum at about the junction of the conducting airways and the gas exchange region of the lungs in humans, rats, guinea pigs, and rabbits. While, the tracheal absorption of gas extensively differs across species. Considering that in a wide array of species, the area expected to receive maximum dose directs the lesion induced by nitrogen dioxide. This framework also helps in the determination of alter in tidal volume in humans with exercise as well as quantity of dose responsible for gas exchange in tissues (Miller, Overton, Myers, & Graham, 1982; Overton Jr, 1984; Overton, Barnett, & Graham, 1987).

Several experimental reports show that nitrogen dioxide or its chemical products subsist in the lungs for comparatively longer interim of time, which has been suggested by radioactively labeled nitrogen. The associated studies have also remarkedly noticed the presence of nitric and nitrous acid or their salts in humans exposed to nitrogen dioxide (EPA, 1993; Berglund et al., 1993; Jarvis et al., 2010).

22.5 Uses

It is primarily used in the grain storage silos and welding operations, as well as in the combustion of fuels or nitrogen-containing materials, the production and use of nitrate explosives. It is used in the preparation of nitric and sulfuric acid.

Some of its major applications include its use as an oxidant in chemical reactions, as a catalyst in certain oxidation reactions, as an inhibitor to prevent polymerization of acrylates during distillation, as a nitrating agent for organic compounds, and as a flour bleaching agent and in increasing the wet strength of paper.

22.6 Sources of contaminants

Natural sources include intrusion of stratospheric nitrogen oxides, bacterial and volcanic action, and lightning. The prime sources of nitrogen dioxide in the air are emission from motor vehicles and energy production. Due to its presence in gaseous form, inhalation is the significant route of contact with humans whether indoor or outdoor. The extent of individual exposure is majorly dependent on the outdoor concentrations in the ambient air, which can be manipulated on the level of indoor sources like tobacco smoking and unvented cooking or heating appliances using gas.

While at the industrial scale the exposure is limited to certain industrial processes, ignition/combustion is a part of the process. Substantially, the industry exposure-related risks are relatively low when compared to external and domestic risks (WHO Air Quality Guidelines for Europe, 2000).

Particularly, the population residing near heavy traffic roads are prone to NO_2 pollution at a larger scale. Studies show that short-term exposure origins for the optimal enhancement in respiratory allergic reactions. Also, there are communications discussing the correlation between NO_2 exposure and mortality, which according to current evidence is not sufficient to conclude the effect of NO_2 on mortality attributed to long-time exposure.

22.7 Risk assessment

The health effects of NO_2 have been extensively assessed in a very large number of studies. Research shows that concentrations exceeding 200 µg/m^3 may have significant health effects as well as short-term peak exposure can increase allergic respiratory reactions. Furthermore, a number of short-term experimental human toxicology studies have reported adverse health effects on consequent exposure to 500 μg/m^3 NO_2 concentrations for an hour. Some studies have demonstrated associations between the exposure and risks of mortality from NO_2, while some contradictory contemplation concludes that prolonged exposure to NO_2 is responsible for increasing the death rate. The nitrogen dioxide exposure study is aimed to determine the relative contribution of indoor and outdoor sources as well as the factor patterns related to daily activity and human exposure to NO_2 (Monn, 2001) have also been conducted.

Hazard identification: Personal exposure measurements can be performed directly or indirectly. In the direct approach, exposure levels are determined on an individual; in the indirect approach, exposure levels are either measured stationarily or determined by models (Monn, 2001).

Risk analysis: The risk evaluation can be considered on the basis of method inherent criteria. These criteria are sensitivity, precision, accuracy, selectivity, and detection limit. Besides these criteria, cost and applicability play a major role in the choice of a particular method. WHO brochure summarizes quality assurance and quality control protocols, which are also important to warrant high-quality measurements (Monn, 2001).

Risk evaluation: The release of NO_2 by combustion of high nitrogen content such as coal by the ratio of pure stoichiometric can be evaluated. The release of NO_2 by combustion of low nitrogen fuels can be seen as a poor versus rich fuel/air ratio. Either way, this technique avoids the ideal stoichiometric ratio because it is the ratio that produces high temperatures at high concentrations of thermal NO_2.

This approach uses nonthermal plasma by injecting gas to ionize ammonia, urea, hexane, methane, or other reducing agents. The temperature, combined with the effect of nonthermal plasma, reduces the plasma ionizing agent that reacts with nitrogen oxide to achieve 94% DRE. This reduces the amount of reducing chemicals, which "slip" through unreacted. Nonthermal plasma act as a catalyst for ionizing pollutants and controlling NO_2 in diesel (United States Environmental Protection Agency, 1999).

22.8 Environmental effects

Nitrogen dioxide plays an important role in the formation of smog, which often produces brown mist in the atmosphere, especially during summer. When exposed to UV rays, NOx molecules break down to form ozone (O_3). This problem exacerbates the presence of volatile organic compounds (VOCs) in the atmosphere, which has an affinity toward NOx to form harmful molecules. Ozone at the ground level is a serious pollutant (bad ozone), unlike the protective ozone layer much higher up in the stratosphere (good ozone). The majority of tropospheric ozone formation occurs when nitrogen oxides (NOx), carbon monoxide (CO), and VOCs react in the atmosphere in the presence of sunlight, specifically the UV spectrum, since NOx, CO, and VOCs are considered ozone precursors (United States Environmental Protection Agency, 1999).

More environmental problems are caused by NO_2 pollution. In the presence of rain, nitrogen oxides form nitric acid, contributing to the acid rain problem. Additionally, NOx deposition in the oceans provides phytoplankton with nutrients, worsening the problem of red tides and other harmful algae blooms.

High levels of NO_2 can have a negative effect on vegetation, including leaf damage and reduced growth, damage to foliage or crop yield. It can make vegetation more susceptible to disease and frost damage. A study of the effect of nitrogen dioxide and ammonia on the habitat of Epping Forest has revealed that pollution is likely to be significantly influencing ecosystem health in the forest. The study demonstrated that local traffic emissions contribute substantially to exceeding the critical levels and critical loads in the area. The critical level for the protection of vegetation is $30\,\mu g/m^3$ measured as an annual average. NOx also reacts with other pollutants in the presence of sunlight to form ozone which can damage vegetation at high concentrations.

22.9 Hazardousness

Nitrogen oxide, nitric acid, and ozone can easily enter the lungs, where they reach and damage delicate lung tissue. Even short-time exposure can damage the lungs of healthy people. In addition, this promotes diseases like asthma. When people with asthma breathe in this pollution for some time, they may need to consult a doctor. Approximately 16% of the US houses and apartments fall within 300 ft. of a major street, increasing risk of exposure to dangerous NO_2 and its derivatives.

This air pollution especially affects the young and the elderly causing respiratory diseases like emphysema and bronchitis. Pollution caused by NO_2 gas can be dangerous to humans with asthma and heart disease and their chances of premature death increase. Living in high levels of nitrogen dioxide for a long time can cause lung diseases. The senses can be affected by this, for example, by reducing a person's ability to smell. Some adverse health effects are briefly discussed in Fig. 22.2.

The effects of toxicity on health were examined in in-person interviews to understand the association of asthma with NO_2. The effect of indoor air pollutants is significant in health as 80% of the world population spend their time indoors. The time spent indoors depends on many factors including geographical area, job activities, and gender among other

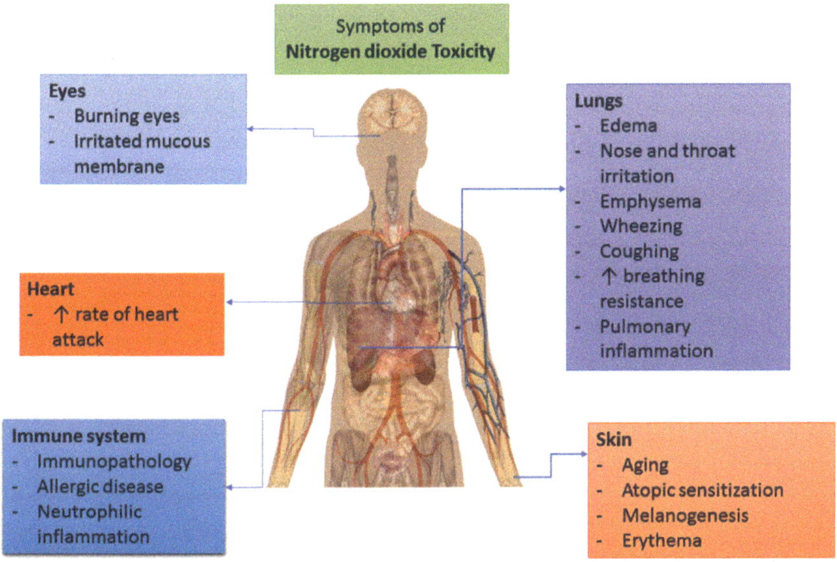

FIG. 22.2 Adverse health effects after exposure to nitrogen dioxide. *Credit: Jaspal Singh.*

variables. Additionally, household insulation is improving; this can result in greater retention of indoor air pollutants such as NO_2.

The prevalence of asthma ranges from 2% to 20% with respect to geographical areas. This may be the result of a "hygiene hypothesis" or "western lifestyle," which captures perception of households that are well and with fewer residents (Heinrich, 2011). One study examined the relationship between nitrogen oxides exposure in residential areas and respiratory symptoms and reported a statistical ratio of 2.23 (95% CI: 1.06, 4.72) among those with a medical diagnosis of asthma and gas stove exposure (Garrett, Hooper, Hooper, & Abramson, 1998).

- **Ocular exposure:** It can cause redness, pain, and severe burning sensation in eyes due to the reaction of nitrogen dioxide and water to form nitric acid.
- **Dermal exposure:** It can cause different risks related to burning, redness, and pain to skin.
- **Inhalation exposure:** Inhalation of nitrogen dioxide gas can cause swelling and slight pain in the lungs, which can go unnoticed. Inhalation of gas may cause cough, indigestion, fever, sputum of the mucosa, decreased pulmonary function, bronchitis, chest pain, pulmonary edema, cyanosis, tachycardia, and tachypnea.
- **Ingestion exposure:** Because nitrogen dioxide is a gas under ambient conditions, oral exposure normally does not occur.

Acute harm due to NO_2 exposure is only likely to arise in occupational settings. Direct exposure to the skin can cause irritation and burns. Only very high concentrations of the gaseous form cause immediate distress: 10–20 ppm can cause mild irritation of the nose and throat, 25–50 ppm can cause edema leading to bronchitis or pneumonia, and levels above 100 ppm can cause death due to asphyxiation from fluid in the lungs. There are often no symptoms at the time of exposure other than a transient cough, fatigue, or nausea, but over an hour's inflammation in the lungs causes edema (NIOSH, 1976).

22.10 Treatment

Just like the toxicity of chlorine and carbon monoxide, nitrogen dioxide gas is harmful to all forms of life. It passes easily through the lungs into the body and its inhalation can lead to heart failure and sometimes death in severe cases. Individuals and races may differ in nitrogen dioxide tolerance levels and varying tolerance levels for gas, such as metabolic rate, barometric pressure, and hematological disorders. But significant risk can result in a fatal condition leading to younger age for heart failure.

The patient should be kept under observation for at least 12 h for hypoxemia if initial symptoms are not present. If the gas exchange is compromised, keep the patient hospitalized for 12–24 h or more. Noncardiogenic pulmonary edema may take up to 48 h to develop. Educate the patient on possible symptoms and instruct the patient to return if symptoms develop (Kamanga, 2017).

For skin or eye exposure, the affected area should be flushed with saline. For inhalation, oxygen is administered, bronchodilators may be administered, and if there are signs of methemoglobinemia, a condition that arises when nitrogen-based compounds affect the hemoglobin in red blood cells, methylene blue may be administered (Agency for Toxic Substances and Disease Registry, 2016).

22.11 Risk management

Nitrogen dioxide (NO_2) is one of the highly reactive gases known as oxides of nitrogen or nitrogen oxide (NOx). NO_2 primarily gets in the air from the burning of fuel, emissions for automobiles, power plants, stoves, and off-road equipment. Birthing air with a high concentration of NO_2 can irritate airways in the human respiratory system, create other health issues, and generate many chemical reactions in the atmosphere to form acid rain. It also takes part in climatic change and ozone depletion. Nitrous oxide, akin to CFCs, is stable at ground level but breaks down on reaching to the stratosphere to form other gases, i.e., nitrogen oxides, that activates ozone-destroying reactions (United States Environmental Protection Agency, 1999).

Therefore, taking into consideration the above key points, various Environment Protection Agencies (EPAs), national and international, and even regional agencies, have designed rules and standards to reduce the emission of NO_2 and NOx. One of them is the National Ambient Air Quality Standards (NAAQS), which specifies the limiting of NO_2 present in outdoor air, inside air, protects human health, and the environment.

The Clean Air Act (CAA) involves the EPA for the management of National Ambient Air Quality Standards (NAAQS) for maintenance of nitrogen oxides level (as one of the six criteria pollutants). The NAAQS for oxides of nitrogen is presently set using NO_2 as the indicator of the larger group of nitrogen oxides. The act also gave EPA the responsibility to review and improve standards from time to time to ensure that they provide the necessary amount of health and environmental protection and even update those standards as crucial.

A lot of documents published throughout the review process include multiple drafts of plans and assessments, reports from the Clean Air Scientific Advisory Committee (CASAC) and Federal Register notices from time to time. Sections 108 and 109 of the Clean Air Act (CAA) govern the establishment, review, and revision, as appropriate, of the NAAQS for each criterion air pollutant regarding NO_2 to give protection for the nation's public health and the environment.

22.12 Current research

Bayesian geostatistical regression (GR) models have recently estimated air pollution (NO_2) exposure at the high spatial resolution, quantified the prediction uncertainty, and recommended probabilistic inference on the exceedance of air quality thresholds (Beloconi & Vounatsou, 2020). They applied these models to estimate annually averaged NO_2 concentrations at 1 km^2 spatial resolution across 44 European countries, integrating information from in situ monitoring stations, satellites, and chemical transport model (CTM) simulations. Besides this, in Ontario, Canada, research has evaluated the potential of nature-based solutions to reduce ozone, nitrogen dioxide, and carbon dioxide through a multitype green infrastructure study.

Furthermore, the selective catalytic reduction of NO + NO_2 (NOx) at low temperature (180–230 °C) with NH_3 has been investigated with Cu—Ni and vanadium oxides supported on titania and alumina monoliths by Blanco et al. They have concluded that the catalytic activity of the system was high with both NO and with NO + NO_2 gas mixtures.

The presence of NO_2 increases the quantity of adsorbed species. These species were seemed to be responsible for the N_2O formation. When NO is also present, it effectively competes with this NO_2 reaction path, and N_2O formation is not observed.

Wang et al. investigated the sensing properties of the P_2C_2 monolayer with eight gas molecules including NO_2 using first-principle calculations. They demonstrated that the adsorption of other gaseous molecules on the P_2C_2 monolayer occurs through a physisorption process with weak Van der Waals interactions, while NO_2 can undergo stable chemisorption with the P_2C_2 monolayer (Wang et al., 2018).

In accordance with a current analysis by NASA researchers, discussed on NASA's Earth Observatory, and compared nitrogen dioxide (NO_2) values detected by the ozone monitoring instrument (OMI) on NASA's Aura satellite in 2020 with the average amounts detected at this time of year from 2005 to 2019 (Lamsal et al., 2020). In early 2020, nitrogen dioxide values in eastern and central China were 10–30% lower than what is normally observed for this time period. There is corroborating evidence that the change is at least partly related to the economic slowdown following the outbreak of coronavirus (COVID-19). Prolonged exposure to nitrogen dioxide may be one of the most important contributors to the lethal events caused by the COVID-19 pandemic in these regions and perhaps throughout the world (Ogen, 2020).

22.13 Nitrogen dioxide toxicity: Case study

Incidental risks in NO_2 have been reported from agriculture (1502,000 ppm), mining explosions (500 ppm), space exploration (250 ppm), military activities (500 ppm), and the burning of nitrogen-containing propellants (Mohsenin, 1994). Due to the pungent, ammonia-like odor of NO_2, individuals experience discomfort and suffocation. Such accidental-risk data, together with relevant animal studies, are useful in establishing emergency risk thresholds. The intensity and duration of NO_2 exposure are directly dependent on the variety of clinical reactions (Lowry & Schuman, 1956; Jones, Proudfoot, & Hall, 1973). Immediate death may also occur upon exposure (Mohsenin, 1994). Exposure above 150 ppm in 30–60 min can result in rapid death, fatal pulmonary edema, or asphyxia. Exposure to NO_2 at concentrations of 150–300 ppm can lead to bronchiolitis fibrosa obliteration accompanied by restrictive and obstructive ventilatory defects that might lead to death in 2–3 weeks (Lowry & Schuman, 1956; NRC, 1977). Such exposure can cause permanent injury to people. The LC_{50} has been estimated to be 174 ppm for a 1-h exposure to humans (Book, 1982).

Four farmers were exposed to high concentrations of NO_2, i.e., estimated between 200 and 4000 ppm when they entered the freshly filled silos. Two individuals died instantly due to high concentrations, and others experienced cough, indigestion, and fever, which recovered after several days but reappeared after about 3 weeks (Lowry & Schuman, 1956; EPA, 1982).

At lower exposure concentrations, a variety of nonlethal effects have been observed. From 50 to 100 ppm for a 30-min exposure, pulmonary edema and bronchiolitis with focal pneumonitis are likely to develop and last from 6 to 8 weeks; recovery is often spontaneous. Individuals exposed for 30 min at concentrations between 25 and 75 ppm might develop bronchial pneumonia, acute bronchitis, dyspnea, cyanosis, chest pain, rales, headaches, eye irritation, a dry nonproductive cough, and vomiting. Such effects usually are resolved in hours but sometimes are followed by relapse with shortness of breath, cough, cyanosis, and fever. In addition to its effect on the lung, NO_2 can transform hemoglobin to Meta-Hb (Lowry & Schuman, 1956; Grayson, 1956; Stern, 1968; Milne, 1969; EPA, 1993). Welders exposed to NO_2 at 3.9–5.4 ppm exhibited 2.3–2.6% MetHb in their blood (Patty, 1963).

22.14 Safety measures

- If preparing gas in test-tube reactions, use small amounts, and take steps to avoid suck-back such as using a bomb valve. Use a fume hood to hold the gas test tube.
- The operations must be performed in isolated and ventilated areas. In case of unavailability of local exhaust and isolation area in the premises, the use of respirators should be given priority.
- The premises with the presence of NO_2 should be equipped with gas detection alarms.
- If there are chances for exposure of more than 1 ppm, then workers need to wear MSHA/NIOSH-approved supplied-air respirator with a full facepiece.
- Chemical-resistant, impervious gloves complying with an approved standard should be worn at all times.
- The specified approved personal protective equipment for the body should be worn before handling this product.

22.15 Handling and storage

The handling of NO_2 should be done in the presence of well-trained workers. The chemicals such as trimethyltin, triethylamine, ammonia, alcohol, acetonitrile, indium, methyl sulfoxide, hydrazines, aluminum powder, and carbon disulfide should be kept away from NO_2, since their presence may cause explosion in the vicinity. NO_2 should not be kept in the proximity of chlorinated hydrocarbons, formaldehyde, fluorine, petroleum, nitrobenzene, copper, copper alloys, and reducing materials. The specified equipment that are compatible with NO_2 must be utilized. Keep equipment free from oil and grease. The container valve must be slowly opened. It is recommended to install cross-purge assemble between the container and the regulator. While handling the containers, the good industrial hygiene and safety procedures should be followed. The containers must be handled with care; do not drag, roll, slide, or drop. The proper labeling should be done at primary stage. The movement of containers should be done by using trolley, hand truck, fork truck, etc. The storage area must be provided with local exhaust and temperature maintained below 50 °C. Stored containers should be periodically checked for leaks. Container valve guards or caps must be kept in place. Containers should be stored in a location free from fire hazard and away from sources of heat and ignition. Store in accordance with local/regional/national/international regulations. Also, study suggests that storage and release of NO_2 on alumina supported barium oxide (Cant & Patterson, 2002).

22.16 Precautions

- Avoid eye and skin contact with NO_2.
- Always use appropriate eye protection when working with NO_2 in the laboratory.
- Wear acid-resistant gloves and clothing.
- Do not eat, smoke, and drink at the place where NO_2 is handled, processed, or stored.
- Wear a face shield along with goggles.
- Perform good work practice to reduce the hazards.
- Handle with care.

References

Advisory Group on The Medical Aspects of Air Pollution Episodes. (1993). *Third report: Oxides of nitrogen*. London: H.M. Stationery Office.

Agency for Toxic Substances and Disease Registry. (2016). *Guidelines for Nitrogen oxides*.

Bauer, M. A., Utell, M. J., Morrow, P. E., Speers, D. M., & Gibb, F. R. (1986). Inhalation of 0.30 ppm nitrogen dioxide potentiates exercise-induced bronchospasm in asthmatics. *American Review of Respiratory Disease*, *134*(5), 1203–1208.

Beloconi, A., & Vounatsou, P. (2020). Bayesian geostatistical modelling of high-resolution NO2 exposure in Europe combining data from monitors, satellites and chemical transport models. *Environment International*, *138*, 105578. https://doi.org/10.1016/j.envint.2020.105578.

Berglund, M., Boström, C. E., Bylin, G., Ewetz, L., Gustafsson, L., Moldéus, P., et al. (1993). Health risk evaluation of nitrogen oxides. *Scandinavian Journal of Work, Environment & Health*, 1–72.

Book, S. A. (1982). Scaling toxicity from laboratory animals to people: An example with nitrogen dioxide. *Journal of Toxicology and Environmental Health, Part A Current Issues*, *9*(5–6), 719–725.

Budavari, S., O'Neil, M. J., Smith, A., & Heckelman, P. E. (1989). *The Merck index*. Rahway, NJ: Merck & Co., Inc.

Cant, N. W., & Patterson, M. J. (2002). The storage of nitrogen oxides on alumina-supported barium oxide. *Catalysis Today*, *73*(3–4), 271–278.

Depayras, S., Kondakova, T., Heipieper, H. J., Feuilloley, M. G., Orange, N., & Duclairoir-Poc, C. (2018). The hidden face of nitrogen oxides species: From toxic effects to potential cure? In S. Soloneski, & M. L. Larramendy (Eds.), *Emerging pollutants: Some strategies for the quality preservation of our environment* (pp. 19–44). United Kingdom: IntechOpen.

Dunlea, E. J., Herndon, S. C., Nelson, D. D., Volkamer, R. M., San Martini, F., Sheehy, P. M., et al. (2007). Evaluation of nitrogen dioxide chemiluminescence monitors in a polluted urban environment. *Atmospheric Chemistry and Physics*, *7*, 2691–2704.

EPA (U.S. Environmental Protection Agency). (1982). *Air quality criteria for oxides of nitrogen, EPA/600/8-82-026*. Research Triangle Park, N.C: U.S. Environmental Protection Agency, Office of Health and Environmental Assessment, Environmental Criteria and Assessment Office. Available from NTIS, Springfield, Va., Doc. No. PB83-131011.

EPA (U.S. Environmental Protection Agency). (1993). *Air quality criteria for oxides of nitrogen. vol. 3*. Washington, DC: U.S. Environmental Protection Agency, Office of Research and Development. EPA/600/8-91/049cF.

Faustini, A., Rapp, R., & Forastiere, F. (2014). Nitrogen dioxide and mortality: Review and meta-analysis of long-term studies. *European Respiratory Journal*, *44*(3), 744–753. https://doi.org/10.1183/09031936.00114713.

Garrett, M. H., Hooper, M. A., Hooper, B. M., & Abramson, M. J. (1998). Respiratory symptoms in children and indoor exposure to nitrogen dioxide and gas stoves. *American Journal of Respiratory and Critical Care Medicine*, *158*(3), 891–895.

Grayson, R. R. (1956). Silage gas poisoning: Nitrogen dioxide pneumonia, a new disease in agricultural workers. *Annals of Internal Medicine*, *45*(3), 393–408.

Heinrich, J. (2011). Influence of indoor factors in dwellings on the development of childhood asthma. *International Journal of Hygiene and Environmental Health*, *214*(1), 1–25.

Holleman, A. F., & Wiberg, E. (2002). Inorganic chemistry. *Journal of Chemical Education*, *79*(8), 944.

Jarvis, D. J., Adamkiewicz, G., Heroux, M. E., Rapp, R., & Kelly, F. J. (2010). Nitrogen dioxide. In *WHO guidelines for indoor air quality: Selected pollutants* World Health Organization.

Jones, G. R., Proudfoot, A. T., & Hall, J. I. (1973). Pulmonary effects of acute exposure to nitrous fumes. *Thorax*, *28*(1), 61–65.

Kamanga, A. (2017). *Nitrogen dioxide toxicity treatment and management*. Available from https://emedicine.medscape.com/article/302133-treatment.

Kley, D., & McFarland, M. (1980). Chemiluminescence detector for NO and NO_2. *Atmospheric Technology*, *12*, 63–69.

Lamsal, L. N., Krotkov, N. A., Vasilkov, A., Marchenko, S., Qin, W., Yang, E. S., et al. (2020). OMI/Aura nitrogen dioxide standard product with improved surface and cloud treatments. *Atmospheric Measurement Techniques Discussions*, 1–56.

Li, Y. Q., Demerjian, K. L., Zahniser, M. S., Nelson, D. D., McManus, J. B., & Herndon, S. C. (2004). Measurement of formaldehyde, nitrogen dioxide, and sulfur dioxide at Whiteface Mountain using a dual tunable diode laser system. *Journal of Geophysical Research-Atmospheres*, *109*(D16).

Lowry, T., & Schuman, L. M. (1956). Silo-filler's disease—A syndrome caused by nitrogen dioxide. *Journal of the American Medical Association*, *162*(3), 153–160.

Marley, N. A., Gaffney, J. S., White, R. V., Rodriguez-Cuadra, L., Herndon, S. E., Dunlea, E., et al. (2004). Fast gas chromatography with luminol chemiluminescence detection for the simultaneous determination of nitrogen dioxide and peroxyacetyl nitrate in the atmosphere. *Review of Scientific Instruments, 75*(11), 4595–4605.

Miller, F. J., Overton, J. H., Myers, E. T., & Graham, J. A. (1982). Pulmonary dosimetry of nitrogen dioxide in animals and man. In *Vol. 21. Studies in Environmental Science* (pp. 377–386). Elsevier.

Milne, J. E. (1969). Nitrogen dioxide inhalation and bronchiolitis obliterans. A review of the literature and report of a case. *Journal of Occupational Medicine: Official Publication of the Industrial Medical Association, 11*(10), 538–547.

Mohsenin, V. (1994). Human exposure to oxides of nitrogen at ambient and supra-ambient concentrations. *Toxicology, 89*(3), 301–312.

Monn, C. (2001). Exposure assessment of air pollutants: A review on spatial heterogeneity and indoor/outdoor/personal exposure to suspended particulate matter, nitrogen dioxide, and ozone. *Atmospheric Environment, 35*(1), 1–32. https://doi.org/10.1016/s1352-2310(00)00330-7.

NIOSH (National Institute for Occupational Safety and Health). (1976). Occupational exposure to oxides of nitrogen (Nitrogen dioxide and nitric oxide). In *Criteria for a recommended standard* (pp. 76–149).

NRC (National Research Council). (1977). *Nitrogen oxides*. Washington, DC: National Academy Press.

Ogen, Y. (2020). Assessing nitrogen dioxide (NO_2) levels as a contributing factor to coronavirus (COVID-19) fatality. *Science of the Total Environment, 726*, 138605. https://doi.org/10.1016/j.scitotenv.2020.138605.

Osthoff, H. D., Brown, S. S., Ryerson, T. B., Fortin, T. J., Lerner, B. M., Williams, E. J., et al. (2006). Measurement of atmospheric NO_2 by pulsed cavity ring-down spectroscopy. *Journal of Geophysical Research-Atmospheres, 111*(D12).

Overton, J. H., Barnett, A. E., & Graham, R. C. (1987). *Significances of the variability of airway paths and their air-flow rates to dosimetry model predictions of the absorption of gases* (No. PB-88-147814/XAB; EPA-600/D-87/364). Research Triangle Park, NC (USA): Northrop Services, Inc.

Overton, J. H., Jr. (1984). Physicochemical processes and the formulation of dosimetry models. *Journal of Toxicology and Environmental Health, Part A Current Issues, 13*(2–3), 273–294.

Patty, F. A. (1963). Inorganic compounds of oxygen, nitrogen, and carbon. *Industrial Hygiene and Toxicology, 2*, 919–923.

Platt, U., & Perner, D. (1980). Direct measurements of atmospheric CH_2O, HNO_2, O_3, NO_2, and SO_2 by differential optical absorption in the near UV. *Journal of Geophysical Research, Oceans, 85*(C12), 7453–7458.

Stern, A. C. (Ed.). (1968). *Vol. 1. Air pollution and its effects* (2nd ed.). Elsevier.

Thornton, J. A., Wooldridge, P. J., & Cohen, R. C. (2000). Atmospheric NO_2: In situ laser-induced fluorescence detection at parts per trillion mixing ratios. *Analytical Chemistry, 72*(3), 528–539.

United States Environmental Protection Agency. (1999). *Nitrogen oxides (NOx), why and how they are controlled*. https://www3.epa.gov/ttncatc1/dir1/fnoxdoc.pdf. (Accessed 20 September 2020).

Wagner, H. M. (1970). Absorption von NO and NO_2 in MIK-und MAK-Konzentrationenbei der Inhalation. *Staub - Reinhaltung der Luft, 30*, 380–381.

Wang, J., Lei, J. M., Yang, G. F., Xue, J. J., Cai, Q., Chen, D. J., et al. (2018). An ultra-sensitive and selective nitrogen dioxide sensor based on a novel P_2C_2 monolayer from a theoretical perspective. *Nanoscale, 10*(46), 21936–21943.

WHO Air Quality Guidelines for Europe. (2000). *Nitrogen dioxide* (pp. 1–33). Copenhagen, Denmark: WHO Regional Office for Europe.

CHAPTER

23

Oxygen: Risk assessment, environmental, and health hazard

Jaspal Singh

Department of Chemistry, Gurukula Kangri (Deemed to be University), Haridwar, Uttarakhand, India

Abbreviations

ARDS	acute respiratory distress syndrome
ATA	atmosphere absolute
CNS	central nervous system
DNA	deoxyribonucleic acid
FMEA	failure modes and effects analysis
ROS	reactive oxygen species

23.1 Introduction

The ubiquitous abundance of elemental oxygen prevails from the atmosphere (20.9%), Earth's crust (45%), and water on Earth (89%). Oxygen is distinctly reactive nonmetal, an oxidizing agent, and a member of the chalcogen group on the periodic table. This diatomic colorless and odorless gas constitutes around 28% of the Earth's atmosphere at standard temperature and pressure. In 1604, Michael Sendivogius isolated oxygen which was believed to be discovered in 1773 by Carl Wilhelm Scheele and in 1774 by Joseph Priestley. However, the term was coined by Antoine Lavoisier in 1777, who gave recognition to oxygen for combustion.

Some of the intermittent usages of oxygen involve steel, plastic and textile production, brazing, welding and steel cutting, propellants of rockets, and oxygen therapy life support systems at various oxygen-deficient places (Chawla & Lavania, 2001). The deficiency of oxygen at higher altitudes in the air can cause hyperoxia and oxygen toxicity (Hafner, Beloncle,

Hazardous Gases
https://doi.org/10.1016/B978-0-323-89857-7.00013-X

Koch, Radermacher, & Asfar, 2015). The clinical manifestations include high oxygen exposure of patients for a short time period or low oxygen exposure of patients for a longer time. Cell damage and death may even occur due to its potential to affect the central nervous system (CNS) and pulmonary system. There is a greater risk of oxygen poisoning in those exposed to prolonged oxygen areas, hyperbaric oxygen therapy patients, prematurely born infants, and underwater divers (Camporesi, 2014).

23.2 Historical background

In the 2nd century BCE, the correlation between combustion and air was entrenched by Philo of Byzantium in his work entitled *Pneumatica*. On account of his work further, Leonardo da Vinci explained the role of air at the time of combustion and respiration. In the 17th century, Robert Boyle confirmed the need for air during combustion. In the later centuries, John Mayow refined the work by depicting that only a portion of air is responsible for the combustion that he called spiritus nitroaereus. Finally, he concluded the requirement of the consumption of nitroaereus for both respiration and combustion.

Oxygen poisoning was first reported by Paul Bert, who in 1878 exhibited convulsions in larks exposed to 15–20 ATA air. Therefore, the after effect on the central nervous system is known as the Bert effect (Rusoke-Dierich, 2018). In 1899, J. Lorain Smith extensively studied the Bert effect and concluded the occurrence of fatal pneumonia on exposure to 73% oxygen at 1 ATA (atmosphere absolute). He unknowingly led to the discovery of the Smith Effect (Chawla & Lavania, 2001; Donald, 1947). Though in early 1954, it was Gerschman and team that suggested the lethal effects of oxygen (Chawla & Lavania, 2001; Cheeseman & Slater, 1993).

23.3 Physicochemical properties

Oxygen is a clear, highly reactive, electronegative, strong oxidizing agent with temperature-dependent solubility. It condenses and freezes at 90.20 and 54.36 K, respectively. Some of the physicochemical properties are enlisted in Table 23.1 (National Center for Biotechnology Information, 2020).

23.4 Physiological role

Hyperoxia, a state with excessive oxygen supply, occurs on the increment of alveolar partial pressure under normal conditions. The oxygen overexposure results in balance disruption between oxidants and antioxidants that damages cells and tissues (Ciencewicki, Trivedi, &

TABLE 23.1 Some physicochemical properties of oxygen.

Name	Oxygen
Symbol	O
Atomic number	8
Atomic weight	15.99 amu
Standard state	Gas at 298 K
Group in the periodic table	16
Group name	Chalcogen
Period in periodic table	2
Block in periodic table	p-Block
Color	Colorless as a gas, liquid is pale blue
Classification	Nonmetallic
Melting point	54.36 K (−218.79°C or −361.82 °F)
Boiling point	90.20 K (−182.95°C or −297.31 °F)
Density	1.429 g/L at 0°C
Vapor density	1.43
Vapor pressure	1 kPa at −211.9°C
Log *P*	0.65
Viscosity	101.325 kPa at 25°C
Heat of vaporization	50.9 cal/g at −183°C
Surface tension	Liquid-surface tension: 13.47 dynes/cm = 0.01347 N/m at −183°C
Triple point temperature	54.4 K
Triple point pressure	0.0015 atm
Heat of fusion	3.3 cal/g

Kleeberger, 2008; Clanton, 2007; Rahman, Biswas, & Kode, 2006; Yee et al., 2006). The extent of toxicity is dependent on the extent of exposure, atmospheric pressure, and inspired oxygen. With high FIO_2 (>0.60), overextended exposure may cause O_2 poisoning, pulmonary toxicity, or the Lorraine Smith effect, lung congestion, pulmonary edema, atelectasis, the shortness of breath, burning of the throat and chest, and painful breathing (Mason, Murray, Nadel, & Gotway, 2015).

The hemoglobin-O_2 buffer system is the main carrier of oxygen at normal partial pressure to body tissues (Dash & Bassingthwaighte, 2006; Jones, 2008; Zhou, Saidel, & Cabrera, 2007). In case of high ATA and short FIO_2 exposure time or the Paul Bert effect, oxygen poisoning can happen, causing toxicity to the CNS (Allen, Demchenko, & Piantadosi, 2009; Guyton & Hall, 2006; Halliwell, 2006; Hedley-Whyte, 2008; Joiner, 2001; Zhou et al., 2007).

Pulmonary capillary endothelial and alveolar epithelial cells are targeted for ROS, resulting in injury-induced lung inflammation, alveolar flooding, bleeding, and collagen, elastin, and hyaline membranes (Kang, Lee, & Kim, 2005; Pagano & Barazzone-argiroffo, 2006; Sies & Jones, 2020; Yee et al., 2006). Above a critical P_AO_2, the hemoglobin-O_2 buffering mechanism fails to cause cell death (Guyton & Hall, 2006; Mantell, Horowitz, Davis, & Kazzaz, 1999).

Oxygen toxicity caused by ROS occurs in overlapping stages, i.e., initiation, inflammation, proliferation, and fibrosis depending on the degree of severity and the reversibility of the injury. Initially, ROS increase and antioxidant levels decrease and the lungs fail to clean the mucosa (Crapo, 1986; Crapo & Tierney, 1974; Crapo, Barry, Chang, & Mercer, 1984).

The enhanced oxygen tension and progressive hypoxemia may occur due to the presence of an augmented level of oxygen in the blood. Air pressure within the attached lung cavity and ventilator-induced injury may change and be indistinguishable from O_2 toxicity. Oxygen toxicity can be reduced by keeping P_AO_2 less than 80 mmHg or FIO_2 less than 0.40–0.50 (Mason et al., 2015). The alveolar–capillary barrier damages due to pulmonary epithelial inflammation initiating the impaired exchange of gas and edema in the lungs (Kannan, Pang, Foster, Rao, & Wu, 2006; Romashko et al., 2003).

23.5 Uses

It is one of the inevitable components used for the generation of energy in all aerobic organisms, playing an important role in the immune system. Some of its uses are discussed below in detail.

Oxygen therapy: Oxygen therapy has found its application in emphysema, pneumonia, congestive heart failure, and diseases with impaired pulmonary artery pressure. The flexibility of therapy made it accessible at hospitals as well as the patient's home.

Hyperbaric medicine: At the time of carbon monoxide poisoning, gas gangrene, and decompression illness (the "bends"), the specially designed oxygen chambers are used. The procedure involves the replacement of CO by oxygen. Oxygen gas is deadly to anaerobic bacteria that causes gangrene, so increasing its partial pressure helps kill them. Chiefly, this medical condition prevails in divers, as they utilize a combination of nitrogen and helium for respiration. Hence, this results in the formation of bubbles of inert gas, where an elevated level of O_2 is part of the treatment. Also, oxygen is used clinically for patients who require mechanical ventilation.

Use in space suits and scuba diving suits: Oxygen in pressurized form is notably used in modern spacesuits. The presence of a high concentration of oxygen at low pressure makes the space suits flexible.

Scuba divers and submariners also rely on artificial oxygen and/or mixtures of oxygen and air. The management of oxygen at higher than sea level pressure limits to rebreathing, decompression, or emergency treatment, whereas deeper diving requires oxygen along with gases such as nitrogen or helium, for the prevention of oxygen poisoning. People flying in mountains and nonpressurized fixed-wing aircraft also carry a complimentary supply of O_2.

Pressurized commercial airplanes: The lack of oxygen at higher altitudes such as in airplanes activates chemical oxygen generators causing oxygen masks to drop. Further inhalation from the mask leads to the release of oxygen by the exothermic reaction.

23.6 Sources of contaminants

It is primarily found in natural minerals like sulfates, phosphates, carbonates, and oxides. Some of the sources of contamination of oxygen include the leakage from the sample chamber at a large or small scale. The leakage can be temperature dependent. Broadly oxygen behaves as a source of nourishment for various physiological processes, for instance, oxidative phosphorylation, respiration, and likewise. However, at normal levels, oxygen is a vital part of metabolism but prove to be hazardous in high concentrations. The vital factors for the toxicity focus on partial pressure and exposure time (Rusoke-Dierich, 2018).

23.7 Risk assessment

The oxygen-enhanced combustor and similar analogous system substantiate the need to be analyzed for any probable risk. Here, risk assessment relies positively on the estimation and computation of risk (Patel, 2016). The oxygen handling premises extensively use frameworks like structural model-based FMEA and fuzzy fault tree analysis for the conductance of effective and efficient risk assessment (Chen, Wu, & Qin, 2014). The steps involved in assessing the risk are enlisted below:

1. *System analysis*: The risk analysis is mainly based on the determination of a number of factors of significant importance. Some of the essential factors include engineering controls, production organization, and protection management. Additionally, the collective data obtained from previous hazards should also be utilized periodically to assess the risks.
2. *Risk assessment and improvement*: In accordance with the conclusion of the fuzzy fault tree, the risk assessment for different accident forms can be proposed and some measures can be erected to improve the safety of the system (Chen et al., 2014).

23.8 Environmental effects

Oxidation is an important process that occurs throughout the environment and plays an important role in unrelated events such as corrosion, combustion, and metabolism of carbohydrates that provide energy to our bodies. In addition, oxygen also exists in the form of ozone in ambient air that is a major component of photochemical smog. Hence, acts as a pollutant responsible for environmental hazard. The presence of dissolved oxygen in water plays a vital role in the existence of marine life, whereas concentrations below 3 ppm could be lethal to aquatic organisms and a further decrease in oxygen level may even direct for death.

23.9 Hazardousness

Generally, ambient air comprises around 21% oxygen, which is essentially nontoxic, although the inspiration of 1 atm of oxygen for more than 12 h can cause respiratory tract irritation, progressive decrease in vital capacity, cough, runny nose, sore throat, and chest pain. In addition, some of the earlier signs of toxicity are chiefly concentrated around the impairment of the lungs (Cooper, Phuyal, & Shah, 2020).

The long-term exposure of oxygen >0.5 ATA could cause lung injury resulting in tracheobronchitis, ARDS, or pulmonary interstitial fibrosis. The 2 ATA of oxygen can characteristically result in mild-to-severe carinal irritation, finally causing chest pain and dyspnea (Clark & Lambertsen, 1967). The oxygen pressures of >3 ATA may cause CNS toxicity with variable signs (Clark, 1982; De Martino, Luchetti, & De Rosa, 1996; Edmonds, Lowry, & Pennefather, 1992; Lambertsen, 1968). Oxygen toxicity may also result in the inhabitation of enzymes with SH groups, DNA damage, and destruction of lipid membranes via lipoperoxide production (Rusoke-Dierich, 2018). Taking oxygen in a higher amount shows some symptoms in the human body, as shown in Fig. 23.1.

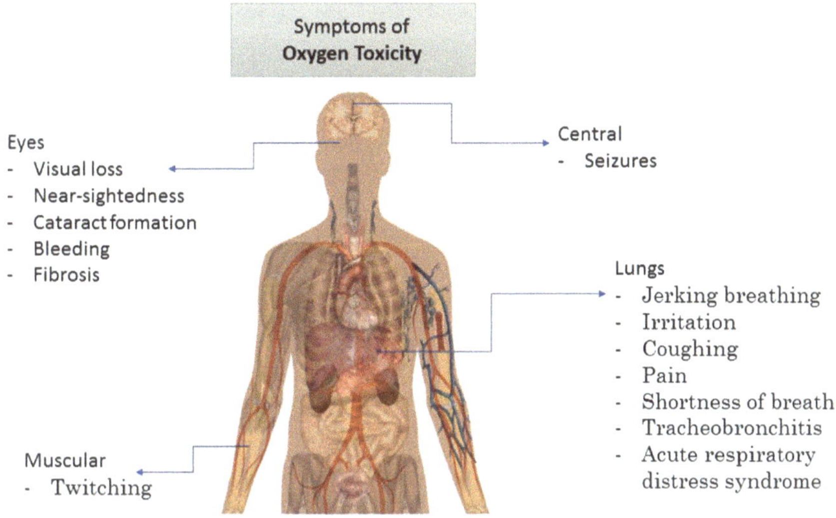

FIG. 23.1 Oxygen toxicity symptoms in the human body. *Credit: Jaspal Singh.*

23.10 Treatment

Oxygen therapy is a treatment that provides a sufficient amount of oxygen to breathe. This is also known as supplemental oxygen, which is only accessible through a health-care provider's prescription. The therapy may be given for a short time in the hospital or at home. Oxygen is stored as gas or liquid in special tanks. These tanks are easily transported to your home and contain a certain amount of oxygen, which needs a refill. Oxygen has a higher risk of fire, so smoke or flammable substances should never be used while using oxygen (National Heart, Lung, and Blood Institute, 2019).

23.11 Risk management

Oxygen toxicity can be managed by reducing exposure to increased oxygen levels. In hospitals, the medical gas pipeline system acts as a major source of oxygen supply within the premises. Any particular damage or leakage in the pipeline may be consequently devastating. Here, risk management plays a crucial role. In order to minimize the related risks, hospital facilities mainly follow the aforementioned points:

- Utilizes the principle of single fault failure in the design
- Risk of failure minimized by using risk management
- Security of supply potentially improved by using a ring main system
- Locate possible areas to minimize the risk of mechanical damage

- Failure of supply potentially controlled by use of strategically located isolation valves
- Ensure a safe system—even under extreme conditions (such as a fire)
- Uses alarm systems to provide early warnings of system faults
- Conduct periodic testing after any work to ensure an end-to-end integrity
- Control of the reinstatement of systems following any work to ensure systems are operating correctly
- Design files updated to reflect all changes to the system
- Training is essential in the mitigation of some of the risks of using medical oxygen. Everyone involved in the process of utilizing oxygen needs suitable training.

23.12 Current research

Oxygen-sensing mechanisms in different eukaryotic kingdoms have the advantages of comparing molecular strategies. A similar work directs the uncontrolled growth in a variety of conditions and stresses during tumorigenesis (Hammarlund, Flashman, Mohlin, & Licausi, 2020).

Another recent research focuses on the sustainable conversion, storage, and transport of hydrogen energy with the help of electrochemical water-splitting technology. The electrolysis of water conducted in the presence of hydrogen/oxygen evolution reaction (HER/OER) in place of noble-metal-based catalysts plays paramount importance (Sun et al., 2020).

Additionally, Kuznetsov et al. recently explored the oxygen evolution reaction (OER) activity in acidic conditions and surface oxygen vacancies. The computed work utilized the DFT calculations, which further introduces a novel catalyst with improved OER performance, i.e., $Y_{1.8}Cu_{0.2}Ru_2O_{7-\delta}$ (Kuznetsov et al., 2020).

23.13 Oxygen toxicity: Case study

Oxygen toxicity is a major concern for underwater divers. By far a considerable number of resembling cases of diving fatality have been acknowledged. Generally, it has been found that divers use a 50% oxygen/nitrogen mixture for most of the dive. As a result, the reports have likely revealed that the affected person gets exposed to a partial pressure of oxygen of 291 kPa (2.9 atm), possibly for as long as 19 min, while it is recommended to use this gas mixture at depths less than 14–18 m (depending on the duration of exposure).

23.14 Safety measures

- The use of oil and grease for lubrication of oxygen or enriched-air equipment should be prohibited as they can ignite spontaneously with explosive violence.
- Smoking must be forbidden in oxygen-enriched atmospheres.
- The equipment carrying oxygen should be checked for any leakage by a timed gas pressure drop test or by using a solution of 1% Teepol in demineralized water.
- Only trained personnel should be allowed to handle the oxygen-carrying types of equipment.
- Employees should be required to use impervious clothing, face shields, gloves, and other appropriate protective clothing necessary to prevent any possibility of skin contact with oxygen.
- Employees handling oxygen should wash their hands thoroughly with soap or mild detergent and water before eating, smoking, and using toilet facilities.
- Employees should use dust and splash-proof safety goggles.

23.15 Handling and storage

- Oxygen is preferably stored in tanks with high pressure, cryogenics, and chemical compounds.
- For economic transport, oxygen is compressed in liquid form and transported in specially insulated tankers.
- The liquefied gas is converted to gaseous form before usage.
- Gas cylinders in use should be protected against being knocked or dropped.
- The oxygen cylinders must be kept free from dust.
- Oxygen concentrator must be stored in a well-ventilated area devoid of curtains or drapes.
- Conduct periodic inspections of oxygen cylinders.
- The filled and empty cylinders must be stacked properly.

23.16 Precautions

- Do not smoke in the proximity of oxygen.
- Avoid usage of any combustible nearby oxygen cylinders.
- Ensure availability of smoke alarms on the premises.
- Use the correct pressure gauge and regulator.

References

Allen, B. W., Demchenko, I. T., & Piantadosi, C. A. (2009). Two faces of nitric oxide: Implications for cellular mechanisms of oxygen toxicity. *Journal of Applied Physiology, 106*(2), 662–667.

Camporesi, E. M. (2014). Side effects of hyperbaric oxygen therapy. *Undersea & Hyperbaric Medicine, 41*(3), 253–257.

Chawla, A., & Lavania, A. (2001). Oxygen toxicity. *Medical Journal Armed Forces India, 57*(2), 131–133. https://doi.org/10.1016/s0377-1237(01)80133-7.

Cheeseman, K. H., & Slater, T. F. (1993). An introduction to free radical biochemistry. *British Medical Bulletin, 49*(3), 481–493. https://doi.org/10.1093/oxfordjournals.bmb.a072625.

Chen, Z., Wu, X., & Qin, J. (2014). Risk assessment of an oxygen-enhanced combustor using a structural model based on the FMEA and fuzzy fault tree. *Journal of Loss Prevention in the Process Industries, 32*, 349–357. https://doi.org/10.1016/j.jlp.2014.10.004.

Ciencewicki, J., Trivedi, S., & Kleeberger, S. R. (2008). Oxidants and the pathogenesis of lung diseases. *Journal of Allergy and Clinical Immunology, 122*(3), 456–468.

Clanton, T. L. (2007). Hypoxia-induced reactive oxygen species formation in skeletal muscle. *Journal of Applied Physiology, 102*(6), 2379–2388.

Clark, J. M. (1982). Oxygen toxicity. In P. B. Bennitt, & D. H. Elliot (Eds.), *The physiology and medicine of diving* (pp. 200–238). London: Bailliere-Tindall.

Clark, J. M., & Lambertsen, C. J. (1967). Pulmonary oxygen tolerance and the rate of development of pulmonary oxygen toxicity in man at two atmospheres inspired oxygen tension. In *Underwater physiology; proceedings of the third symposium* (pp. 439–451). Baltimore Williams and Wilkins.

Cooper, J. S., Phuyal, P., & Shah, N. (2020). Oxygen toxicity. In *StatPearls*. StatPearls Publishing.

Crapo, J. D. (1986). Morphologic changes in pulmonary oxygen toxicity. *Annual Review of Physiology, 48*(1), 721–731.

Crapo, J. D., & Tierney, D. F. (1974). Superoxide dismutase and pulmonary oxygen toxicity. *American Journal of Physiology-Legacy Content, 226*(6), 1401–1407.

Crapo, J. D., Barry, B. E., Chang, L. Y., & Mercer, R. R. (1984). Alterations in lung structure caused by inhalation of oxidants. *Journal of Toxicology and Environmental Health, Part A Current Issues, 13*(2–3), 301–321.

Dash, R. K., & Bassingthwaighte, J. B. (2006). Simultaneous blood–tissue exchange of oxygen, carbon dioxide, bicarbonate, and hydrogen ion. *Annals of Biomedical Engineering, 34*(7), 1129–1148.

De Martino, G., Luchetti, & De Rosa, R. C. (1996). Toxic effects of oxygen. In M. Michael, A. Marroni, & C. Longoni (Eds.), *Handbook of Hyperbaric Medicine* (pp. 59–68). New York: Springer.

Donald, K. W. (1947). Oxygen poisoning in man. *British Medical Journal, 1*(4506), 667–672.

Edmonds, C., Lowry, C., & Pennefather, J. (1992). Oxygen toxicity. In C. Edmonds, C. Lowry, & J. Pennefather (Eds.), *Diving and subaquatic medicine* (pp. 241–256). Oxford: Butterworth-Heinemann.

Guyton, A. C., & Hall, J. E. (2006). *Textbook of medical physiology* (pp. 764–765). Philadelphia, PA: Elsevier Saunders.

Hafner, S., Beloncle, F., Koch, A., Radermacher, P., & Asfar, P. (2015). Hyperoxia in intensive care, emergency, and peri-operative medicine: Dr. Jekyll or Mr. Hyde? *Annals of Intensive Care*, 5(1), 42.

Halliwell, B. (2006). Reactive species and antioxidants. Redox biology is a fundamental theme of aerobic life. *Plant Physiology, 141*(2), 312–322.

Hammarlund, E. U., Flashman, E., Mohlin, S., & Licausi, F. (2020). Oxygen-sensing mechanisms across eukaryotic kingdoms and their roles in complex multicellularity. *Science, 370*(6515).

Hedley-Whyte, J. (2008). Pulmonary oxygen toxicity: Investigation and mentoring. *The Ulster Medical Journal, 77*(1), 39.

Joiner, J. T. (Ed.). (2001). *NOAA diving manual: Diving for science and technology*. Best Pub Co.

Jones, N. L. (2008). An obsession with CO_2. *Applied Physiology, Nutrition, and Metabolism, 33*(4), 641–650.

Kang, K. W., Lee, S. J., & Kim, S. G. (2005). Molecular mechanism of nrf2 activation by oxidative stress. *Antioxidants and Redox Signaling, 7*(11 – 12), 1664–1673.

Kannan, S., Pang, H., Foster, D. C., Rao, Z., & Wu, M. (2006). Human 8-oxoguanine DNA glycosylase increases resistance to hyperoxic cytotoxicity in lung epithelial cells and involvement with altered MAPK activity. *Cell Death and Differentiation, 13*(2), 311.

Kuznetsov, D. A., Naeem, M. A., Kumar, P. V., Abdala, P. M., Fedorov, A., & Müller, C. R. (2020). Tailoring lattice oxygen binding in ruthenium pyrochlores to enhance oxygen evolution activity. *Journal of the American Chemical Society, 142*(17), 7883–7888.

Lambertsen, C. J. (1968). Effects of excessive pressures of oxygen, nitrogen, carbon dioxide, and carbon monoxide: Implications in aerospace and undersea environments. *Medical Physiology*, 836–867.

Mantell, L. L., Horowitz, S., Davis, J. M., & Kazzaz, J. A. (1999). Hyperoxia-induced cell death in the lung-the correlation of apoptosis, necrosis, and inflammation. *Annals of the New York Academy of Sciences, 887*(1), 171–180.

Mason, R. C., Murray, J. F., Nadel, J. A., & Gotway, M. (2015). *Textbook of respiratory medicine.* Elsevier Health Sciences.

National Center for Biotechnology Information. (2020). *PubChem Compound Summary for CID 977, Oxygen.* Available from https://pubchem.ncbi.nlm.nih.gov/compound/Oxygen.

National Heart, Lung, and Blood Institute. (2019). *Oxygen therapy.* Available from https://www.nhlbi.nih.gov/health-topics/oxygen-therapy.

Pagano, A., & Barazzone-argiroffo, C. (2006). Alveolar cell death in hyperoxia-induced lung injury. *Annals of the New York Academy of Sciences, 1010*(1), 405–416.

Patel, I. (2016). *Oxygen Therapy: Risk Assessment.* Available from https://www.networks.nhs.uk/nhs-networks/london-lungs/documents/oxygen-risk-assessment-sel-oxygen-study-day.

Rahman, I., Biswas, S. K., & Kode, A. (2006). Oxidant and antioxidant balance in the airways and airway diseases. *European Journal of Pharmacology, 533*(1–3), 222–239.

Romashko, J., III, Horowitz, S., Franek, W. R., Palaia, T., Miller, E. J., Lin, A. , et al. (2003). MAPK pathways mediate hyperoxia-induced oncotic cell death in lung epithelial cells. *Free Radical Biology and Medicine, 35*(8), 978–993.

Rusoke-Dierich, O. (2018). Intoxication of breathing gases during diving. *Diving Medicine*, 283–306. https://doi.org/10.1007/978-3-319-73836-9_26.

Sies, H., & Jones, D. P. (2020). Reactive oxygen species (ROS) as pleiotropic physiological signalling agents. *Nature Reviews Molecular Cell Biology*, 1–21.

Sun, H., Yan, Z., Liu, F., Xu, W., Cheng, F., & Chen, J. (2020). Self-supported transition-metal-based electrocatalysts for hydrogen and oxygen evolution. *Advanced Materials, 32*(3), 1806326.

Yee, M., Vitiello, P. F., Roper, J. M., Staversky, R. J., Wright, T. W., McGrath-Morrow, S. A., et al. (2006). Type II epithelial cells are critical target for hyperoxia-mediated impairment of postnatal lung development. *American Journal of Physiology-Lung Cellular and Molecular Physiology, 291*(5), L1101–L1111.

Zhou, H., Saidel, G. M., & Cabrera, M. E. (2007). Multi-organ system model of O_2 and CO_2 transport during isocapnic and poikilocapnic hypoxia. *Respiratory Physiology and Neurobiology, 156*(3), 320–330.

CHAPTER

24

Ozone: Risk assessment, environmental, and health hazard

Kavita Rawat[a] *and Gagan Matta*[b]

[a]Department of Chemistry, H.N.B. Garhwal University, Badshahithaul Campus, Tehri Garhwal, Uttarakhand, India

[b]Department of Zoology & Environmental Science, Gurukula Kangri (Deemed to be University), Haridwar, Uttarakhand, India

Abbreviations

APEX Air Pollution Exposure Model
CFC chlorofluorocarbon
VOC volatile organic compound

24.1 Introduction

Ozone is a strong oxidant and short-lived air pollutant that adversely affect human health, environment, and ecological system. The meaning of ozone is "to smell" taken from the Greek language. The presence of ozone in photochemical fog promotes oxygenated organic species and cells by reacting with several compounds. Its presence is an alternative to the stratosphere, but undesirable in the troposphere (Finlaysin-Pitts & Pitts, 1993). Ozone is produced by various processes and is a unique substance, formed by the reaction of UV light with NOx in the presence of volatile organic compounds. Ozone can be recognized as the most damaging air pollutant in the environment due to its phytotoxicity and can be produced by carbon monoxide (CO), methane (CH_4), non-methane hydrocarbons, or some other organic compounds (Ainsworth, Yendrek, Sitch, Collins, & Emberson, 2012; Atkinson & Arey, 2003; Fowler et al., 1999, 1999). After a very long time, Ozone has been described

Hazardous Gases
https://doi.org/10.1016/B978-0-323-89857-7.00019-0

as a trimer of oxygen accompanied by an odor like electrical wires (Anel, 2016).

24.2 Occurrence

Ozone is mainly found in the upper layer of the atmosphere, i.e., stratosphere; hence, it is also known as stratospheric ozone. In the stratospheric region, air oxygen reacts with ultraviolet radiations to produce ozone. Halogens show reactivity toward the water to form ozone in the seawater.

24.3 Historical background

Stratospheric ozone is an inorganic molecule firstly recognized sensitively by Van Marum. Schonbein gave the perception of ozone in 1839, but the molecular structure of ozone was described by Jean-Luis in 1860. Although Schonbein could not describe the structure of ozone, he did much research in this field and suggested correlation with the high concentrations of ground-level ozone as well as its severe effect on human health. Schonbein gave a significant contribution to the research field of ozone. He suggested that ozone can be produced through electrical processes, which were later found to be correct, due to its presence in smog. Van Marum continued his research and suggested that only 0.003% ozone is found in the atmosphere along with other gases, whereas most of it is spread in the stratosphere (Iribarne & Godson, 1996; Wayne, 1985).

24.4 Physicochemical properties

- Ozone absorption occurs between 220 and 290 nm in the ultraviolet region of the electromagnetic spectrum. It blocks the harmful ultraviolet radiation of the sun and protects the earth as well as its inhabitants. In the absence of the ozone layer, more harmful radiations reach into the earth's surface and damage plants, animals, and affect the human body.
- The color of pure ozone gas is blue with a pleasant smell. It has a strong irritating odor, which can cause headache and nausea postinhalation.
- The solubility of ozone is more in water than oxygen, i.e., 49% by volume at 0°C.
- On cooling in the liquid air, it liquefies with a deep blue color.
- It boils at 161.2 K temperature and violet-black crystals form after crystallization and get its molten form at 80.6 K.
- Turpentine oil and acetic acid are the best solvents to dissolve ozone.

As ozone is a highly unstable compound and decomposes at room temperature explosively, while at 25°C, the ozonized oxygen decomposes very slowly; ozone decomposes into oxygen very rapidly at 573 K (Eq. 24.1).

$$2O_3 \xrightarrow{573\,K} 3O_2 \quad (24.1)$$

The nascent oxygen produced in the reaction is responsible for the oxidation of many substances as shown in the following reactions:

$$O_3 \rightarrow O_2 + [O] \quad (24.2)$$

Ozone oxidizes PbS (lead sulfide) to give lead sulfate (Eq. 24.3).

$$4O_3 + PbS \rightarrow 4O_2 + PbSO_4 \quad (24.3)$$

Iodine is released from a solution of potassium iodide by the decomposition of O_3 (Eq. 24.4).

$$O_3 + 2KI + H_2O \rightarrow 2KOH + O_2 + I_2 \quad (24.4)$$

The reaction between hydrochloric acid and ozone produces chlorine gas (Eq. 24.5).

$$2HCl + O_3 \rightarrow O_2 + H_2O + Cl_2 \quad (24.5)$$

Sulfur trioxide forms by the reaction of ozone and sulfur dioxide (Eq. 24.6).

$$3SO_2 + O_3 \rightarrow 3SO_3 \quad (24.6)$$

Iodic acid forms by the reaction of ozone and moist iodine (Eq. 24.7).

$$5O_3 + I_2 + H_2O \rightarrow 2HIO_3 + 5O_2 \quad (24.7)$$

It oxidizes $K_4[Fe(CN)_6]$ to give $K_3[Fe(CN)_6]$, Eq. (24.8).

$$O_3 + 2K_4[Fe(CN)] + H_2O \rightarrow 2K_3[Fe(CN)_6] + 2KOH + O_2 \quad (24.8)$$

Ozone oxidizes stannous chloride to stannic chloride in the acidic medium (Eq. 24.9).

$$O_3 + 3SnCl_2 + 6HCl \rightarrow 3SnCl_4 + 3H_2O \quad (24.9)$$

When silver metal is heated with ozone, silver oxide forms in the initial stages of the reaction, and due to the reduction of this oxide, silver metal turns black, as shown in Eqs. (24.10) and (24.11).

$$2Ag + O_3 \rightarrow Ag_2O + O_2 \quad (24.10)$$

$$Ag_2O + O_3 \rightarrow 2Ag + 2O_2 \quad (24.11)$$

Mercurous oxide forms by passing ozone through mercury, and due to this, mercury loses its meniscus level and clings to the wall of glass, also called mercury tail, Eq. (24.12), though it can be restored after shaking with water.

$$O_3 + 2Hg \rightarrow Hg_2O + O_2 \quad (24.12)$$

Ozone reduces peroxide to give oxygen (Eqs. 24.13 and 24.14).

$$H_2O_2 + O_3 \rightarrow H_2O + 2O_2 \quad (24.13)$$

$$BaO_2 + O_3 \rightarrow BaO + 2O_2 \quad (24.14)$$

Ozonides are additional products, formed by the reaction between unsaturated organic compounds and ozone (Eq. 24.15).

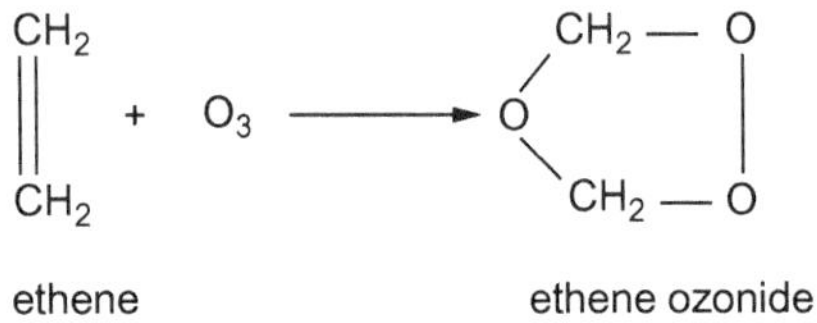

(24.15)

These ozonides decompose to give aldehydes and hydrogen peroxide in most cases. This reaction is known as "ozonolysis" (Eq. 24.16).

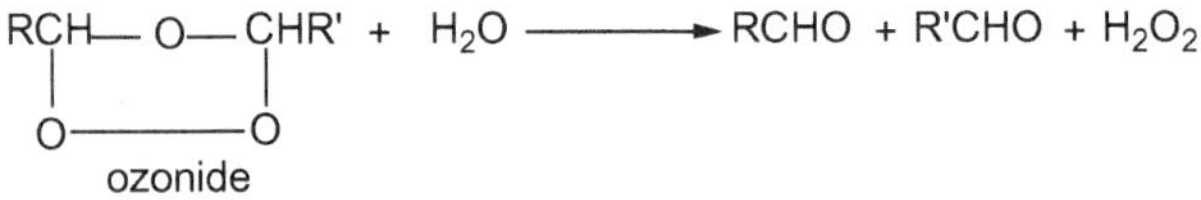

(24.16)

24.5 Physiological role

This section briefly portrays the reason beneath the harmful effect after the long-term inhalation of ozone. This gas tends to progressively affect the whole-body system, by disturbing cellular physiology in several ways. Oxygen delivery is disrupted by oxidative stress, due to which aging in red blood cells induces. Superoxide is a metabolite that hinders growth factor receptors derived by platelets and responsible for the growth factor signal transduction (Mohanty, Nagababu, & Rifkind, 2014).

Hydrogen peroxide, a fundamental reactive oxygenated species, has many biological and therapeutic effects. Hydrogen peroxide concentration is responsible for the determination of cell proliferation and it can be produced by the reaction of ozone with unsaturated fatty acids (Antunes & Cadenas, 2000; Forman, 2008; Halliwell, Clement, & Long, 2000; Halliwell, Clement, Ramalingam, & Long, 2000). Hydrogen peroxide concentration is essential as it regulates signal transduction in physiological quantities (Baeuerle & Henkel, 1994; Bocci, 2006; Stone & Yang, 2006).

24.6 Uses

Ozone can be used in the treatment of wounds, and mainly it is used in air gangrene caused by an infection in the absence of the air associated with *Clostridium*.

Topical ozone therapy: Ozone therapy has been used and widely studied for decades, due to its documented effectiveness with minimal side effects. Ozone therapy is often regarded as a global treatment for several diseases including infections. Creative ways to use this therapy, including herniated disks and pelvic vascular therapy, are promising and, in some cases, are even less aggressive alternative to surgery.

Bacterial and viral therapy: Ozone is capable of inactivating different kinds of bacteria and viruses in the water. Different microbes as parasitic spores and oocytes require more ozone than other microorganisms.

Other uses

- Ozone is used to treat infections, injuries, multiple diseases and is also used to disinfect drinking water (McLean, 2009).
- Ozone has hemodynamic, antibacterial, and anti-inflammatory properties and therefore, can be used for the treatment of many diseases (Stoker, 1902).
- Ozone is the most potent oxidizer among other oxidizers and due to ease of availability, it can be used for industrial or residential purposes.
- Oxidation of iron and manganese is easily carried out by ozone; therefore, it can be used for the removal of these transition metal ions in drinking water systems.

24.7 Sources of contamination

Ozone can be formed by the reaction of nitrogen oxide and volatile organic compounds (VOCs) in the presence of UV light. Volatile organic compounds are mainly produced by vehicle emissions, fuel-burning processes, and evaporation from a large number of fixed and portable sources, which tend to remain in the air until the sun blends them into ozone. The main sources of the NOx emission are high traffic density and thermal power generation plants in the urban areas. The emission of ozone in the atmosphere is maintained by controlling the production and transport of ozone into the region. Furthermore, ozone is additionally transferred into the lower atmosphere from the stratosphere through the stratosphere-troposphere exchange process. Carbon monoxide emission that acts as VOCs in the air during peak rush hours increases the atmospheric ozone production capacity (refer Chapter 8). It is usually produced on hot sunny days when there is less air present in the atmosphere (*U.S.* EPA, 2013).

24.8 Risk assessment

Ozone is used in the purification of drinking water, sanitizing air, wastewater treatment, and other chemical industries. People working in areas with a high level of ozone are at risk of being exposed to ozone such as outdoor workers in the paper and mash factories, wastewater treatment plant workers, etc. (Jhun, Fann, Zanobetti, & Hubbell, 2014; Slapar, Velders, & Matthijsen, 1998). The studies in urban areas illustrating the public health impact of this pollutant and quantitative risk and exposure assessment had been conducted to estimate the risk of various health effects associated with ambient ozone exposure (U.S. EPA, 2007a, 2007b).

APEX model: The modeling of human exposure and risk assessment based on the published data from human exposure studies can be implemented using air pollution exposure model. APEX model is based on the probabilistic exposure model and is used to estimate the human exposure and risk associated with the toxic exposure of ozone at local, metropolitan, or provincial scales using a stochastic microenvironment approach. APEX model randomly collects the data of a sample of an individual based on census data and simulates each individual's movement for their demographics. The activity pattern or long-term activity sequence for each simulated individual is constructed for a day or longer up to a year and breathing rates of each event are assigned by the APEX model on the basis of the demographics of the simulated person. APEX model then calculates the concentration of each pollutant in the microenvironment associated with the simulated individual and the activity pattern. All this information is used by the APEX model to obtain a series of time-averaged exposures spanning the time period of simulation. APEX model is also very useful in the estimation of dose for the simulated individual based on the exposure and breathing rates for each individual (U.S. EPA, 2012ba, b).

24.9 Environmental Effect

Ozone is a solid oxidant that is harmful to the farming, environment, and consequently can decrease crop yields and harvest quality. Being a photochemical pollutant, it can harm plant tissue structure, biochemical activity, and hinder plant development. Chronic exposure to O_3 stress decreases the chlorophyll content as this stress is harmful to plant cell pigment and the activity of chlorophyll enzyme increases (Wilkinson, Mills, Illidge, & Davies, 2012). Ozone stress is mainly responsible for the reduction in plant photosynthesis activities, hence causing the plant death (Ainsworth & Mcgrath, 2010; Arneth et al., 2010; Feng, Kobayashi, & Ainsworth, 2008; Fuhrer, 2009).

Though the ozone level in the troposphere is low, still it is high enough to threaten crops and flora, causing damage to crops (Fiscus, Booker, & Burkey, 2005; Fuhrer & Booker, 2003; Weschler, 2006). Tropospheric ozone puts an adverse effect on the ecosystem, agriculture, environment, and human health. There are two main effects of ozone exposure on the ecosystem, i.e., biomass loss and visible foliar injury that have negative impacts on the vegetation. A high concentration of ozone is responsible for the reduction in crop yield and growth productivity of castor oilseed plants. These negative impacts can be reduced by selecting suitable cultivars and demonstrated ozone resistance, as ozone stress is directly proportional to the oxidative potential of plants, which provides adaptability to the plant seeds toward tropospheric ozone pollution (Rathore & Chaudhary, 2018).

The physiological and developmental processes of flora as plant growth and the ability of adaptation of increased levels of UVB are directly affected by the UVB radiations. The foundation of aquatic food cycles was formed by the phytoplankton and its productivity limited to the euphotic area. Orientation and mobility in phytoplankton are affected by the solar UVB radiation, and due to this, survival rates of these organisms reduce. A small increase in UVB exposure results in a decrease in populations for whole marine organisms. The rise of UVB radiation affects terrestrial and aquatic biochemical cycles.

Tropospheric ozone is phytotoxic and increased concentration of ozone is a major problem in the urban and nonurban ecosystems due to the casual ecological effects of ozone. High exposure to ozone is mainly responsible for the decrease in crop growth rate and reduction in crop yields. In the case study, the mean concentrations of ozone were measured for 6 h and checked the responses of plants in terms of crop growth, yield, pigment, and other physiological properties. It was observed that these parameters of crops get reduced in mung bean (*Vigna radiata*), spinach, wheat, and mustard, planted in the urban region of Varanasi, India. Therefore, it concluded that long-term exposure to ozone may damage vegetation with a reduction in crop growth and quality. The exposure of ozone is very high during the daytime in the month of April and May in urban areas (Agrawal, Singh, Rajput, Marshall, & Bell, 2003).

24.10 Hazardousness

- The main component of the fog is ozone, a strong breathing irritant, and the most widespread secondary air pollutant. Ozone is very toxic as it affects the respiratory tract, reduces pulmonary function, and induces airway inflammation. It can cause health problems and humans may have variable sensitivity to ozone exposure. A high level of ozone

exposure has been linked to various health problems and other cardiopulmonary complications (Weinhold, 2008).
- The effect of ozone exposure reduces the breath count and patients feel difficulty to breathe normally.
- Ozone aggravates asthma. Due to the high level of ozone, most people suffer from asthma attacks and they essentially need the doctor's vision or extra oculars.
- Ozone cells redden and damage the cells that line the lungs and cause lung diseases. Long-term ozone exposure reduces immunity and increases chance of bacterial infections.

24.11 Treatment

The reaction of unsaturated fatty acids with ozone produces free radicals, and if an individual is exposed to ozone, no particular treatment is available, although it is recommended that vitamin "E" intake can prevent these types of chemical reactions (Pryor, Prier, & Church, 1983; Pryor, Stanley, Blair, & Cullen, 1976; Roehm, Hadley, & Menzel, 1971). Dietary intake of vitamin "E" can considerably protect humans and animals from the noxious effects of ozone.

24.12 Risk management

The following are ways to reduce ozone exposure in schools and office settings:

- Minimize the penetration of outside gas on the premises.
- Control the gaseous reactions by restricting the utilization of materials and items that react with ozone.
- Limit the utilization of printers, scanners, and different gadgets and apparatus that produce ozone either deliberately or inadvertently. There are some ways by which the indoor ozone concentration can be maintained below the given WHO guidelines. These include keeping the doors and windows shut when outside ozone exposure is very high as well as utilization of suitable material coatings for an inactive decrease of ozone (Salonena, Salthammer, & Morawska, 2018).

24.13 Current research

Pig-pen effect: Mixing of skin oil and ozone delivers a haze of individual contamination. Scientists revealed in their work that skin oils contain fixings, for example, squalene, unsaturated fats, and wax esters. Wearing clothes more than a day without washing too long is the main cause of individual pollution, e.g., the clothes are more saturated with oils and react with unstable gas ozone.

The mechanism of ozone gas generation can be understood with the help of a novel ozone gas prediction model within the corona discharge region and in that ionizing region, oxygen is converted into O_3. A coupled equation was derived between the electrostatic field and O_3 concentration, which gives the information related to influencing factors of different parameters like polarity, voltage, current, gas velocity, and free diffusion. A pin cylindrical corona discharge device provides the corona discharge status and ozone gas concentration in positive and negative polarity conditions. These negative and positive polarity conditions show the different status of corona discharge suggesting that the distribution of ozone concentration is not uniform. This prediction model is a perfect way to predict the ozone gas generation and distribution within the domain range (Zhu, Chen, Shi, & Shangguan, 2020).

24.14 Ozone toxicity

CFCs and other ozone emitting substances have adverse effects on the human body and this recognition led to the Montreal Protocol on substances that deplete the ozone layer in 1987, a landmark agreement approved by United Nation to eliminate ozone-depleting substances from the atmosphere. Without the agreement, the US would have seen an additional 280 million cases of skin malignancy and 1.5 million skin cancer deaths. Over 30 years after the Montreal Protocol, NASA researchers reported the direct confirmation that CFC's emission decreases resulting in the recovery of Antarctic ozone, and depletion in the region declined 20% since 2005. Furthermore, United Nation confirmed in a scientific assessment that the ozone layer recovering process is in progress and anticipated that it would heal completely in the North equator regions by the 2030s, followed by the south equator regions in the 2050s and polar regions by 2060s.

24.15 Safety measures

- In order to minimize health risk, adequate training and supervision must be established.
- Ozone plants must be maintained with satisfactory safe working systems.
- Cleaning materials must be oil-free in order to clean the ozone units and pipework.
- It is essential to calibrate all the sensor equipment.

- Ozone input control is important for the process.
- Activated carbon filters should not expose to ozone concentrations greater than 20 g/m^3 because the reaction is automated and an explosion may occur.
- In regions of extreme focus, gas covers with appropriate canister and full facepiece or goggles or the utilization of provided air respirators are suggested.
- Representatives should wash quickly with a cleanser when the skin is wet or debased.

24.16 Storage and handling

- There are some practical disadvantages of the liquid ozone storage and the mixture can be compressed and maintained at 70 atm. The slow thermal decomposition of ozone can be reduced by cooling to −78°C temperature and most of the gas is stored in suitable ducts with minimal loss of ozone.
- Ozone storage is important to avoid contact with organics due to the occurrence of violent reactions.
- Severe explosive risk is there if liquefied ozone is exposed to heat or inflammation.

24.17 Precautions

- Wear defensive eye security, gloves, and dress to forestall any sensible likelihood of skin or eye to eye connection. Well-being gear providers/producers can give suggestions on the most defensive glove/garments material for your activity.
- All defensive garments (suits, gloves, footwear, and headgear) ought to be perfect, accessible every day, and put on before work.
- Wear gas verification compound goggles and face shield except if full facepiece respiratory assurance is worn.

References

Agrawal, M., Singh, B., Rajput, M., Marshall, F., & Bell, N. B. J. (2003). Effect of air pollution on peri-urban agriculture: A case study. *Environmental Pollution, 126*, 323–329.

Ainsworth, E. A., & Mcgrath, J. M. (2010). Direct effects of rising atmospheric carbon and ozone on crop yields. *Climate Change and Food Security, 37*, 109–130.

Ainsworth, E. A., Yendrek, C. R., Sitch, S., Collins, W. J., & Emberson, L. D. (2012). The effects of tropospheric ozone on net primary productivity and implications for climate change. *Annual Review of Plant Biology, 63*, 637–661.

Anel, J. A. (2016). The stratosphere: History and future a century after it's discovery: A review of the physics of the stratosphere, by goody. *Contemporary Physics*, *57*(2), 230–233.

Antunes, F., & Cadenas, E. (2000). Estimation of H_2O_2 gradients across biomembranes. *FEBS Letters*, *475*(2), 121–126.

Arneth, A., Harrison, S. P., Tsigaridis, K., Menon, S., Bartlein, P. J., Feichter, H., et al. (2010). Terrestrial biogeochemical feedbacks in the climate system. *Nature Geoscience*, *3*, 525–532.

Atkinson, R., & Arey, J. (2003). Atmospheric degradation of volatile organic compounds. *Chemical Reviews*, *103*(12), 4605–4638.

Baeuerle, P. A., & Henkel, T. (1994). Function and activation of NF-kappa B in the immune system. *Annual Review of Immunology*, *12*(1), 141–179.

Bocci, V. A. (2006). Scientific and medical aspects of ozone therapy. State of the art. *Archives of Medical Research*, *37*(4), 425–435.

Feng, Z., Kobayashi, K., & Ainsworth, E. A. (2008). Impact of elevated ozone concentration on growth, physiology and yield of wheat (Triticum aestivum, L.): A meta analysis. *Global Change Biology*, *14*, 2696–2708.

Finlaysin-Pitts, B. J., & Pitts, J. N., Jr. (1993). Atmospheric chemistry of tropospheric ozone formation: Scientific and regulatory implications. *Air and Waste*, *43*(8), 1091–1100.

Fiscus, E. L., Booker, F. L., & Burkey, K. O. (2005). Crop responses to ozone: Uptake, modes of action, carbon assimilation and partitioning. *Plant, Cell and Environment*, *28*(8), 997–1011.

Forman, H. J. (2008). Hydrogen peroxide: The good, the bad, and the ugly. In *Oxidants in biology* (pp. 1–17). Dordrecht: Springer.

Fowler, D., Cape, J. N., Coyle, M., Flechard, C., Kuylenstierna, J., Hicks, K., et al. (1999). The global exposure of forests to air pollutants. *Water, Air, and Soil Pollution*, *116*(1–2), 5–32.

Fowler, D., Cape, J. N., Coyle, M., Smith, R. I., Hjellbrekke, A. G., Simpson, D., et al. (1999). Modeling photochemical oxidant formation, transport, deposition and exposure of terrestrial ecosystems. *Environmental Pollution*, *100*(1–3), 43–55.

Fuhrer, J. (2009). Ozone risk for crops and pastures in present and future climates. *The Science of Nature*, *96*, 173–194.

Fuhrer, J., & Booker, F. (2003). Ecological issues related to ozone: Agricultural issues. *Environment International*, *29*(2–3), 141–154.

Halliwell, B., Clement, M. V., & Long, L. H. (2000). Hydrogen peroxide in the human body. *FEBS Letters*, *486*(1), 10–13.

Halliwell, B., Clement, M. V., Ramalingam, J., & Long, L. H. (2000). Hydrogen peroxide. Ubiquitous in cell culture and in vivo. *IUBMB Life*, *50*, 251–257.

Iribarne, J. V., & Godson, W. L. (1996). *Thermodynamics of the atmosphere*. Ministry of the Environment: Publications Center.

Jhun, I., Fann, N., Zanobetti, A., & Hubbell, B. (2014). Effect modification of ozone related mortality risks by temperature in 97 US cities. *Environment International*, *73*, 128–134.

McLean, L. (2009). *The miracle of ozone therapy*.

Mohanty, J., Nagababu, E., & Rifkind, J. M. (2014). Red blood cell oxidative stress impairs oxygen delivery and induces red blood cell aging. *Frontiers in Physiology*, *5*, 84.

Pryor, W. A., Prier, D. G., & Church, D. F. (1983). Detection of free radicals from low temperature ozone-olefin reactions by ESR spin trapping: Evidence that the radical precursor is a trioxide. *Journal of American Chemical Society*, *105*, 2883–2888.

Pryor, W. A., Stanley, J. P., Blair, E., & Cullen, G. B. (1976). Autoxidation of polyunsaturated fatty acids. Part I. effect of ozone on the autoxidation of neat methyl linoleate and methyl linoleate. *Archives of Environmental Health*, *3l*, 201–210.

Rathore, D., & Chaudhary, I. J. (2018). Ozone risk assessment of castor (Ricinus communis L.) cultivars using open top chamber and ethylenediurea (EDU). *Environmental Pollution*. https://doi.org/10.1016/j.envpol.2018.10.036.

Roehm, I. N., Hadley, J. C., & Menzel, D. B. (1971). Oxidation of unsaturated fatty acids by ozone and nitrogen dioxide: A common mechanism of action. *Archives of Environmental Health*, *23*, 142.

Salonena, H., Salthammer, T., & Morawska, L. (2018). Human exposure to ozone in school and office indoor environments. *Environment International, 119*, 503–514.
Slapar, H., Velders, G., & Matthijsen, J. (1998). Ozone depletion and skin cancer incidence: A source risk approach. *Journal of Hazardous Materials, 61*, 77–84.
Stoker, G. (1902). Ozone in chronic middle-ear deafness. *The Lancet, 160*(4131), 1187–1188.
Stone, J. R., & Yang, S. (2006). Hydrogen peroxide: A signaling messenger. *Antioxidants and Redox Signaling, 8*(3–4), 243–270.
U.S. EPA. (2007a). *O_3 population exposure analysis for selected urban areas*. Research Triangle Park, NC: Office of Air Quality Planning and Standards (EPA document number EPA-452/R-07-010).
U.S. EPA. (2007b). *O_3 health risk assessment for selected urban areas*. Research Triangle Park, NC: Office of Air Quality Planning and Standards (EPA document number EPA 452/R-07-009).
U.S. EPA. (2012a). *Total risk integrated methodology (TRIM) - air pollutants exposure model documentation (TRIM. Expo/APEX, version 4.4) volume II: Technical support document*. Research Triangle Park, NC: Office of Air Quality Planning and Standards, U.S. Environmental Protection Agency. EPA-452/B-12-001b.
U.S. EPA. (2012b). *Total risk integrated methodology (TRIM) - air pollutants exposure model documentation (TRIM. Expo/APEX, version 4.4) volume I: User's guide*. Research Triangle Park, NC: Office of Air Quality Planning and Standards, U.S. Environmental Protection Agency. EPA-452/B-12-001a.
U.S. EPA. (2013). *Integrated science assessment of ozone and related photochemical oxidants (final report)*. Research Triangle Park, NC: EPA Office of Research and Development (EPA document number EPA/600/R-10/076F).
Wayne, R. P. (1985). *Chemistry of atmospheres*. Clarendon Press.
Weinhold, B. (2008). *Ozone nation: EPA standard panned by the people*.
Weschler, C. J. (2006). Ozone's impact on public health: Contributions from indoor exposures to ozone and products of ozone-initiated chemistry. *Environmental Health Perspectives, 114*(10), 1489–1496.
Wilkinson, S., Mills, G., Illidge, R., & Davies, W. J. (2012). How is ozone pollution reducing our food supply? *Journal of Experimental Botany, 63*, 527–536.
Zhu, Y., Chen, C., Shi, J., & Shangguan, W. (2020). A novel simulation method for predicting ozone generation in corona discharge region. *Chemical Engineering Science, 227*, 115910.

CHAPTER

25

Phosgene: Risk assessment, environmental, and health hazard

Ashok Kumar Sharma[a] *and Nitish Kumar*[b]

[a]Department of Chemistry, Deen Bandhu Chhotu Ram University of Science and Technology, Murthal, Haryana, India

[b]Department of Chemistry, Gurukula Kangri (Deemed to be University), Haridwar, Uttarakhand, India

Abbreviations

ABG	arterial blood gas
ACGIH TLV-TWA	American Conference of Governmental Industrial Hygienists, Threshold Limit Value–Time Weighted Average
AEGL	Acute Exposure Guideline Levels
AIE	aggregation-induced emission
CalEPA	California Environmental Protection Agency
CHC	chlorinated hydrocarbon
DATPES	3,6-bis(1,2,2-triphenylvinyl)benzene-1,2-diamine-based sensor
DFT	density functional theory
EEGL	emergency exposure guidance levels
EPA	Environmental Protection Agency
FMEA	Failure Mode and Effect Analysis
HAZOP	hazard and operability study
HIRA	hazard identification and risk assessment
LOPA	layer of protection analysis
MAC	maximal accepted concentration
MAK	maximum workplace concentration
NFPA	National Fire Protection Association
NIOSH REL-STEL	Recommended Exposure Limits-Short Term Exposure Limit
NIOSH	National Institute of Occupational Safety and Health
OSHA PEL-TWA	Occupational Health and Safety Administration Permissible Exposure Limits-Time Weighted Average
QRA	quantitative risk assessment

Hazardous Gases
https://doi.org/10.1016/B978-0-323-89857-7.00016-5

25.1 Introduction

Phosgene is a colorless, volatile liquid (b.p. 8.3°C, 48 °F) with an odor of new-mown hay or green corn, also called carbonyl chloride; used as poison gas, in organic synthesis and in making dye. It is prepared by the reaction of carbon monoxide with chlorine in the presence of activated charcoal. It has a trigonal planar structure as shown in Fig. 25.1.

$COCl_2$, first developed in 1812, was most probably best known as the warfare specialist that made the most concoction in the First World War. It remains, in any case, an important, poisonous modern part of the world, which was manufactured in megatons by various industrial synthetic processes. Although the supply and development of phosgene are currently restricted, their widespread use as a chemical intermediate ensures their continued industrial presence. Because of its high poisonous characteristic quality, it is a major subject of worry as the potential for inadvertent ecological or environmental accident proceeds (Jugg, 2016).

25.2 Occurrence

Phosgene is not freely present in the natural environment and can be generated in high-temperature areas by exposing the hydrocarbon compounds of chlorine. Chlorinated hydrocarbon (CHC) is sometimes used or generated in manufacturing in materials that include the elements of chlorine, hydrogen, and carbon. $COCl_2$ is believed to be produced by vapors of chlorinated solvents exposed to high temperatures. Chlorinated solvents are products containing chlorine that are typically used for dissolving or cleaning certain substances such as paint stripping, metal cleaning, and dry cleaning in industrial processes. For low-lying countries, it is easily available because of its lighter density than air.

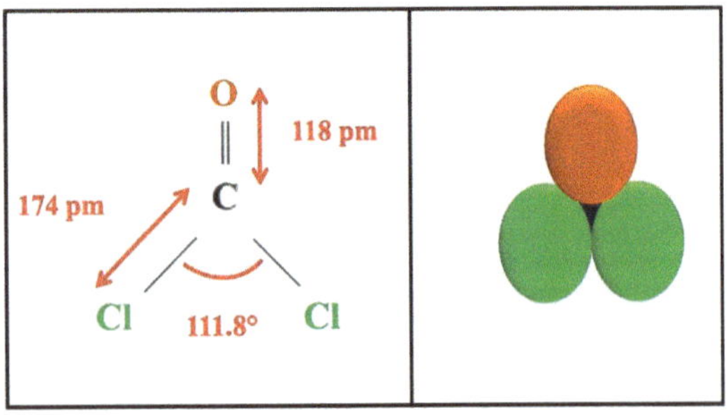

FIG. 25.1 Structural formula of phosgene. *Credit: Nitish Kumar.*

25.3 Historical background

In 1812, the first $COCl_2$ gas was synthesized by a 22-year-old British amateur chemist John Davy. It does not have phosphorus; however, its name was derived from the Greek words phos (light) and genesis (born). In Belgium on December 19, 1915, 88 tons of phosgene was released during the trench war in France that killed 69 people and badly wounded more than 1200, presumably supposed to be implemented with a more sensible approach. German people were satisfied with the results, so they quickly began fighting with grenades filled with phosgene. The combining of chlorine and mustard gas accounted for more than 60% of all deaths in the chemical war during the First World War. During the Second World War, soldiers were well prepared and ready for the possible use of this deadly gas. However, at least 10,000 Soviet soldiers were allegedly injured by the release of phosgene during the 1942 Nazi Germany Kerch Battle.

25.4 Physicochemical properties

$COCl_2$ is a colorless, flammable gas at room temperature that has a suffocating smell like new-mown hay. Odor, however, does not warn sufficiently of dangerous levels. It is mildly irritating at high levels. It is a colorless fluid that fumes at temperature less than 47°F; contact with the fluid may cause frostbite. The liquids or gas slowly spills into hydrochloric acid, which can irritate and damage the cells in the presence of water (sweat, saliva, tears). Some of the physical properties of $COCl_2$ are shown in Table 25.1.

Moisture (water or alcohol) reacts with phosgene. In the presence of water, HCl slowly disintegrates to give CO_2. It produces toxic and corrosive smoke when heated to decomposition. $COCl_2$ is strongly responsive to various chemicals (e.g., alkalis, ammonia, copper, amines, aluminum); in the presence of water; attacks many metals, plastics, and rubber too.

25.5 Physiological role

Although poorly soluble in water, $COCl_2$ interact with water at high enough concentration, similar to the chemical interaction in which chlorine produces hydrochloric acid in mucous membranes to liberate hydrochloric acid. The major amount of inhaled $COCl_2$ enters the distal airways because of their poor water solubility. The phosgenous (very reactive free radical) carbonyl group in the distal respiratory tract damages alveolar

TABLE 25.1 Physicochemical properties of phosgene.

Physicochemical properties of phosgene	
Formula	$COCl_2$
Odor	Suffocating, like musty hay
Molecular weight	98.9 Da
Molecular structure	Trigonal, planar
Boiling point (760 mmHg)	47 °F (8°C)
Melting point (760 mmHg)	−118°C
Freezing point	−198 °F (−127°C)
Specific gravity	1.43 (liquid at 32°)
Vapor pressure	1215 mmHg at 68 °F (20°C)
Gas density	3.48 (air = 1)
Water solubility	Slight
Dipole moment	1.17 D

capillary membrane and causes pulmonary edema that is not cardiogenic. It is very important that this procedure is delayed and that a severe lung injury (Fig. 25.2), possibly life-threatening, occurs without significant symptoms, even after an initial presentation is delayed.

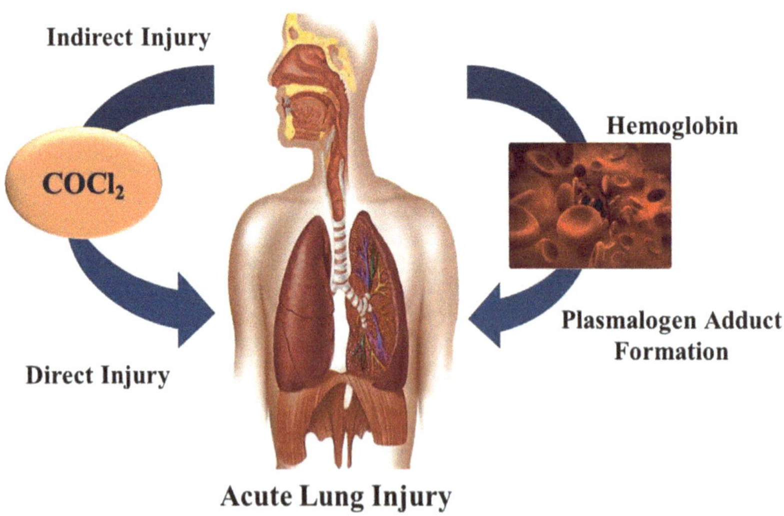

FIG. 25.2 Systematic representation of phosgene inhalation causes acute lung injury. *Credit: Nitish Kumar.*

25.6 Uses

$COCl_2$ is widely used in the production of resins, pesticides, herbicides, insecticides, metallurgy, dyes, polyurethane, and polycarbonate. In various pharmaceutical and organic industries, $COCl_2$ is used primarily as a building block in the production sites of other products. Isocyanate production consumes approximately 85% of the world's production of phosgene. The primary use of $COCl_2$ is in toluene di-isocyanate production, which is a precursor to polyurethane resins that is used to make foams, elastomers, and lacquers. In addition to its production in the industrial sector, potential sources of atmospheric phosgene include transient combustion, thermal decomposition, and photo-oxidation of chlorinated hydrocarbons. $COCl_2$ is a commonly used medium-sized chemical that is primarily produced in many organic chemicals. For the synthesis of polymers based on isocyanates, carbonic acid esters, and acid chlorides, $COCl_2$ is used.

25.7 Sources of contaminants

Many industrial processes, including the processing of direct $COCl_2$ as the intermediate origin of isocyanates, polycarbonates, carbamates, thiocarbamates, and phenyl ureas, are the possible sources of $COCl_2$ emission. Process descriptions and estimates of emissions from hypothetical facilities involved in the production or application of $COCl_2$ are included. Large amounts of phosgene are produced in situ and constitute less than 2% of total production of commercial phosgene. Therefore, $COCl_2$ is usually produced in industries that manufacture isocyanates, polycarbonates, carbonates, etc.

25.8 Risk assessment

The extent of risk depends on how much the people or living things are closer to the exposure site of phosgene gas. Moreover, what type of exposure is also the major concern for avoiding the health hazards; if phosgene is released in the form of gas, then the possibility of exposure is through skin contact, eye contact, or by the breathing air and if it is in the form of liquid and released into water, contaminated water or intake of food acts as phosgene's source of exposure. To prevent any probable chances of risk, many international regulations and guidelines have been proposed to derive the level of phosgene gas; some of them are summarized in Table 25.2.

TABLE 25.2 Inhalation standards and guidelines for phosgene.

Guideline	Specification	References
OSHA PEL-TWA[a]	The airborne exposure limit 0.1ppm with exposure duration 8h	National Institute of Occupational Safety and Health (NIOSH) (2010)
NIOSH REL-STEL[b]	The airborne exposure limit 0.1ppm over 10h and 0.2ppm for 15min	National Institute of Occupational Safety and Health (NIOSH) (2010)
ACGIH TLV-TWA[c]	The airborne exposure limit 0.1ppm with exposure duration 8h	American Conference of Governmental Industrial Hygienists (ACGIH) (2013)
AEGL[d]	Three levels (AEGL-1, AEGL-2, and AEGL-3) are developed for five exposure duration (10 and 30min at 0.60ppm, 1h at 0.30ppm, 4h at 0.08ppm, and 8h at 0.09ppm)	National Research Council (NRC) (2002)
EEGL[e]	The airborne exposure limit 0.2ppm for 1h and 0.02ppm for 24h	National Research Council (NRC) (1985)
MAK[f]	The airborne exposure limit 0.1ppm for 8h	Maximum Workplace Concentration (MAK) (2012)
MAC[g]	The airborne exposure limit 0.1ppm for 8h	Maximal Accepted Concentration (MAC) (2012)

[a] *Occupational Health and Safety Administration Permissible Exposure Limits-Time Weighted Average.*
[b] *Recommended Exposure Limits-Short Term Exposure Limit.*
[c] *American Conference of Governmental Industrial Hygienists, Threshold Limit Value–Time Weighted Average.*
[d] *Acute Exposure Guideline Levels.*
[e] *Emergency Exposure Guidance Levels, National Research Council.*
[f] *Maximale Argeitsplatzkonzentration (maximum workplace concentration).*
[g] *Maximaal Aanvaaarde Concentratie (maximal accepted concentration)*

25.9 Environmental effects

Exposure to high $COCl_2$ levels can severely damage and even kill wildlife. These effects are however only probable when large quantities of $COCl_2$ are released—for instance as a result of an accidental discharge. There are not enough usual environmental levels to cause major effects. $COCl_2$ pollution is not likely to affect the global environment.

25.10 Hazardousness

Acute Effects: Acute exposure to $COCl_2$ inhalation causes severe human respiratory symptoms, including lung edema, pulmonary emphysema, and even death. Acute exposure signs include vomiting, chest compression, coughing, painful respiration, and bloody sputum. Death comes from anoxemia, causing paralysis, asphyxiation, or heart failure of the respiratory center (The Merck Index, 1989; US Department of Health and Human Services, 1993; U.S. Environmental Protection Agency, 1986). Acute $COCl_2$ poisoning can also affect individual's brain, heart, and blood. The results of these effects are primarily due to lack of oxygen and not phosgene's direct action.

Chronic Effects (noncancer): Chronic exposure to low $COCl_2$ levels was observed to cause certain tolerances to the noted acute effects in humans but could also lead to irreversible pulmonary changes in emphysema and fibrosis (Sittig, 1985). In animals, chronic exposure to $COCl_2$ has been found to increase the incidences of chronic pneumonitis along with acute and fibrinous pneumonia. EPA has not established a reference concentration (RfC) or a reference dose (RfD) for phosgene (U.S. Environmental Protection Agency, 1999). A chronic exposure rate reference inhalation of 0.0003 mg/m^3 has been determined for phosgene based on pulmonary toxicity in rats by the California Environmental Protection Agency (CalEPA). The exposure levels at or below that CalEPA refers to, are not expected to cause adverse health effects. It is not a specific hazard estimator, but rather a benchmark for evaluating the possible impacts. The potential for adverse health effects is increasing at lifetime exposures that increase over and above the reference exposure (California Environmental Protection Agency (CalEPA), 1997).

Dermal: If the skin is wet or humid, it can cause skin irritation and redness if the phosgene vapor is in contact. Pressures can lead to frostbite in contact with liquid phosgene. As the surface area (body weight ratio) is higher, therefore children are more prone to skin-affected toxicants.

Ocular: High vapor levels can cause tearing and decrease blood in the eye. Fluid $COCl_2$ contact can result in corneal clouding and delayed perforation.

Hematologic: $COCl_2$ may lead to hemolysis which could lead to pulmonary capitular plugging in serious cases. Standard blood tests may be used to detect most hematologic alterations (e.g., bleeding, methemoglobinemia, suppression of the bone marrow, and anemia).

Hepatic: $COCl_2$ may be cytotoxic to the liver directly in cases of high exposure, resulting in necrosis and function loss.

Renal: $COCl_2$ can be cytotoxic to the kidneys, which results in necrosis and function loss in case of high exposure.

Gastrointestinal: Nausea and vomiting may occur following exposure to $COCl_2$.

25.11 Risk management

This section describes the information for minimizing phosgene-related risk, the composite of good practices including education and training program, safe handling of phosgene gas, and response to emergencies.

The educational training of the workers is very much effective for safe handling and prevents any health risk. The NOISH recommends the occupational safety and health guideline for phosgene to prevent the injuries at workplace (National Institute of Occupational Safety and Health (NIOSH), 2010). The government and nongovernment bodies may conduct an audit of the workplace to maintain specific and good practices. Safety management is another point of view to minimize accidental phenomena. The different workplace (chemical plant, medication industries, fertilizers manufacturing units) have different safety guideline, in which several methods are given for safety reviews like FMEA (failure mode and effect analysis) mostly used in the automotive industry, LOPA (layer of protection analysis), QRA (quantitative risk assessment), HIRA (hazard identification and risk assessment), and HAZOP (hazard and operability study) (Indian Phosgene Council (IPC), 2017). The design of the workplace (building, office, pilot plant) must include some safety options like ventilation system, alarm system, elevator, phosgene emission control, steam ammonia curtain, and phosgene destruction system.

25.12 Current research

Nowadays, the scientific community has developed a number of phosgene gas sensors that are cost-effective, precise, and portable. A ratiometric fluorescent chemosensor was developed by Wang, Zhong, and Song (2017) for selective and visual detection of phosgene in both liquid and gas phases. Another research conducted by Cheng, Yang, Li, Gao, and Wang (2019) revealed that AIE and DATPES fluorescent sensors have the capability to detect the phosgene gas up to 0.1 ppm. A new approach was developed by using the DFT technique for the adsorption of phosgene gas on $(XY)_{12}$ (X = Al, B, and Y = N, P) nanoclusters (Padash, Nasrabadi, Rad, Jesionowski, & Ehrlich, 2019). This technique may be used for the preparation of phosgene destruction system. The new and safe guidelines for sampling and handling of phosgene have also being prepared by the agencies from time to time (National Institute of Occupational Safety and Health (NIOSH), 2010). The mechanism of toxicity, countermeasures, and pharmacological intervention is a debatable topic till now (Pauluhn & Li, 2015); the researchers have tried to solve these mysteries to minimize the health risk.

25.13 Phosgene toxicity: Case study

The exposure of phosgene is associated with different symptoms like respiratory problems and skin disease. Galdston, Filley, and Hopson (1947) checked the residual effects of phosgene poisoning on six patients during acute exposure. Some measurements such as ventilator function test, respiratory function test, circulatory function test, and psychiatric examination were performed to know about the health hazard of the patients. The physical examination showed shallow breathing and some abnormality in lung functioning. Half of the patients have shown a significant reduction in the voluntary breathing capacity. Wyatt and Allister (1995) studied a case of 43-year-old patient who was admitted in hospital by the accidental phosgene respiration occupational exposure. The patient had symptoms like dyspnea (respiratory rate 24 min^{-1}), tachycardia (pulse 100 min^{-1}) and also pharyngeal erythema. The patient was kept under observation for 24 h after he was discharged but the patient complaint about exertional dyspnea for the next 2 weeks.

A 25-year-old male patient was studied by Kumar et al. (2012) after the accidental phosgene inhalation. The patient suffered from acute respiratory distress syndrome and needed ventilator support. The patient did not receive any antidote and any other diagnosis. The positive pressure ventilation achieved only by the timely administered supportive management system helped the patient in recovering. An accidental case of phosgene gas was studied by Vaish et al. (2013). This study involved 10 different patients, who were hospitalized (February 2011, King George's Medical University, Lucknow, India) after an accidental phosgene exposure. The outcomes of this study revealed that 30% of patients suffered from ocular symptoms such as redness and lacrimation after 2–3 h of exposureand around 50% felt diffuse chest pain. Sixty percent of the patients had shown the failure in type 1 respiratory by testing with arterial blood gas (ABG). Noncardiogenic pulmonary edema was observed in the chest X-rays of all the patients. Most of the patients recovered in 1–3 weeks after the exposure.

A 58-year-old man was admitted to the emergency ward of the hospital post an accidental phosgene inhalation exposure. Primarily, the patients have complaints of a pained throat. The preliminary examination showed some vital signs: respirations 12 min^{-1}, blood pressure 175/118 mmHg, heart rate 98 min^{-1}, and oxygen saturation of 93% on room air. After 2.5 h, the patient informed about dyspnea symptoms and his oxygen saturation level reached below 90%. The chest X-ray after 3.5 h of exposure had shown the diffuse alveolar infiltrates. However, no improvement was observed in the patient and then the patient died (Hardison, Wright, & Pizon, 2014).

25.14 Safety measures

- If the release of phosgene happens outdoors, remove the phosgene from the region. Go to the highest terrain as $COCl_2$ is heavier than air and settles into lower areas. Get out of the building if the $COCl_2$ release happens indoors.

- Remove your clothes, immediately clean your entire body with water and soap, and seek medical assistance as soon as possible.
- Removing and disposing of clothing:
  - Take off the clothes quickly that have $COCl_2$ fluid on them. The clothes should not be pulled over your head rather cut off from the body. Seal the clothes in a plastic bag as early as possible. Instead, screen a second bag of plastic with the first one. This is how the fabric is stripped and checked to protect you and other people from any toxins on your clothes.
  - Contact local or state health department or emergency services upon delivery of your items in plastic bags. Don't take plastic bags with you.
  - If you help others remove their clothes, try not to touch any contaminated areas and remove their clothes as quickly as possible.
- Rinse your eyes in plain water for 10–15 min if your eyes are burning or your vision is blurred. If you wear contact lens, remove them. Don't put your contact lens in your face. And if you have eyeglasses, wash them with soap. After you have cleaned them, you should put the glasses back on.
- Do not induce vomiting or drink any liquid if you have eaten (swallowed) $COCl_2$.
- Search for health care immediately.

25.15 Handling and storage

- $COCl_2$ gas can be converted into a liquid by refrigeration and under pressure so that it can be transported and stored. While it can be transported as liquefied, pressurized fuel, 99% of phosgene is used in the place of output.
- Store away from the workplace and away from the sun in tightly closed containers in an isolated area.
- $COCl_2$ shall be kept off the heating and refrigeration lines.
- Leaks should be tested periodically on cylinders.
- $COCl_2$ reacts with water and moisture to form hydrogen chloride gas and carbon dioxide.
- $COCl_2$ is not compatible with alkalis, ammonia, alcohols, oxidizing agents (such as perchlorates, peroxides, permanganates, chlorates, nitrates, chlorine, bromine, and fluorine), metals (such as copper, aluminum, and lithium), and combustibles (such as wood, paper, and oils).

25.16 Precautions

People's health risk against exposure depends upon how close they are to the source of phosgene. The NFPA704 (National Fire Protection Association) is a widely used labeling system for identifying the material properties against fire, spill, and safety purposes like a hazard of acute/chronic exposure (National Fire Protection Association (NFPA), 2010). The label of phosgene as shown in Fig. 25.3 is given by NFPA704.

- **Health (blue)**: 4
- **Flammability (red)**: 0
- **Reactivity (yellow)**: 1
- **Special (white)**: None

Some important precautions and advice are listed below:

- First, go away from the area where the chances of gas release is high.
- At the workplace, you must be careful and follow the instructions, which are given in the material safety data sheet.
- Safety, handling, and training have the capability to prevent accidental phosgene inhalation exposure.
- Follow the safety guideline as given by the different independent agencies like NIOSH, OSHA, AEGL, EEGL, MAK, etc.
- The risk management system has defined the possible issues, human errors that may cause accidental exposure, and hence, people are requested to follow the instruction to minimize the hazard.
- Different safety tools like alarm system, elevator, phosgene emission control, steam ammonia curtain, and phosgene destruction system should be checked and calibrated on a definite time interval.
- Besides these, if individual get exposed to phosgene, first remove their clothes, give a quick bath, and provide medical facility as soon as possible.

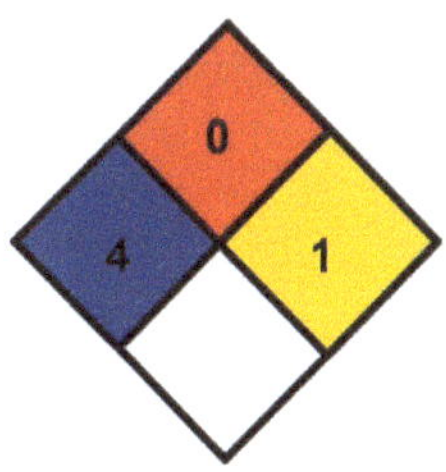

FIG. 25.3 NFPA 704 label for phosgene. *Reproduced with permission of NFPA from NFPA 704,* Standard System for the Identification of the Hazards of Materials for Emergency Response, *2017 edition. Copyright© 2016, National Fire Protection Association.*

- No antidote has been prepared for phosgene till now. Hence, exposed individual should be observed under preliminary examination for up to 48h, because in many cases it had been observed that symptoms develop after a long time.

References

American Conference of Governmental Industrial Hygienists (ACGIH) (2013). *TLVS and BEIs based on the documentation of threshold limit values for chemical substances and physical agents. Phosgene.* Cincinnati, OH: American Conference of Governmental Industrial Hygienists.

California Environmental Protection Agency (CalEPA) (1997). *Technical support document for the determination of noncancer chronic reference exposure levels.* Draft for Public Comment Berkeley, CA: Office of Environmental Health Hazard Assessment.

Cheng, K., Yang, N., Li, Q. Y., Gao, X. W., & Wang, X. J. (2019). Selectively light-up detection of phosgene with an aggregation- induced emission-based fluorescent sensor. *ACS Omega, 4*, 22557–22561.

Galdston, M., Filley, G. L., & Hopson, J. L. (1947). A study of the residual effects of phosgene poisoning in human subjects. I. After acute exposure. *The Journal of Clinical Investigation, 26* (2), 145–168.

Hardison, L. S., Wright, E., & Pizon, A. F. (2014). Phosgene exposure: A case of accidental industrial exposure. *Journal of Medical Toxicology, 10*(1), 51–56.

Indian Phosgene Council (IPC) (2017). *Phosgene handling. General safety guideline, best industrial practices and medical approach, compiled by Indian Phosgene Council.* (2017). http://www.indianchemicalcouncil.com/docs/PHOSGENE-SAFETGUIDELINES.pdf.

Jugg, B. (2016). Toxicology and treatment of phosgene induced lung injury. In *Chemical warfare toxicology* (pp. 117–153).

Kumar, A., Chaudhari, S., Kush, L., Kumar, S., Garg, A., & Shukla, A. (2012). Accidental inhalation injury of phosgene gas leading to acute respiratory distress syndrome. *Indian Journal of Occupational & Environmental Medicine, 16*(2), 88–89.

Maximal Accepted Concentration (MAC) (2012). *SDU Uitgevers, Nationale MAC List.* SDU Uitgevers, The Hague (under the auspices of the Ministry of Social Affairs and Employment) The Netherlands. Sandall, T.E., 1922. The later effects of gas poisoning. *Lancet, 203*, 857–859.

Maximum Workplace Concentration (MAK) (2012). *Deutsche Forschungsgemeinschaft, List of MAK and BAT values*: (p. 114). German Research Association. Report no. 48.

National Fire Protection Association (NFPA) (2010). Fire protection guide to hazardous materials, *NFPA 704 Identification of the Hazards of Materials for Emergency Response (1996 edition)*: (p. 704) (14th ed.).

National Institute of Occupational Safety and Health (NIOSH) (2010). *Phosgene. NIOSH pocket guide to chemical hazards. Online at* (2010). http://www.cdc.gov/niosh/npg/ngpd0504.html.

National Research Council (NRC) (1985). *Emergency and continuous exposure limits for selected airborne contaminants phosgene*: (pp. 69–86). Washington, DC: National Academy Press.

National Research Council (NRC) (2002). *Acute exposure guideline levels for selected airborne chemicals phosgene*: (pp. 15–70). (Vol. 2). Washington, DC: National Academy Press.

Padash, R., Nasrabadi, M. R., Rad, A. S., Jesionowski, T., & Ehrlich, H. (2019). A comparative computational investigation of phosgene adsorption on $(XY)_{12}$ (X = Al, B and Y = N, P) nanoclusters: DFT investigations. *Journal of Cluster Science, 30*, 203–218.

Pauluhn, J., & Li, W. L. (2015). In H. Salem, & S. A. Katz (Eds.), *Mechanisms involved in the inhalation toxicity of phosgene.* (3rd ed.). CRC Press. ISBN 978-1-4665-5273-9.

Sittig, M. (1985). *Handbook of toxic and hazardous chemicals and carcinogens* (2nd ed.). Park Ridge, NJ: Noyes Publications.
The Merck Index (1989). In S. Budavari (Ed.), *An encyclopedia of chemicals, drugs, and biologicals.* (11th ed.). Rahway, NJ: Merck and Co. Inc.
U.S. Department of Health and Human Services (1993). Registry of Toxic Effects of Chemical Substances *(RTECS, online database).* Bethesda, MD: National Toxicology Information, National Library of Medicine.
U.S. Environmental Protection Agency (1986). *Health assessment document for phosgene (external review draft).* EPA/600/8–86/022A Research Triangle Park, NC: Environmental Criteria and Assessment Office, Office of Health and Environmental Assessment, Office of Research and Development.
U.S. Environmental Protection Agency (1999). *Integrated risk information system (IRIS) on phosgene. National center for environmental assessment.* Washington, DC: Office of Research and Development.
Vaish, A. K., Consul, S., Agrawal, A., Chaudhary, S. C., Gutch, M., Jain, N. , et al. (2013). Accidental phosgene gas exposure: A review with background study of 10 cases. *Journal of Emergencies, Trauma, and Shock, 6*(4), 271–275.
Wang, S. L., Zhong, L., & Song, Q. H. (2017). A ratiometric fluorescent chemosensor for selective and visual detection of phosgene in solutions and in the gas phase. *Chemical Communication, 53*, 1530–1533.
Wyatt, J. P., & Allister, C. A. (1995). Occupational phosgene poisoning: A case report and review. *Journal of Accident and Emergency Medicine, 12*, 212–213.

CHAPTER

26

Phosphine: Risk assessment, environmental, and health hazard

Karuna Saini and R. D. Kaushik

Department of Chemistry, Gurukula Kangri (Deemed to be University), Haridwar, Uttarakhand, India

Abbreviations

ACGIH	American Conference of Governmental Industrial Hygienists
ALS	advanced life support
NOEL	nonobservable effect level
ATSDR	Agency for Toxic Substances and Disease Registry
CKMB	creatine kinase myocardial band
CNS	central nervous system
CPK	creatine phosphokinase test
CPR	cardiopulmonary resuscitation
DPR	detailed project report
ECG	electrocardiogram
ERPG	Emergency Response Planning Guideline
GGTP	gamma-glutamyl transpeptidase
GSH	glutathione
LAD	lactic acid dehydrogenase
LDH	lactate dehydrogenase
OECD	Organization for Economic Co-operation and Development
OEHHA	Office of Environmental Health Hazard Assessment
OSHA	Occupational Safety and Health Administration
PEL	permissible exposure limit
ppm	parts per million
SDS	safety data sheet
SGPT	serum glutamic pyruvic transaminase
STEL	short-term exposure limit
$T_{\mathbf{triple}}$	triple point temperature
USDA	United States Department of Agriculture
USEPA	United States Environmental Protection Agency

Hazardous Gases
https://doi.org/10.1016/B978-0-323-89857-7.00012-8

WH&S	Worker Health and Safety Branch
WHO	World Health Organization
Δ_{subH}	enthalpy of sublimation
Δ_{vapH}	enthalpy of vaporization

26.1 Introduction

Phosphine is considered to be an extremely toxic gas that was first discovered by Philippe Gengembre (Fluck, 1973). It belongs to pnictogen hydrides and appears as colorless, spontaneously flammable, and explosive gas at ambient temperature. It is usually formed after the breakdown of organic matter and used to fumigate agricultural products such as tobacco, grain, animal feed, and processed food for storage. In addition, it can also be used in the manufacturing of organophosphines, retardants, as a precursor in the semiconductor industry, and as a doping agent (Bettermann, Krause, Riess, & Hofmann, 2000). Organophosphine compounds are composed of phosphine which is represented by formula R_3P, where R is an organic derivative. It acts as a catalyst and forms complexes with various metal ions (magnesium, aluminum, and zinc) (Oktaviani, 1988).

Phosphine has reducing and nucleophilic properties, hence responsible for kidney and liver damage and anoxia in the brain. Due to this fact, phosphine is referred as highly toxic gas, which negatively affects the ecosystem. Phosphine with its toxic properties can cause serious symptoms even at low concentrations (0.3ppm). Phosphine has the toxicological significance, which somehow is responsible for wildlife poisoning (Oktaviani, 1988).

Phosphine has the trigonal pyramidal geometry (Forsyth, Rankin, & Robertson, 1990; Hansen & Bartell, 1965) (Fig. 26.1) and its chemical formula is PH_3. The geometrical structure of phosphine has one lone pair with a 93.5° bond angle between H-P-H, the bond length of P—H is 1.42Å, and C_{3v} molecular symmetry. The solid aluminum phosphide and magnesium phosphide can produce phosphine on reacting with

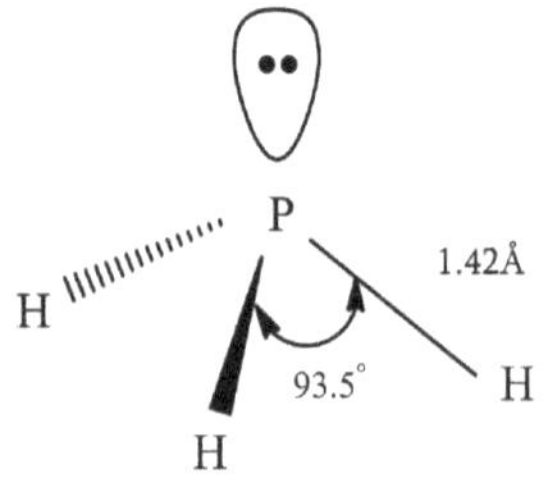

FIG. 26.1 Geometrical structure of phosphine.

water. Zinc phosphide is responsible for the liberation of phosphine gas, by reacting in the stomach and intestine, with water and hydrochloric acid (Dogan et al., 2014).

26.2 Occurrence

Phosphine is available in the atmosphere in very low concentrations with a high degree of variability in concentration (Pasek, Sampson, & Atlas, 2014). It can be found in human tissues, blood, urine, and saliva. The biochemical cycle of phosphorus plays a vital role in its formation. Decaying of organic matter through partial reduction and disproportionation reactions leads to the conversion of phosphate directly into phosphine (Sousa-Silva et al., 2020). Phosphine can also be considered as a biomarker of life due to its origination by anaerobic bacteria. These anaerobic bacteria live in oxygen-sparse environments, viz., sewers, landfills, or even animal guts (Greaves et al., 2020).

26.3 Historical background

In 1783, Philippe Gengembre (1764–1838) discovered phosphine by igniting white phosphorus in the aqueous solution of potassium carbonate. Phosphorus is found in a gaseous state due to its elemental association (Lavoisier, 1789) which, in turn, generates the phosphine gas by combining with hydrogen gas. It is also termed "phosphure d hydrogen" (viz., phosphide of hydrogen).

Diphosphine can also be used to generate phosphine which is produced by isolating the calcium phosphide (Ca_3P_2) using a wet trap (Varma, 2018). The experiment results indicate that PH_3 is generally inflammable. However, it may result in fire due to the presence of an impurity in the air.

26.4 Physicochemical properties

Pure phosphine is extremely noxious, reactive, flammable, and colorless gas. It has the "fishy" or "garlicky" odor and this is due to the presence of substituted phosphines and diphosphines (Oktaviani, 1988). It is pyrophoric at high concentration and generally has a dazzling flame during heating. The gas is sparingly soluble in water (approximately 2.5% v/v at around 20°C). Phosphine has better solubility in most of the organic solvents between 2.5% and 15% v/v. It behaves as highly toxic when present in concentrations up to 282 mg/m^3. Aluminum phosphide (AlP) or magnesium phosphide (Mg_3P_2) generates phosphine under suitable industrial procedures. It is a strong reducing agent and nucleophilic agent (Lam et al., 1991). PH_3 is stable and does not get autoxidized under standard temperature, pressure, and humidity conditions. At high air pressure and the presence of oxygen, autoxidation of phosphine may occur. Other properties are given in Table 26.1 (OEHHA, 2002).

TABLE 26.1 Physical properties of phosphine.

Molecular formula	**PH_3**
Chemical names	Phosphane, phosphorus hydride, phosphoretted hydrogen, phosphorus trihydride
Physical state	Gas
Appearance	Colorless gas
Molecular mass	33.99 g/mol
Critical temperature	324.76 K
Critical pressure	947.90 psia
Boiling point	185.37 K (OEHHA, 2002)
Melting point	139.37 K
Specific gravity	1.203
Density	Absolute 1.529 g/L (0°C) (OEHHA, 2002) Relative to air: 1.17 at 25°C (1 atm)
T_{triple}	39.41 K
Δ_{vap} H	14.6 ± 0.1 kJ/mol
Δ_{sub} H	17.2 kJ/mol
Vapor pressure	3460 kPa
Critical pressure	6540 kPa
Flammability (solid, gas)	1.2–98 vol%
Solubility in organic solvents	Soluble in alcohol, ether, and cyclohexanol (Nayak, Daglish, Phillips, & Ebert, 2020)
Henry's law constant	123.46 atm @ m^3/mol (Wilhelm, Battino, & Wilcock, 1977)
Air half-life	5 h (light); 28 h (dark)
Octanol-water partition coefficient (log K_{ow})	−0.27 (Pepelko et al., 2004)
Solubility in water	2.5 mL gas in 100 mL @ 20°C (3.5 mg/100 mL) (Pepelko et al., 2004)

26.5 Physiological role

Phosphine is a toxidrome that is commonly used as a rodenticide and insecticide. Various dose-dependent actions of phosphine have been described as follows (Cha et al., 2007):

(i) Conversion of ferric iron (Fe^{3+}) to ferrous iron (Fe^{2+}) by reduction,
(ii) Production of iron from horse ferritin,
(iii) On lipid peroxidation of iron release from ferritin.

Hemoglobin: Phosphine-induced hemoglobin denaturation, oxidation to methemoglobin, and formation of an abnormal pigmented form of hemoglobin "Verdichromogen" have also been described (Trimborn & Klimmer, 1962). Studies of purified hemoglobin (Potter, Rong, Griffith, White, & Garry, 1991) showed that with increasing duration of exposure, phosphine in concentrations as low as 0.11 μM steadily leads to the formation of hemichrome pigment. In intact red blood cells, the formation of Heinz bodies (hemoglobin protein aggregates) at PH_3 concentrations as low as 2 μg/mL has also been noticed. The toxicant effects in intact cells as well as in purified hemoglobin were eradicated by incubation in a reduced oxygen atmosphere, which shows an oxygen requirement due to phosphine hemoglobin toxicity (Potter et al., 1991).

Peroxidases, lipid peroxidation: Phosphine acts as a strong reducing agent, so it shows properties such as peroxidation, the formation of peroxides, and their reduction are of mechanistic and therapeutic concern. In animal studies, the inhibition of myeloperoxidase enzyme at concentrations of 8 mg/m^3 (Pazynich, Mazur, Podloznyĭ, Chinchevich, & Mandrichenko, 1984) has been reported. Further, oxidation of phosphine gives rise to reactive phosphorylating species (Lam et al., 1991). As such, effects on cholinesterase are also possible. Significant inhibition of cholinesterase was detected in animals. Studies of grain fumigant applicators (Potter et al., 1991) and in vitro studies in human RBCs explain that significant phosphine-induced inhibition of red cell cholinesterase occurs at concentrations more than 10 μg/mL.

26.6 Uses

Phosphine is considered to be the most significantly used pesticide. However, its usage had been discouraged since 2008. Phosphine is mainly used in commodity fumigation, space fumigation, spot fumigation, and burrowing pest control fumigation. Commodity fumigation includes fumigation of cabbage, barley, and almonds. The space fumigation involves fumigation of commercial building, warehouses, storage, animal husbandry area, and food processing area or handling. Spot fumigation is characterized as fumigation of a food market and processing units, and finally, the burrowing fumigation involves landscape fumigation, animal burrow entrance, and vertebrate pest control (Fonger, 1995).

Apart from fumigation, phosphine is also used in the semiconductor industry (Pepelko et al., 2004) and as a signal to the approaching boats. Its traces are used as a dopant and an originator for the deposition of compound semiconductors. Phosphine is used as a polymerization and condensation catalyst (Arthur, 1990) in the textile industry for the preparation of alkyl and aryl phosphonium halides, and for curing the epoxy resins and phase dopants during the manufacturing of organophosphines and phosphonium halides.

26.7 Sources of contaminants

Phosphine gas is produced from sewage plant landfills, compost, river sediments, and organic material (Dévai, Felföldy, Wittner, & Plósz, 1988; Glindemann, Stottmeister, & Bergmann, 1996) due to bacterial reduction. Generally, metal phosphides, like aluminum phosphide, magnesium phosphide, and zinc phosphide, are the main commercial sources of phosphine. Magnesium phosphides and aluminum phosphide are used to control insects in stored grains and other edible products as fumigants available as tablets, pellets, ropes, or strips, and sachets (Meister, 1999).

26.8 Risk assessment

Several agencies have formed various guidelines and issued "upgrade its risk assessment process as part of the tolerance setting procedures" and "Pesticides in the Diets of Infants and Children" (USEPA, 1997a, 1997b). In order to follow these guidelines, the following available information must be considered: (a) exposure from nonoccupational sources, (b) effect due to cumulative exposure by pesticide and other related substances having similar toxicity mechanisms, (c) the effect in utero exposure, and (d) potential for endocrine-disrupting effects.

Worker Health and Safety Branch (WH&S) formulated a DPR to understand the exposure effects of phosphine under various occupational and bystander scenarios. Metal phosphide tablets may release toxic phosphine gas in contact with moisture (either in fluids or air). Moreover, the inhalation of toxic gas and explosion may cause a serious hazard, which should be considered efficiently.

It has a very less probability that the residue of phosphine remains at the consumption time. USDA's Pesticide Data Program has not considered phosphine assay for the assessment of risk. This signifies the less probability of phosphine under residue detection. In occupational exposure, both occupational and residential bystander exposure have been considered. These exposures were considered under various fumigation operations in different spaces, viz., flat storage facilities, farm bins, grain elevators, shipping containers, warehouses, and rail cars (bulk and box cars). Burrowing pest fumigation and spot fumigation have also been analyzed. Various occupational categories have been assigned with different tasks which also include applicators, viz., spent fumigant retrievers and aerators.

Under the considered assumptions, it has been observed that toxicity by precursor compounds used may result in airborne phosphine exposure. The risk assessment is primary dependent on the available phosphine toxicity data. The assumption considered under this may reduce the chance of toxicity for the defined range.

26.9 Environmental effects

Air: Phosphine reacts with the by-product of a chemical reaction between ozone and water, i.e., hydroxide radical (HOx). The rate of reaction increases with the presence of nitro oxide (NOx) impurities. Under normal concentrations of hydroxide radicals, the half-life of phosphine is 28h. However, this value may decrease to 5h due to an increase in hydroxide radical concentration. The reaction results in the production of phosphorus oxyacids and inorganic phosphate. In a total number of 40days, however, the complete disappearance of phosphine can be observed.

Water and soil: Moisture content plays a crucial role in the reduction of phosphine percentage in soil. The depressed diffusion rate results in the phosphine reduction from the soil matrix. This process requires a total of 18days to completely remove the 1000ppm of phosphine from dry soil and 40days from moisture saturated soil. Further, the soil type plays a vital role in this process. Based on the report, it has been identified that phosphine at normal atmospheric pressure conditions possesses 0.27 (v/v at 17°C) solubility in water (Rubin, 2014).

Wildlife and food crops: The animals fed by phosphine-fumigated food have not been found to exhibit serious effects. WHO report concluded that "it is unlikely that the use of phosphine or phosphides has toxicological significance". Howbeit, the accidental poisoning of wildlife has been known to occur (Oktaviani, 1988).

26.10 Hazardousness

Lungs can be easily affected by phosphine absorbance. It has been found in the literature that the liver and brain are least influenced by the phosphine distribution. Further, the skin is found to be insensitive to metal phosphides, which may result in phosphine generation. Metal phosphide hydrolysis is not responsible for phosphine absorbance, which leads to the presence of traces of aluminum phosphide in blood and liver followed by human ingestion. Phosphine can readily react with heme and copper-containing proteins' components present in the human body, which may lead to an increase in phosphine distribution.

(a) **Acute toxicity**: In human, various reports confirmed that phosphine is lethal and its consumption may cause anorexia, nausea, cough, dyspnea, inappetence and dry mouth, hypotension, cardiac arrhythmia, vomiting, acidosis, epigastric distress, chest tightness, headache, lethargy, irritability, jaundice, pulmonary crepitation, giddiness, numbness/paraesthesia, altered sensoria, hypotension, and cardiovascular collapse. Critical nonobserved effect level (NOEL) of 5 ppm was adopted to test the acute toxicity of phosphine exposure (Schaefer, Newton, Gruebbel, Busey, & Shaheen, 1998). The NOEL signifies the highest dose or exposure level of a substance or material that produces no noticeable toxic effect on tested animals.

(b) **Subchronic toxicity:** This includes to test the potential for subchronic toxicity under phosphine exposure. A critical NOEL of 1 ppm has been set under this procedure.

(c) **Chronic toxicity:** The NOEL under this test is highest among others, viz., 3 ppm, this is considered as the highest dose used in phosphine-related exposure studies (Schaefer et al., 1998).

(d) **Genotoxicity:** In humans, the clastogenic role is found to be consistent with epidemiologic studies. It has been reported in studies that results were inconsistent on laboratory animals; however, evidence for micronucleus induction in mouse splenic lymphocytes was observed on exposure to a period of 13 weeks.

26.11 Treatment

Phosphine comes under the category of multisystem toxicants which may result in pulmonary irritation, cardiovascular collapse, and CNS depression. Till now, no antidote is available for phosphine poisoning. The treatments are usually provided in support of respiratory and cardiovascular functions. Phosphine toxicity is generally caused by inhalation and due to the higher specific weight of phosphine than air, it may cause asphyxiation in low-lying areas and poorly ventilated areas.

The toxicity of phosphine can easily be identified based on various symptoms, viz., tremors, vertigo, diplopia, ataxia, cough, restlessness, irritability, drowsiness, dyspnea, retrosternal discomfort, abdominal pain, and vomiting. Phosphine also interacts with enzymes and protein synthesis, mainly present in heart, mitochondria and cells of lungs. This may cause various related issues, viz., hypotension, reduction in cardiac output, tachycardia, oliguria, anuria, cyanosis, pulmonary edema, tachypnea, jaundice, hepatosplenomegaly, ileus, seizures, and diminished reflexes.

Phosphine causes various serious effects on the human body. During the respiratory problem, secure airway and respiration should be preferred along with endotracheal intubation. In the case of the unsuitability of the procedure, cricothyroidotomy should be performed with care.

The patients having bronchospasm should be treated with aerosolized bronchodilators. Further, additional risks may be caused by multiple chemical exposures to bronchial sensitizing agents in these situations. Cardiac sensitizing agents are good; however, it should be avoided to have exposure to several chemicals which if not avoided may increase the risk of cardiac arrhythmias. Advanced life support (ALS) protocols should be preferred to treat the patients having comatose, hypotensive, or having seizures or cardiac arrhythmias (ATSDR, 2014).

The children who usually develop stridor can be treated with racemic epinephrine aerosol. A dose of 0.25–0.75 mL of 2.25% racemic epinephrine solution in 2.5 cm^3 water should be repeated every 20 min as per requirements. For shock or hypotension, fluid administration should be used.

26.12 Risk management

In order to avoid the risk associated with phosphine toxicity, various levels of toxicity are categorized as follows:

(a) NIOSH Recommended Exposure Level (REL): TWA 0.3 ppm (0.4 mg/m^3)
(b) NIOSH Short Term Exposure Level (STEL): 1 ppm (1.4 mg/m^3)
(c) NIOSH (Immediately Dangerous to Health or Life): 50 ppm (ATSDR, 2014)
(d) OSHA Permissible Exposure Limit (PEL): TWA 0.3 ppm (0.4 mg/m^3)
(e) ACGIH Threshold Limit Value (TLV): TWA 0.3 ppm (0.42 mg/m^3)
(f) ACGIH STEL: 1 ppm (1.4 mg/m^3)
(g) OEHHA chronic reference exposure level: 0.0006 ppm (0.0008 mg/m^3)
(h) ERPG-2 (Emergency Response Planning Guideline): 0.5 ppm

26.13 Current research

The current work related to PH_3 is mainly focused on its biological chemistry and metal complexes. More research is inclined based on its widespread application as a fumigant. Further, the uses of metal phosphine complexes are also the main areas on which current research is focused. Applications such as heart-imaging agents and anticancer drugs are the most common applications of metal complexes. Researches are more focused on diphosphine and related complexes such as gold, silver, and copper responsible for cytotoxicity and anticancer activity (Berners-Price & Sadler, 1988).

Organization for Economic Co-operation and Development (OECD) in vitro toxicology protocol with human donor skin had reported two common and odorous chemicals, hydrogen sulfide and phosphine. Results show that undamaged human skin provides a good barrier to hydrogen sulfide (up to 800 ppm) and phosphine (up to 1000 ppm) penetration for up to 30 min exposures, with little variability in the presence of clothing or elevated temperature and humid conditions. A practical guideline template for skin decontamination has been developed, and implications of the research for first responders have been outlined (OECD, 2015).

The pests present at storage commodities are a major threat throughout the world causing severe economic losses to a variety of food products on a large scale. Phosphine is becoming a major insecticide and fumigant as a grain's protectants against *Tribolium castaneum* and *Rhyzopertha dominica* (Opit, Phillips, Aikins, & Hasan, 2012).

26.14 Phosphine toxicity: Case study

Various cases of phosphine toxicity are available in the literature. The phosphine poisoning in 29 subjects was observed due to grain freighter, yield evidences of occult blood (urinary tract injury), bilirubinuria, and increased SGPT, GGPT, and LDH (liver damage). The effect of acute poisoning (inhalation) in terms of delayed pulmonary edema, nasal mucosa necrosis, and the myocardial injury was also observed (Dan et al., 2007).

O'Malley reported two deaths and three critical illnesses due to phosphine poisoning caused by agricultural products (O'Malley et al., 2013).

In another case study of phosphine lethality, two siblings were found to be dead within 18 h due to 1-h exposure to fumigated wheat (Arthur, 1990). It was found in the postmortem report that death was caused by phosphine fumigation. The lungs of the siblings were found to be congested. Ethanol formation was also detected in the blood and tissues of both siblings that also indicated phosphine poisoning.

26.15 Safety measures

Always follow the manufacturer's guidelines:

(a) A safety data sheet (SDS) should be obtained from a chemical supplier which has the guidelines regarding symptoms and treatment of poisoning, along with the chemical data.
(b) Protective clothing as per SDS and label should be used along with a correct respirator system. Impervious gloves should be necessarily used during phosphine handing (Works, 1998).
(c) An observer should always be there in the premises with access to respiratory protection.
(d) Sign the areas for which fumigation is directed as per the Management Regulation.
(e) Use signs for fumigation information (Grains Research & Development Corporation, 2013).
(f) Phosphine should not be used for the transit under the grain stage (Grains Research & Development Corporation, 2013).
(g) A hand pump and gas detector tube should be used to monitor the atmosphere in the vicinity of fumigation. The concentration results should be less than 0.3 ppm (Oktaviani, 1988).
(h) Proper precautions should be used before the movement of grain just after the fumigation process. It is advised to ensure the proper dispersion of gas fumes (Grains Research & Development Corporation, 2013).
(i) The opening of phosphine containers should be preferred in the open space. The opening should be avoided under shedding or silo.
(j) Containers of phosphine should be stored properly.
(k) Used phosphine tablets should be correctly disposed.
(l) Partially filled containers should not be disposed of by the burial process.
(m) A prescribed dose should be used to kill the insects. Overdose should be prohibited.
(n) Keep phosphine away from the reach of children.

26.16 Handling and storage

(a) Store in a cool, dry, well-ventilated location. Separate from oxidizing materials.

(b) Store separately from acids, alkalies, and halogenated compounds.
(c) Metal phosphide should be kept in a well-ventilated, dry warehouse, and tightly packed into polyethylene-lined containers (Works, 1998).
(d) Formulated commercial metal phosphide preparation should be stored dry and out of reach of children in a locked cupboard.

26.17 Precautions

(a) A full-face covered respiratory system along with the breathing apparatus should be used under the concentrations which are more than 0.3 ppm (Oktaviani, 1988).
(b) The used filter must be validated for phosphine.
(c) The used filter should be destroyed immediately.
(d) Under exposure at higher concentrations, self-contained breathing apparatus should be preferred for performing operations, viz., breakdowns.
(e) Facial hair should be removed as it reduces the sealing of the mask at the skin.
(f) The filter should be used for the prescribed life hours.

References

Arthur, D. (1990). *Phosphine. In National toxicology program executive* Available from https://ntp.niehs.nih.gov/ntp/htdocs/chem_background/exsumpdf/phosphineadlittle_508.pdf.

ATSDR (2014). *Phosphine.* Available from http://www.atsdr.cdc.gov/mmg/mmg.asp?id=1013&tid=214.

Berners-Price, S. J., & Sadler, P. J. (1988). *Phosphines and metal phosphine complexes: Relationship of chemistry to anticancer and other biological activity*: (pp. 27–102). https://doi.org/10.1007/3-540-50130-4_2.

Bettermann, G., Krause, W., Riess, G., & Hofmann, T. (2000). Phosphorus compounds, inorganic. In *Ullmann's encyclopedia of industrial chemistry*. Wiley-VCH Verlag GmbH & Co. KGaA. https://doi.org/10.1002/14356007.a19_527.

Cha, U., Valmas, N., Collins, P. J., Reilly, P. E. B., Hammock, B. D., & Ebert, P. R. (2007). Disruption of Iron homeostasis increases phosphine toxicity in Caenorhabditis elegans. *Toxicological Sciences*, *96*(1), 194–201. https://doi.org/10.1093/toxsci/kfl187.

Dan, P., Saing, D., Lara-subiabre, B. A., Chacón, M., Chacón, A., Díaz Larenas, C. , et al. (2007). *Acute exposure guideline levels for selected airborne chemicals. Estudios pedagógicos (Valdivia)*. Vol. 6. Washington: The National Academies Press Available from unige.ch/fapse/SSE/.../2001_36.html%0Ahttp://www.oei.es/metas2021/APRENDYDESARRPROFESIONAL.pdf%0Ahttp://www.rieoei.org/deloslectores/475Caceres.pdf%0Ahttp://www.ub.edu/obipd/la-profesion-docente-ante-los-desafios-del-presente-y-futuro/%0Ahttp://www.redalyc.

Dévai, I., Felföldy, L., Wittner, I., & Plósz, S. (1988). Detection of phosphine: New aspects of the phosphorus cycle in the hydrosphere. *Nature*, *333*(6171), 343–345. https://doi.org/10.1038/333343a0.

Dogan, E., Güzel, A., Çiftçi, T., Aycan, I., Çelik, F., Çetin, B. , et al. (2014). Case report zinc phosphide poisoning. *Case Reports in Critical Care*, *2014*, 1–3.

Fluck, E. (1973). The chemistry of phosphine. In *Inorganic chemistry* (pp. 1–64). Berlin, Heidelberg: Springer.

Fonger, G. C. (1995). Hazardous substances data bank (HSDB) as a source of environmental fate information on chemicals. *Toxicology*, *103*(2), 137–145. https://doi.org/10.1016/0300-483X(95)03145-6.

Forsyth, G. A., Rankin, D. W. H., & Robertson, H. E. (1990). Determination of the molecular structure of tris(trimethylsilyl)phosphine in the gas phase by electron diffraction, supported by molecular mechanics calculations. *Journal of Molecular Structure*, *239*, 209–217. https://doi.org/10.1016/0022-2860(90)80215-6.

Glindemann, D., Stottmeister, U., & Bergmann, A. (1996). Free phosphine from the anaerobic biosphere. *Environmental Science and Pollution Research*, *3*(1), 17–19. https://doi.org/10.1007/BF02986806.

Grains Research & Development Corporation (2013). *Fumigating with phosphine, other fumigants and controlled atmospheres.* Available from https://grdc.com.au/__data/assets/pdf_file/0025/206791/fumigating-with-phosphine.pdf.pdf.

Greaves, J. S., Richards, A. M. S., Bains, W., Rimmer, P. B., Sagawa, H., Clements, D. L. , et al. (2020). Phosphine gas in the cloud decks of Venus. *Nature Astronomy*, 1–20. https://doi.org/10.1038/s41550-020-1174-4.

Hansen, K. W., & Bartell, L. S. (1965). Electron diffraction study of the structure of PF51a. *Inorganic Chemistry*, *4*(12), 1775–1776. https://doi.org/10.1021/ic50034a023.

Lam, K. S., Salmon, S. E., Hersh, E. M., Hruby, V. J., Kazmierski, W. M., & Knapp, R. J. (1991). A new type of synthetic peptide library for identifying ligand-binding activity. *Nature*, *354*, 82–84.

Lavoisier, A. (1789). *The french revolution and Lavoisier's execution. In Encyclopædia Britannica* Available from https://www.britannica.com/biography/Antoine-Lavoisier.

Meister, R. T. (1999). *Farm chemicals handbook '99.* OH: Willoughby.

Nayak, M. K., Daglish, G. J., Phillips, T. W., & Ebert, P. R. (2020). Resistance to the fumigant phosphine and its management in insect pests of stored products: A global perspective. *Annual Review of Entomology*, *65*, 333–350. https://doi.org/10.1146/annurev-ento-011019-025047.

O'Malley, M., Fong, H., Sánchez, M., Roisman, R., Nonato, Y., & Mehler, L. (2013). Inhalation of phosphine gas following a fire associated with fumigation of processed pistachio nuts. *Journal of Agromedicine*, *18*, 151–173. https://doi.org/10.1080/1059924X.2013.768135.

OECD. (2015). Genetic Toxicology Guidance Document. Available from: https://www.oecd.org/chemicalsafety/testing/Genetic%20Toxicology%20Guidance%20Document%20Aug%2031%202015.pdf.

OEHHA (2002). *Phosphine.* Available from https://oehha.ca.gov/media/downloads/air/document/7803512.pdf.

Oktaviani, J. (1988). Environmental health criteria 73. *United nations environment programme, international labour organisation, and the world health organization world.* Vol. 51. Geneva: World Health Organization.

Opit, G., Phillips, T., Aikins, M., & Hasan, M. M. (2012). Phosphine resistance in *Tribolium castaneum* and *Rhyzopertha dominica* from stored wheat in Oklahoma. *Journal of Economic Entomology*, 105.

Pasek, M. A., Sampson, J. M., & Atlas, Z. (2014). Redox chemistry in the phosphorus biogeochemical cycle. *Proceedings of the National Academy of Sciences of the United States of America*, *111*(43), 15468–15473. https://doi.org/10.1073/pnas.1408134111.

Pazynich, V. M., Mazur, I. A., Podloznyĭ, A. V., Chinchevich, V. I., & Mandrichenko, B. E. (1984). Experimental substantiation and prediction of time-related maximum permissible concentration of phosphine in the air. *Gigiena i Sanitariia, 1*, 13–15.

Pepelko, B., Seckar, J., Harp, P. R., Kim, J. H., Gray, D., & Anderson, E. L. (2004). Worker exposure standard for phosphine gas. *Risk Analysis, 24*(5), 1201–1213. https://doi.org/10.1111/j.0272-4332.2004.00519.x.

Potter, W. T., Rong, S., Griffith, J., White, J., & Garry, V. F. (1991). Phosphine-mediated Heinz body formation and hemoglobin oxidation in human erythrocytes. *Toxicology Letters, 57*(1), 37–45. https://doi.org/10.1016/0378-4274(91)90117-O.

Rubin, A. L. (2014). *Phosphine risk characterization document.* Available from http://www.cdpr.ca.gov/docs/risk/rcd/phosphine.pdf.

Schaefer, G. J., Newton, P. E., Gruebbel, M. M., Busey, W. M., & Shaheen, D. G. (1998). Acute and subchronic inhalation neurotoxicity of phosphine in the rat. *Inhalation Toxicology, 10*(4), 293–320. https://doi.org/10.1080/089583798197691.

Sousa-Silva, C., Seager, S., Ranjan, S., Petkowski, J. J., Zhan, Z., Hu, R. , et al. (2020). Phosphine as a biosignature gas in Exoplanet Atmospheres. *Astrobiology, 20*(2), 235–268. https://doi.org/10.1089/ast.2018.1954.

Trimborn, H., & Klimmer, O. R. (1962). Experimental investigations of the chemical hemoglobin changes caused by hydrogen phosphide in vitro. *Archives Internationales de Pharmacodynamie et de Thérapie, 137*, 331–347.

USEPA (1997a). *The federal insecticide, fungicide and rodenticide act (FIFRA) and federal food, drug and cosmetic act (FFDCA) as amended by the food quality protection act (FQPA) of August 3, 1996. Doc. #730L97001, March 1997.* Office of Pesticide Programs, United States Environmental Protection Agency.

USEPA. (1997b). Raw and processed food schedule for pesticide tolerance reassessment. Federal Register 62(149): 42020–42030. United States Environmental Protection Agency.

Varma, P. (2018). *Phosphine.* Available from https://alchetron.com/Phosphine.

Wilhelm, E., Battino, R., & Wilcock, R. J. (1977). Low-pressure solubility of gases in liquid water. *Chemical Reviews, 77*(2), 219–262. https://doi.org/10.1021/cr60306a003.

Works, H. (1998). *Pesticide fact sheet.* Available from http://www.sierraclub.ca/national/programs/health-environment/pesticides/imidacloprid-fact-sheet.shtml.

CHAPTER

27

Phosphorus pentafluoride: Risk assessment, environmental, and health hazard

Sanjeev Maken[a] *and Karuna Saini*[b]

[a]Department of Chemistry, Deenbandhu Chhotu Ram University of Science and Technology, Murthal, Haryana, India

[b]Department of Chemistry, Gurukula Kangri (Deemed to be University), Haridwar, Uttarakhand, India

Abbreviations

CWA chemical warfare agent
IMS ion mobility spectrometer
MSHA Mine Safety and Health Administration
NIOSH National Institute for Occupational Safety and Health
OSHA Occupational Safety and Health Administration
SEI solid electrolyte interphase
TIC toxic industrial chemical

27.1 Introduction

Phosphorus pentafluoride (PF_5) was discovered in 1876 by Thorpe, who prepared it by reacting AsF_3 and PCl_5. It comes under the category of halophosphine. The chemical equation for the phosphorus pentafluoride is given in the following equation (Thorpe, 1876):

$$5AsF_3 + 3PCl_5 \rightarrow 3PF_5 + 5AsCl_3 \qquad (27.1)$$

It is a colorless, toxic, flame extinguisher, pungent smelling, and incombustible gas (Thorpe, 1876). A single-crystal X-ray of phosphorus pentafluoride illustrates a trigonal bipyramidal geometry in which the

Hazardous Gases
https://doi.org/10.1016/B978-0-323-89857-7.00022-0

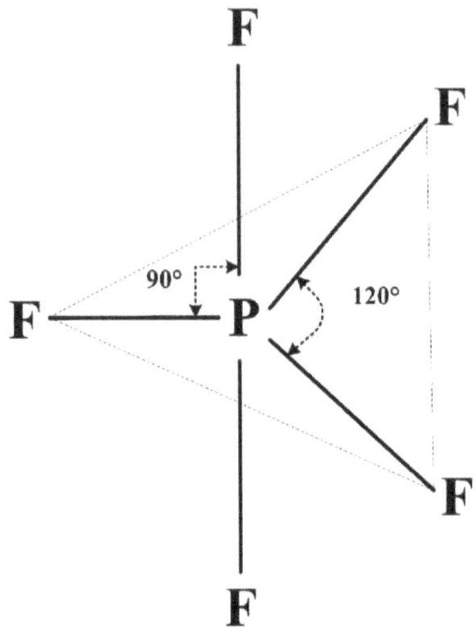

FIG. 27.1 Trigonal bipyramidal structure of phosphorus pentafluoride.

P—F bond is equal by length and the equatorial bond is shorter than the axial P—F bond as shown in Fig. 27.1 (Hansen & Bartell, 1965; Mootz & Wiebcke, 1987). The two different type of bonds between P and F are due to the presence of pseudo berry rotation. Phosphorus pentafluoride acts as Lewis acid that has a property to hydrolyze readily. This gas creates extreme irritation to the skin, mucous membranes, eyes, and even cause pneumonitis, and pulmonary edema, or severe burns.

It produces corrosive fumes on reacting with water. It is extremely harmful to inhale as it binds with iron in blood hemoglobin further causing toxicity. Long-term vulnerability to heat may cause explosive rupturing and rocketing. It is commonly used as a polymerization catalyst, as a source of ion implantation for semiconductors, and fluorinating agent in organosilicon chemistry (Kifer & Van Dyke, 1972). Hydrofluoric acid and phosphoric acid are formed with its rapid decomposition.

27.2 Occurrence

The major possible source is the partial reduction and disproportionation of phosphate available in decaying organic matter. Synthesis of phosphorus pentafluoride is usually performed under different methods to obtain the phosphorus pentafluoride (Aroutiounian, 2018).

1. Phosphorus trifluoride reacts with bromine followed by heating, as shown in Eqs. (27.2), (27.3):

$$PF_3 + Br_2 \rightarrow PF_3Br_2 \quad (27.2)$$

$$5PF_3Br_2 \rightarrow 3PF_5 + 2PBr_5 \quad (27.3)$$

2. Phosphorus pentachloride reacts with arsenic trifluoride on heating (Eq. 27.4):

$$PCl_5 + 5AsF_3 \rightarrow 3PF_5 + 5AsCl_3 \quad (27.4)$$

3. Phosphorus trifluoride treated to an electric spark in the absence of air (Eq. 27.5):

$$5PF_3 \rightarrow 3PF_5 + 2P \quad (27.5)$$

4. Phosphorus pentaoxide reacts with calcium fluoride on heating (Eq. 27.6):

$$P_2O_5 + 5CaF_2 \rightarrow 2PF_5 + 5CaO \quad (27.6)$$

5. Phosphorus oxyfluoride reacts with hydrogen fluoride and sulfur trioxide on heating (Eq. 27.7):

$$POF_2 + 2HF + SO_3 \rightarrow PF_5 + H_2SO_4 \quad (27.7)$$

27.3 Historical background

Phosphorus pentafluoride (PCl_5) was first prepared in 1876 by Thorpe. It is a strong nonoxidizing acid that is known as a halogenating agent. It reacts with water and air and further breakdown to form hydrofluoric acid (HF) and phosphoric acid (H_3PO_4).

27.4 Physicochemical properties

Phosphorus pentafluoride (PF_5) is colorless and odorless under the normal boiling point, i.e., 84.5°C, and fusing point 93.78°C. As a kind of fluorinating agent, PF_5 can carry out ion-transfer and can be used widely in fields such as battery manufacturing, electronic industry, catalyzer, and macromolecular material. Lewis acid property of the phosphorus pentafluoride

TABLE 27.1 Physical properties of phosphorus pentafluoride.

Properties	Value
IUPAC name	Phosphorus pentafluoride
Chemical formula	PF_5
Density	5.526 kg/m^3 (g/l)
Odor	Unpleasant
Appearance	Colorless gas
Molecular shape	Trigonal bi-pyramidal
Dipole moment	0 D
Molar mass	125.966 g/mol
Boiling point	−84.50°C (−120.30°F)
Melting point	−93.79°C (−136.80°F)
Solubility in water	Hydrolysis
Specific heat	20.26 Btu/lbmol°F
Specific gravity	4.31
Specific volume	3.10 ft^3/lb
Critical temperature	66.1 °F
Gas density	0.323 lb./ft^3
Critical pressure	492.3 psia

makes it relevant to readily hydrolyzed. It is about $4\frac{1}{2}$ times heavier than air; it may be collected by downward displacement and may be poured from vessel to vessel. Some of the physical properties of phosphorus pentafluoride are given in Table 27.1 (Synquest Laboratories, 2012).

27.5 Physiological role

In lungs, there are capillaries, the blood vessels in the walls of alveoli (small sacs) transporting red blood cells. In alveoli, red blood cells release CO_2 and absorb oxygen and then circulate oxygen to the entire body. Inhalation of phosphorus pentafluoride damages liver capillaries and also lung capillaries become enlarged or dilated. Once these capillaries dilate, red blood cells are unable to absorb the required oxygen owing to decrease in size of alveoli. This results in low oxygen level and patient feels breathlessness or shortness of breath known as dyspnea.

27.6 Uses

Phosphorus pentafluoride has a variety of applications and it is most commonly used as the source of phosphorous in the semiconductor industry and as a catalyst in the ionic polymerization (Aroutiounian, 2018). In the case of the organosilicon industry, phosphorus pentafluoride acts as a fluorinating agent. This can also be employed in different materials used for drying having the nonreacting property and for moisture absorption in a vacuum system with different roles such as dehydrating agent, a condensing agent in organic synthesis and catalyst in asphalt air-blowing and sugar industry. Thermal degradation of solid electrolyte interphase (SEI) can be done by attacking the phosphorus pentafluoride (Jongjung et al., 2016). Phosphorus pentafluoride is also useful for the synthesis of phosphorus oxychloride, metaphosphoric acid, phosphate esters, and ammonium polyphosphate. The oximes of phosphorus pentafluoride can also be used as herbicides in its complex form.

27.7 Sources of contaminants

1. Industrial processes are the sources that release phosphorus pentafluoride and use this gas as an intermediate in the production of polycarbonates, phenyl ureas carbamates, isocyanates thiocarbamates.
2. Lithium-ion batteries generate a large amount of heat and a significant amounts of phosphorus pentafluoride gas and smoke (Larsson, Andersson, Blomqvist, & Mellander, 2017).

27.8 Risk assessment

The toxic industrial phosphorus pentafluoride and other chemicals may get induced in human milk which is a major occupational risk. The women that are in their reproductive age (13–55 years) working at workplaces handling PF_5 are more vulnerable than men. The effect of the exposure may hinder the health of the newborn, which is mainly due to breastfeeding (Wolff, 1983).

However, during breastfeeding, the risk is also taken to the home and the infant is passively infected. Irrespective of acute and chronic hazards, phosphorus pentafluoride may cause other hazards such as the storage gas cylinder may explode on heating. Hence, it is necessary to conduct risk assessment.

In the workplace, various engineering controls and practices are the most effective way to reduce the risk associated with phosphorus pentafluoride. Other than that, the effective way of reducing the risk at the workplace is to provide local exhaust ventilation at the site. Isolation of operations may also reduce the risk of exposure. Further, the use of protective equipment and respirators can be useful to control the risk related to exposure.

27.9 Environmental effects

a. Exposure to a high level of phosphorus pentafluoride may harm and even kill wildlife. However, these effects are seen only when large amounts of phosphorus pentafluoride are evolved.
b. The use of lithium-ion batteries evolves heat and intense amount of phosphorus pentafluoride gas and smoke which may result in atmosphere toxicity (Larsson et al., 2017).
c. Phosphorus pentafluoride fumes on reacting with the moisture present in the air form the traces of POF_3 and HF. The byproduct of the reaction is corrosive in nature.

27.10 Hazardousness

Phosphorus pentafluoride hazardous effects are broadly classified into two different categories, viz., acute health effects and chronic health effects.

i. Acute health effects

Acute (short-term) health effects may be observed immediately or shortly just after the phosphorous pentafluoride exposure. Under acute health effects, severe damage to the skin can be observed in terms of irritation and burn. The damage to eye can also happen. The inhalation of gas may lead to irritation in the nose and throat along with lung irritation which may result in coughing and breathing shortness. Higher exposure of phosphorous pentafluoride may build-up fluid in the lungs (pulmonary edema) and displace the oxygen which may lead to suffocation.

ii. Chronic health effects

Various chronic hazards may be caused by phosphorous pentafluoride exposure. These are as follows:

a. Cancer hazard

Phosphorus pentafluoride has not yet been examined in terms of cancer-causing agent in animal.

TABLE 27.2 Symptoms of phosphorus pentafluoride.

Effects on potential health	Symptoms of short-term exposure	Symptoms of long-term exposure
By inhalation	• Dizziness • Nausea • Difficulty in breathing • Irritation • Headache • Bluish skin color • Kidney damage • Lung congestion	• Tooth decay • Digestive disorders
By skin contact	• Burns • Nausea	• Burns • Nausea
By eye contact	• Burns • Tearing • Blindness	• Burns • Tearing • Blindness

b. Reproductive hazard

Phosphorus pentafluoride has not yet been evaluated in terms of its effects on the reproduction system.

The exposure of phosphorus pentafluoride may also damage the liver. The strong fume of phosphorus pentafluoride in air severely affects the eyes, liver, lungs, skin, and irritates the nose and throat. However, no exposure limits (occupational) have been established for phosphorus pentafluoride. The risk and various symptoms of phosphorus pentafluoride are given in Table 27.2.

27.11 Treatment

Under overexposure conditions, chest X-ray and liver function tests should be considered . Medical testing does not control the exposure; these are mainly done to examine the damage caused by exposure. The mixed exposure damages by phosphorus pentafluoride are caused due to alcohol consumption which may lead to liver damage.

27.12 Risk management

In order to avoid the risk caused due to the poisonous effect of the phosphorus pentafluoride, the management of the instigated risk is a necessary step that needs to be understood. In this way, the sensitive sensors used for the toxicity detection can be used to detect even low concentration of PF_5. The toxic and chemical warfare agents (CWA) can be detected using the sensor devices and semiconductor gas sensor technique (Aroutiounian, 2018).

There are various techniques that can be used to detect the toxic concentrations of phosphorus pentafluoride. Raman spectroscopy, Fourier transform infrared spectrophotometry, ion mobility spectrometer (IMS), photoionization, colorimetric indicators, and mass spectrometers combined with gas chromatography are among the most used and latest technologies for the detection of CWAs and toxic industrial chemicals (TICs). The advantages of the aforementioned techniques have been summarized in Table 27.3. Ion mobility spectrometer (IMS), semiconductors, and SAW-type techniques are the preferred methodologies for the detection of gas toxicity.

Some other measures include:

a. *Fire fighting measures for fire and explosion hazards*

Availability of extinguishing media containing carbon dioxide, which reduces the risk of explosion.

b. *Exposure controls, personal protection exposure limits*

i. Ventilation

A local exhaust or process enclosure ventilation system should be used. Under explosive concentrations, explosion-resistant ventilation equipment should be preferred.

ii. Personal protective equipment

Personal protective equipment is found to be suitable for various specific site conditions such as open space work, confined workplace,

TABLE 27.3 Various detection techniques along with their advantages.

Techniques	Advantages
Photoionization	Rapid response, sensitive to organics
Paper tape	Very specific to certain
Fourier transform FT-IR	More sensitive than standard IR, rapid response
Gas chromatography	Applicable to a wide variety of gases, good sensitivity
Ion mobility spectrometer (IMS)	Very good sensitivity, very specific to certain
Semiconductor	Low cost, fast response
SAW type	Portability, high sensitivity, low cost

and very few instances of site work. OSHA 1910.132 standard defines the requirement to determine the suitable personal protective equipment for each category of hazard and to formulate the instruction for the employees about the guideline for their use.

Other ways to reduce the risk are given below:

Skin clothing protection: Avoid skin contact with phosphorus pentafluoride. Always wear protective gloves and clothing which should be cleaned after each usage.

Eye protection: To reduce the impact of phosphorus pentafluoride, properly design nonvented, and impact-resistant goggles should be used in the fume and vapor environment of gas. A face shield protect the eyes from the corrosive and toxic environment. A contact lens should be avoided.

Respiratory protection: Proper guidelines should be followed to use respiratory protection. The training should be done by explaining each and every aspect of use. Medical tests as explained in OSHA 1910.134 should be followed. MSHA/NIOSH-approved air respirator system should be preferred.

27.13 Current research

Nowadays, most of the research on phosphorus pentafluoride is focused to enhance its application for its safe use and under other similar areas to increase the life of human beings. In this way, for storage battery applications, it is observed that the thermal gradient of phosphorus pentafluoride under elevated temperature (60–130°C) is lower than the other similar products. It is due to the fact that Solid Electronic Interface (SEI) decomposes at temperature above 90°C. The graphite electrode is usually degraded by the attack of phosphorus pentafluoride, which leads to the generation of lithium hexafluorophosphate ($LiPF_6$). $LiPF_6$ is thermally unstable, which makes the decomposition product (PF_5) highly reactive Lewis acid. Based on the results, it was concluded that careful X-ray photoelectron spectroscopy is essential to understand the change in chemical compositions and readiness of thickness of the SEI layer under overexposure conditions of phosphorus pentafluoride. Further, the damage caused due to the impact of gas on SEI needs to be investigated in future studies.

27.14 Phosphorous pentafluoride: Case study

Ninothypsthys, suffered the tragic devastation caused by this gas which previously, was a resourceful and fertile land. Ninothypsthys atmosphere got completely filled with phosphorus pentafluoride, nitrogen dioxide, methane, and phosgene. It has been found in various government studies that exposure in such an environment may even lead to death. Even after 15 years, the vulnerability of the incident may cause the degeneration with irreversible internal decay. Smoking sludge and slag rolls along with sluggish tar are some of the symptomatic harms caused by the incident (Saftner, 2014).

27.15 Safety measures

a. Skin contact should be avoided with phosphorus pentafluoride by wearing safety gloves and clothing, which should be cleaned daily (Synquest Laboratories, 2012).
b. Eyes should be properly protected using nonvented goggles (impact-resistant) or face shields.
c. Contact lenses should not be worn during working with PF_5.

Various first aid measures against phosphorus pentafluoride poisoning are given in Table 27.4.

TABLE 27.4 First aid measure against phosphorus pentafluoride poisoning.

First-aid measures in general	a. In the case of unwell feeling, concern for immediate medical advice (National Center for Biotechnology Information, 2020) b. Always keep the distance from the affected person under the contaminated area
First-aid measures after inhalation	a. Move the person to breathe pure oxygen, if this is not possible, give artificial respiration b. Get immediate medical services
First-aid measures after skin contact	a. Lukewarm water should be used to wash the affected parts b. Do not touch or rub the affected area c. Keep out contaminated clothing and shoes
First-aid measures after eye contact	a. Contact lenses should be removed b. Flush eyes thoroughly with water for few minutes
First-aid measures after ingestion	a. The affected person should not be given anything via the mouth b. Rinse the mouth with water c. The immediate medical service should be given

27.16 Handling and storage

a. Store in a well-ventilated, cool, and dry place and away from products that oxidize (National Center for Biotechnology Information, 2020).
b. The container should be tightly closed.
c. It should be kept away from water or steam gases, acids, alkalies, ammonia, amines, and halogenated compounds.
d. Only a trained person should be allowed to enter the premises containing the gas.
e. Piercing or burning should be avoided. The valve should be closed after each use or when empty.
f. All safety precautions should be clearly understood before use.
g. Protective equipment should be used to reduce the risk via breathing fumes, gas, mist, spray, and vapors.
h. Contact with skin and eyes should be avoided.
i. The gas receptacle should be handled safely.
j. Good industrial hygiene and safety procedures should be followed.
k. Hands should be washed after each handling of the product.
l. Storage criterion: Avoid sunlight. Temperatures should not exceed 50 °C.

27.17 Precautions

Various measures need to be undertaken to evaluate the risk at the workplace, viz.,

a. Automation in the transfer of phosphorus pentafluoride from cylinder to other storage containers,
b. Good work practices such as contaminated clothes should be changed with clean clothes, the workplace should be equipped with eyewash fountains for emergency uses, the facility of the emergency shower should be provided to reduce the risk caused by skin exposure,
c. Further, the instruction of not to eat, smoke, and drink should be carefully followed along with hand wash after each shift.

References

Aroutiounian, V. M. (2018). Semiconductor gas sensors for detection of chemical warfare agents and toxic industrial chemicals. *Alternative Energy and Ecology (ISJAEE)*, *03*(1–3), 37–48. https://doi.org/10.15518/isjaee.2018.01-03.037-048.

Hansen, K. W., & Bartell, L. S. (1965). Electron diffraction study of the structure of PF51a. *Inorganic Chemistry*, *4*(12), 1775–1776. https://doi.org/10.1021/ic50034a023.

Jongjung, K., Hyun-seung, K., Jae Gil, L., Hosang, P., Jeong Beom, L., Jiyong, S., et al. (2016). Degradation of solid electrolyte interphase by phosphorus pentafluoride generated by thermal decomposition of lithium hexafluorophosphate. *ECS Meeting Abstracts. MA2016-03*, https://doi.org/10.1149/ma2016-03/2/646.

Kifer, E. W., & Van Dyke, C. H. (1972). Use of phosphorus pentafluoride as a fluorinating agent in organosilicon chemistry. III. Synthesis and characterization of some new fluorosiloxanes. *Inorganic Chemistry*, *11*(2), 404–408. https://doi.org/10.1021/ic50108a040.

Larsson, F., Andersson, P., Blomqvist, P., & Mellander, B. E. (2017). Toxic fluoride gas emissions from lithium-ion battery fires. *Scientific Reports*, *7*(1), 1–13. https://doi.org/10.1038/s41598-017-09784-z.

Mootz, D., & Wiebcke, M. (1987). The crystal structure of phosphorus pentafluoride. *Zeitschrift für Anorganische und Allgemeine Chemie*, *545*, 39–42.

National Center for Biotechnology Information (2020). *PubChem compound summary for CID 24295, phosphorus pentafluoride.* https://pubchem.ncbi.nlm.nih.gov/compound/Phosphorus-pentafluoride. (Accessed 10 November 2020).

Saftner, B. (2014). Ninothypsthys. In *Punctuated Equilibrium*. Xlibris.

Synquest Laboratories (2012). Phosphorus pentafluoride safety data sheet M015202. *Synquest Laboratories*, *77*(58), 1–8. Available from http://synquestlabs.com/msds/M000/M015-2-02.pdf.

Thorpe, T. E. (1876). Phosphorus pentafluoride. *Proceedings of the Royal Society of London*, 122–123.

Wolff, M. S. (1983). Occupationally derived chemicals in breast milk. *American Journal of Industrial Medicine*, *4*(1–2), 259–281.

CHAPTER

28

Silane: Risk assessment, environmental, and health hazard

Shuvasish Choudhury[a], *Ajendra Kumar*[b], *and Neeraj Kumar*[c]

[a]Central Instrumentation Laboratory, Assam University, Silchar, Assam, India

[b]Department of Mathematics and Statistics, Gurukula Kangri (Deemed to be University), Haridwar, Uttarakhand, India

[c]Building Energy Efficiency Group, CSIR-Central Building Research Institute, Roorkee, Uttarakhand, India

Abbreviations

CPR	cardio-pulmonary resuscitation
CVD	chemical vapor deposition
NIOSH	National Institute for Occupational Safety and Health
SCBA	self-contained breathing apparatus

28.1 Introduction

Silane (SiH_4) is a group 14 hydride, flammable and colorless gas with an unpleasant smell of vinegar (acetic acid). The presence of $Si(OR)_3$ and organofunctional (vinyl-, amino-, epoxy-, etc.) groups is practically responsible for its use as a precursor to elemental silicon and adhesion mechanism. Silane can also be applied as a pretreatment to the inorganic materials such as glass fibers or added to the resin. It functions as a mediator for the efficient promotion of adhesion among dissimilar, inorganic, and organic matrices (Matinlinna, Lassila, Özcan, Yli-Urpo, & Vallittu, 2004). As per their diverse function and substrates, they are known as primers or coupling agents.

https://doi.org/10.1016/B978-0-323-89857-7.00025-6

28.2 Occurrence

Silane is a saturated hydrosilicon compound with the general formula Si_nH_{2n+2} in which there is the presence of a single bond between hydrogen and silicon atoms. Similar to alkanes (saturated hydrocarbons), silane can either be linear with each silicon atom bound to at most two others or branched in which the silicon backbone bifurcates or trifurcates at silicon atoms bound to three or four others. Additionally, the term silanes can also be applied for cyclosilanes (general formula: Si_nH_{2n}, $n > 2$). Moreover, the silanes in which few or all hydrogen atoms have been substituted by other atoms or groups are also called silanes. The additional atoms or groups may be the same or different, reactive or nonreactive, and if reactive, these possess either inorganic or organic reactivity. These can be classified based on atoms or groups that are bonded to the silicon. The most frequently used silanes are silyl hydrides (Si—H). Other examples include silyl amines ($Si\text{–}NR_mH_{2\text{-}m}$, $m = 0, 1, 2$), halosilanes (Si–X, X is a halogen), and organosilanes (Si—C). Siloxanes and polysiloxanes are saturated silicon–oxygen hybrids with branched or unbranched backbones and can be considered silanes (Lopez, Bergueiro, & Fidalgo, 2014).

28.3 Historical background

In the mid-19th century, silane was first produced by German chemists, Heinrich Buff and Friedrich Woehler. Around 1968 the silane found its use in the semiconductor industry. The manufacturing of silane involves a batch reaction of trichlorosilane or silicon tetrachloride in a molten lithium aluminum hydride salt bath. For additional or specific characteristics, the nonreactive groups such as methyl can also be incorporated. Methylsilanes act as raw materials for manufacturing silicones forming the basis for all silicone chemistry.

28.4 Physicochemical properties

Silane is rapidly ignited in air, lighter than air, toxic and strong irritant to skin, eyes, and mucous membranes. Containers with silane may rupture under prolonged exposure to fire or heat. Some of the properties of silane are discussed in Table 28.1.

TABLE 28.1 Physicochemical properties of silane.

Property	Value
Molecular weight	32.11
Gas density of 0.083 lb./ft^3	1.33 g/L
Boiling point, 1 atm	−169.6 °F
Freezing point, 1 atm	−299 °F
Critical temperature	19.90 (−7°C)
Critical pressure	632 psia (43 atm)
Thermal stability	Silane is thermally stable up to 788°°F
Water solubility	Silane is slightly soluble in water

28.5 Physiological role

Silane acts as a coupling agent in the interphase region, the area between an inorganic substrate and an organic substrate, and acts as a bonding, or bridging agent to improve the adhesion between the two dissimilar materials. The alkoxy groups on silicon can hydrolyze to silanols, either with added water or from residual water on the inorganic surface, and create a siloxane bond. The acidity of the silanol groups depends upon the organofunctional group of the silane. On combination with an acidic primer, this gas increases the microshear bonds significantly. With ceramics that are not etchable to form a microretentive surface (e.g., alumina, zirconia), optimization of silane-promoted adhesion may play a great role in the future.

28.6 Uses

Silane has found application in various areas that are discussed in the below section:

1. **Adhesion promoters:** Silane contributes toward the enhancement of adhesion by decreasing the invasion of moisture at the interface. This eventually leads to improved moisture resistance, temperature resistance, and chemical resistance.
2. **Crosslinking agents:** For improved tear resistance, elongation at the break, tear propagation resistance, and abrasion resistance, silanes can be used to form crosslink polymers such as polyesters.

3. **Water scavengers:** They can also be costumed to absorb excess moisture due to their fast reaction with water preventing premature cure during compounding, enhance uniform curing, and improve package stability.
4. **Coupling agents:** Silane can react efficiently with both the reinforcement and the resin matrix of complex material. For the further advancement of the stronger bond at the interface, it may bond inorganic fillers or fibers to organic resins by forming a chemical bridge between them. The bond results in enhanced mixing, greater soldering of pigment or fillers to resins, escalate matrix robustness, diminish water intake of composite, and minimal wear.
5. **Other uses:** Silanization of orthodontic brackets (Eliades, Lekka, Eliades, & Brantley, 1994; Winchester, 1991) and silanized ceramic root canal posts in endodontics (Purton, Love, & Chandler, 2000) has been favored. In a denture, the long-range intraoral stain protection can be accomplished by surface modification with a fluorocarbon chain containing silane. The fluoroalkylethyl silane is used in dental ceramics for reduced stress corrosion and 3,3,3-trifluoro-propyl trimethoxysilane in filler surface treatment.

Application of silane in dentistry:

a. **Ceramic restorations and repairs:** Silanes have improved the bond between gold alloys and porcelain by about 25%, which was earlier done by micromechanical attachment techniques (Shell & Nielsen, 1962). Silanes do not form a bond with metal surfaces as they do with ceramic (Eames, Rogers, Feller, & Price, 1977). Siloxane bonds between the resin and ceramic matrix are moisture sensitive (Tjan & Nemetz, 1988) and expected to deteriorate by hydrolysis, over time. Studies have also substantiated that usage of silane augment bond strength of resin composite (Hayakawa et al., 1992). The upcoming ceramic repair systems (Matsumura, Kawahara, Tanaka, & Atsuta, 1989) are used in combination with silica coating/silicatization (Ozcan, Alkumru, & Gemalmaz, 2001) (i.e., airborne SiO_2 particle conditioning, use of acidic primer) (Kato, Matsumura, Ide, & Atsuta, 2001) (HF acid etching), or heat treatment of silanized substrate due to its capability of forming bonds with both inorganic and organic surfaces. The reason behind the enhanced bond strength is the mechanical interlocking of the resin composite and conformation of covalent siloxane bonds via silane treatment.
b. **Silanes on filler materials:** For enhanced physical properties of the composite, silanization is done. This procedure of silanization interprets an improved wetting for filler particles and diminished viscosity

among filler and fluid resin. In denture material, the process of silanization of E-glass and aramid fibers aggravates the adherence between fibers and organic acrylic resin (Vallittu, 1993).

c. **Base and noble metal alloys:** Silane is preferred for significant adhesion of resin composite to ceramic and efficient bonding of resin to feldspathic porcelain and metal.

28.7 Sources of contaminants

Silane is a gaseous material at atmospheric conditions, the only means of contamination is the impurity of the other gaseous materials, which inhibit the properties of the silane. Moisture is considered the most difficult way of contamination to control. If the moisture contaminates, particles will be formed downstream of the filter due to silicon dioxide nuclei formation, which can be led by nuclei condensing and particle formation. The clustering of siloxane molecules can also lead to particle formation. It has been hypothesized that siloxane molecules increase the particle formation downstream of the filter due to collision of the siloxane molecules formed by moisture in the gas on reacting with silane (Gotlinsky, O'Sullivan, Horikoshi, & Babasaki, 2000).

28.8 Risk assessment

It has been found that the overpressure and flame acceleration dictate mostly through silane concentration, followed by silane cloud size. It is suggested to incorporate (Dorofeev, 1996) model for detonation at the worst situation modeling in risk assessment for explosion through accidental silane release into the air without comprehensive information of silane flame propagation or concentration (Tsai et al., 2017). The acoustic theory for expanding flame (Thomas & Williams, 1966) and unconfined gas explosion models (Dorofeev, 1996) have been recommended for evaluating risk in case of silane release and explosion (Tsai et al., 2017).

The ignition performance of silane on contact with air is not always predictable. So, operation, safe design, and maintenance of silane disposal system are prime important to regulate the risk of the silane and its operational life-cycle. Failure to appropriately maintain vent source characteristics increases the chances of an accidental reaction or energy release that can cause personal, equipment, building, and environmental risk (Vanommeren, 2011).

28.9 Hazardousness

Silane is not known to sensitize human upon repeated or prolonged contact, and no synergistic effects have been reported. The major hazards containing silane are explained in the previous literature due to pyrophoric behavior demonstrated by silane. It has been well documented that it instantly ignites when released into the atmosphere. It has a huge broad flammable range in air of 1.4%–96% by volume at normal pressure and temperature. However, the silane testing experiments show the explosions or fires that can occur in silane/air mixtures of the concentrations between 1.4% and 4.1% (by volume) (Tamanini, Chaffee, & Jambor, 1998). Similar to other significantly hazardous industrial materials, it can be pretty manageable if treated and handled properly with respect. If accidentally released or mishandled, it can cause serious fire or explosion hazards (Ngai et al., 2007; Thomas, Bambrey, & Oakley, 2018).

28.10 Treatment

Silane has moderate acute peroral and percutaneous toxicity, severe primary skin and eye irritant, an aspiration hazard, and high intrinsic acute inhalation toxicity, but in moist air conditions lethal toxicity may be reduced. Therefore, treatment of silane toxicity is based on two factors, i.e., lethal dose and exposure time. Instead of that, first aid should be given according to the patient's symptoms.

Eye contact: Immediately flush with a large amount of water for at least 15 min, lifting upper and lower lids. Remove contact lenses.

Skin contact: Quickly remove contaminated clothing, immediately wash the contaminated body with soap and water.

Inhalation: Begin rescue breathing, if breathing has stopped and (cardiopulmonary resuscitation) CPR if heart action is stopped.

28.11 Risk management

Silane is a pyrophoric gas that requires special equipment and handling precautions. Hazardous gas management systems generally require continuous monitoring. The SiH_4 treatment areas, i.e., purification, manufacturing, filling, delivery, and supply system, require periodic routes for letting out gas during normal procedures. Also, accidental venting may occur from certain systems depending on their design and significant upset scenarios. Silane disposal can be handled safely by an initial understanding of all hazards, identifying the most essentially safe disposal type in conjunction with regulatory and facility requirements. Silane must be appropriately handled as part of the complete safety risk management for the facility (Vanommeren, 2011).

28.12 Current research

Silane pretreatment had been applied in steel manufacturing to inhibit corrosion. The toxic traditionally and environmentally harmful applied chromates could be better substituted by certain silanes (Matinlinna et al., 2004; Sundararajan & Van, 2000). Silanes are mostly used as a coupling agent in dentistry (Sundararajan & Van, 2000; Zakir et al., 2016).

28.13 Silane toxicity: Case study

- **Taiwan, Nov. 23, 2005**: While changing the cylinder silane was released, this ignited and fatally injured the operator rupturing the cabinet.
- **Silane Gas Explosion at Osaka University, Oct. 2, 1991**: Around 4pm on October 2, 1991, at the School of Engineering Science, Osaka University, a student closed a valve of the plasma CVD system. This triggered an explosion of a silane container supplying gas to the CVD system killing two students and igniting the city gas and organochlorine solvents to start a fire.

28.14 Safety measures

The safety measures taken while handling silane gas are discussed in Table 28.2.

28.15 Handling and storage

- Keep apart from burning surfaces, sparks, open blazes, and other ignition origins.
- Use only nonsparking and eruption-proof equipment.
- Use safety clothing when handling cylinders.
- Protect cylinders from exterior injury by avoiding dragging, rolling, sliding, or dropping.
- While shifting the cylinder, always use the cart and keep the removable valve shield integrated.
- The lifting of the cylinder should never be attempted by its cap.

TABLE 28.2 Safety measures taken while handling silane gas.

Appropriate engineering controls	Usage of an explosion-proof local exhaust system, equipment, and lighting; automatic gas panels, flow valves, and gas monitors
Eye protection	Wear safety glasses, vapor-proof goggles, or a face shield during cylinder change out
Skin and body protection	Wear metatarsal shoes, work gloves, protective clothing for cylinder handling
Respiratory protection	Application of an air-supplied or air-purifying cartridge if the activity level is exceeded. For emergencies or cases with unknown exposure levels, use a self-contained breathing apparatus (SCBA)
Thermal hazard protection	Wear cold insulating gloves when trans-filling or breaking transfer connections

- Never induce any object into cap openings as it may create a source for leakage.
- For the removal of overtight or rusted caps, an adjustable strap wrench should be used.
- Keep the valve of the cylinder closed after usage and slowly open the valve for subsequent use.
- Store at a temperature near 125 °F (52°C).

28.16 Precautions

Silane has a wide flammable range of 1.4%–96%; however, its release into the air may not always ignite instantly or sometimes not at all. Several serious industrial incidents have reported burning and detonation of leaked silane in the air (Chen et al., 2006). The US NIOSH has fixed an exposure limit of 5 ppm (over 8-h time-weight average) as occupational exposure of silane to workers (NIOSH, 2011). The general precautions are listed as follows:

- Avoid direct physical contact. Use appropriate, approved safety equipment. Untrained individuals should not handle silane or its container. Handling should occur in a chemical fume hood. Protect containers against physical damage.
- Silane should be stored in locked buildings and out of reach of children and unauthorized personnel.
- All efforts should be made to control exposure through the enclosure of medical or adhesive applications, exhaust ventilation, and good housekeeping.

References

Chen, J. R., Tsai, H. Y., Chen, S. K., Pan, H. R., Hu, S. C., Shen, C.C., et al. (2006). Analysis of silane explosion in a photovoltaic fabrication plant. *Process Safety Progress*, *25*(3), 237–244.

Dorofeev, S. B. (1996). Blast effect of confined and unconfined explosions. In B. Sturtevant, J. Shepherd, & H. Hornung (Eds.), *Vol. 1. Proceedings of the 20th ISSW* (pp. 77–86). Singapore: Scientific Publishing Co.

Eames, W. B., Rogers, L. B., Feller, P. R., & Price, W. R. (1977). Bonding agents for repairing porcelain and gold: An evaluation. *Operative Dentistry*, *2*(3), 118–124.

Eliades, T., Lekka, M., Eliades, G., & Brantley, W. A. (1994). Surface characterization of ceramic brackets: A multi-technique approach. *American Journal of Orthodontics and Dentofacial Orthopedics*, *105*(1), 10–18.

Gotlinsky, B., O'Sullivan, J., Horikoshi, M., & Babasaki, S. (2000). Eliminating siloxane impurities from silane process gas using next generation purification. *Ultrapure Gas Delivery*, 1–4.

Hayakawa, T., Horie, K., Aida, M., Kanaya, H., Kobayashi, T., & Murata, Y. (1992). The influence of surface conditions and silane agents on the bond of resin to dental porcelain. *Dental Materials*, *8*(4), 238–240.

Kato, H., Matsumura, H., Ide, T., & Atsuta, M. (2001). Improved bonding of adhesive resin to sintered porcelain with the combination of acid etching and a two-liquid silane conditioner. *Journal of Oral Rehabilitation*, *28*(1), 102–108.

Lopez, S., Bergueiro, J., & Fidalgo, J. (2014). Silane. *Encyclopedia of Toxicology*, *4*, 261–263.

Matinlinna, J. P., Lassila, L. V., Özcan, M., Yli-Urpo, A., & Vallittu, P. K. (2004). An introduction to silanes and their clinical applications in dentistry. *International Journal of Prosthodontics*, *17*(2).

Matsumura, H., Kawahara, M., Tanaka, T., & Atsuta, M. (1989). A new porcelain repair system with a silane coupler, ferric chloride, and adhesive opaque resin. *Journal of Dental Research*, *68*(5), 813–818.

Ngai, E. Y., Huang, K. P. P., Chen, J. R., Shen, C. C., Tsai, H. Y., Chen, S. K., et al. (2007). Field tests of release, ignition, and explosion from Silane cylinder valve and gas cabinet. *Process Safety Progress*, *26*(4), 265–282.

NIOSH. (2011). *Pocket guide to chemical hazards.* Silicon Tetrahydride, Centers for Disease Control and Prevention.

Ozcan, M., Alkumru, H. N., & Gemalmaz, D. (2001). The effect of surface treatment on the shear bond strength of luting cement to a glass-infiltrated alumina ceramic. *International Journal of Prosthodontics*, *14*(4).

Purton, D. G., Love, R. M., & Chandler, N. P. (2000). Rigidity and retention of ceramic root canal posts. *Operative Dentistry*, *25*(3), 223–227.

Shell, J. S., & Nielsen, J. P. (1962). Study of the bond between gold alloys and porcelain. *Journal of Dental Research*, *41*(6), 1424–1437.

Sundararajan, G. P., & Van, O. W. J. (2000). Silane based pretreatments for automotive steels. *Surface Engineering*, *16*(4), 315–320.

Tamanini, F., Chaffee, J. L., & Jambor, R. L. (1998). Reactivity and ignition characteristics of Sflane/air mixtures. *Process Safety Progress*, *17*(4), 243–258.

Thomas, G. O., Bambrey, R. J., & Oakley, G. L. (2018). A study of flame acceleration and the possibility of detonation with silane mixtures. *Process Safety and Environmental Protection*, *117*, 278–285.

Thomas, A., & Williams, G. I. (1966). Flame noise: Sound emission from spark-ignited bubbles of combustible gas. *Proceeding of the Royal Society A*, *294*, 449–466.

Tjan, A. H., & Nemetz, H. (1988). A comparison of the shear bond strength between two composite resins and two etched ceramic materials. *International Journal of Prosthodontics*, *1*(1).

Tsai, H.-Y., Lin, Y.-J., Chang, Y.-C., Lin, J.-S., Chen, J.-R., & Ngai, E. Y. (2017). Unconfined silane-air explosions. *Journal of Loss Prevention in the Process Industries*, *49*, 700–710.

Vallittu, P. K. (1993). Comparison of two different silane compounds used for improving adhesion between fibers and acrylic denture base material. *Journal of Oral Rehabilitation, 20*(5), 533–539.

Vanommeren, J. (2011). Silane gas venting/treatment lessons learned and best practices. *Process Safety Progress, 30*(2), 104–114.

Winchester, L. (1991). Direct orthodontic bonding to porcelain: An in vitro study. *British Journal of Orthodontics, 18*(4), 299–308.

Zakir, M., Ashraf, U., Tian, T., Han, A., Qiao, W., Jin, X., et al. (2016). The role of Silane coupling agents and universal primers in durable adhesion to dental restorative materials—A review. *Current Oral Health Report, 3*(3), 1–10.

CHAPTER

29

Styrene: Risk assessment, environmental, and health hazard

Ritu Arora[a] *and Manila*[b]

[a]Department of Mathematics and Statistics, Gurukula Kangri (Deemed to be University), Haridwar, Uttarakhand, India

[b]Department of Chemistry, Gurukula Kangri (Deemed to be University), Haridwar, Uttarakhand, India

List of abbreviations

°C	degree Celsius
°F	degree Fahrenheit
g/cm³	gram per cubic centimeter
km	kilometer
mg/m³	milligram per cubic meter
mi	miles
min	minutes
ppm	parts per million
µg/kg	microgram per kilogram
µg/m³	microgram per cubic meter

29.1 Introduction

Styrene ($C_6H_5CH{=}CH_2$) is a colorless, oily, and viscous and easily vaporable liquid. It is also known as vinylbenzene, ethenylbenzene, cinnamene, or phenylethylene. The pure samples of styrene are colorless but the old samples can turn up yellowish. It evaporates nimbly and thus has to be stored at a temperature below 68°F (20°C). The pure and natural form of styrene has a sweet smell while synthetically prepared samples contain aldehydes which impart an unpleasant odor to it. At high concentrations, styrene is pungent in odor.

Hazardous Gases
https://doi.org/10.1016/B978-0-323-89857-7.00015-3

Styrene is an important monomer compound that is widely used in the manufacture of plastics, synthetic rubbers, polyesters, latex paints, and coatings (Collins & Richey, 1992; Miller, 1994). It is also a precursor to an extensively used polymer called polystyrene having a broad spectrum of applications in various fields.

Recently, an infamous incident (Vizag gas leak) related to styrene gas leakage was reported at LG Polymers India Private Limited located in Vishakhapatnam, India, on May 07, 2020, leaving 11 people dead, besides 1000 sick on exposure to the gas as per NDRF (National Disaster Response Force). This plant was involved in the manufacturing of polystyrene, high-impact polystyrene, expandable polystyrene, and other engineering plastic compounds.

29.2 Occurrence

Natural occurrence: Styrene derived its name from *Storax balsam*, the resin obtained from liquidambar trees of the family Altingiaceae. Styrene naturally occurs in minute quantities in plants and foods like coffee beans, balsamic trees, and peanuts and the highest proportion being found in cinnamon (Some Industrial Chemicals, 1994; Steele et al., 1994). It is formed as a degradation product in plants having cinnamic acid such as balsamic trees (Duke, 1985). This chemical occurs naturally as a by-product of fungal and microbial metabolism (Shimada et al., 1992; Shirai & Hisatsuka, 1979). It is also found in coal-tar.

Sequential occurrence: Styrene can be found in air, water, and soil, although it is not a formal part of the composition of gases in our atmosphere. It is in consequence of its release from the various industrial units involved in the production and use of styrene-based products (Agency for Toxic Substances and Disease Registry, 2012). Other sources that give rise to styrene concentration in the environment are cigarette smoke, vehicle exhaust, and incineration of styrene-based polymers (World Health Organization, 1983). It is also found to be released during the use of photocopiers (Agency for Toxic Substances and Disease Registry, 2012). A negligible amount of styrene may be liberated by recently manufactured plastic wares also (World Health Organization, 2000).

29.3 Historical background

Styrene is a naturally occurring compound with the captivating property of autopolymerization prompting various scientists from time to time to explore more regarding its fascinating properties. The records of research in styrene dated back to 1839, when a German apothecary

Eduard Simon isolated a volatile liquid from the American sweetgum tree (*Liquidambar styraciflua*) resin. He named this liquid as "styrol" (nowadays, styrene) (Semmler, 1906; Simon, 1839). He also observed the gradual conversion of styrol into a hard, rubber-like substance when it comes in contact with air, light, or heat. He named this compound as "styrol oxide" (Simon, 1839). In 1845, the German chemist August Hofmann and his student John Blyth gave the empirical formula of styrene: C_8H_8 (Blyth & Hofmann, 1845). They had also discovered Simon's "styrol oxide" (whom they named "metastyrol") composed of similar empirical formula as that of styrene (Blyth & Hofmann, 1845). In 1865, another German chemist Emil Erlenmeyer found that styrene had the property to form a dimer (Erlenmeyer, 1865). In the year 1866, French chemist Marcelin Berthelot stated that the "metastyrol" was a polymeric form of styrene (Berthelot, 1866). Simultaneous, investigations on another component of storax, i.e., cinnamic acid, were being carried out and it was determined that decarboxylation of cinnamic acid yields "cinnamene" (or "cinnamol") which appeared to be styrene. In 1866, Erlenmeyer suggested that both styrene and cinnamol might be vinylbenzene (Erlenmeyer, 1866). However, it was found that styrene obtained from distilling storax resin was optically active, whereas the styrene obtained from cinnamic acid was optically inactive (Berthelot, 1867). This ambiguity was resolved by the Dutch chemist Van't Hoff in 1876 that a contaminant was responsible for displaying optical activity of the styrene obtained from distilling storax resin (Hoff van't, 1876).

29.4 Physicochemical properties

Styrene is a flammable, toxic, probably carcinogenic with chemical formula $C_6H_5CH{=}CH_2$. It is a colorless, viscous liquid in pure form and a derivative of benzene. Its IUPAC name is ethenylbenzene. Other names include cinnamene, styrol, and vinylbenzene. It evaporates easily above the temperature of 68°F (20°C). It is moderately soluble in water, soluble in ethanol, and highly soluble in benzene (World Health Organization, 2000), Table 29.1.

Styrene spontaneously undergoes polymerization to give polystyrene without any external initiators. This phenomenon is known as autopolymerization which takes place at ambient temperature and above. This reaction is exothermic and involves a real risk of thermal rampant and outburst. One such incident was reported on September 28, 2019, in a ship docked in Ulsan, Republic of Korea. In this incident, 5250 metric tons of styrene monomer triggered an explosion of the tanker Stolt Groenland (Marine Accident Investigation Branch, 2019). The autopolymerization

TABLE 29.1 Physicochemical properties of styrene.

Formula	$C_6H_5CH{=}CH_2$
Molecular weight	104.15
Density (g/cm^3)	0.909
Melting point (°F)	−22
Boiling point (°F)	293
Flashpoint (°F)	88
Viscosity (cP@68 °F)	0.762
Refractive index	1.5469
Vapor pressure (mmHg@68°F)	5
Dipole moment (D)	0.13

can be kept under control by a continuous supply of polymerization inhibitors like 4-*tert*-butyl catechol in trace amounts.

29.5 Physiological role

Styrene may enter the human body by inhalation, ingestion, or dermal contact (Agency for Toxic Substances and Disease Registry, 2012). Once entered into the body through any pathway, styrene is widely distributed throughout the fatty tissues of the body which is followed by the slow process of its elimination through urine. It is an indicator of accumulation of styrene in the body, especially in cases where daily exposure of a person to a higher concentration is evident as in the case of the workers employed in styrene-based industrial units.

Biotransformation of styrene: Styrene is oxidized to styrene-7,8-oxide, which on enzymatic hydrolysis yields phenylethylene glycol which gives mandelic acid and phenyl glyoxylic acid on further oxidation; both of these contribute to being the major metabolites of styrene in humans (Some Industrial Chemicals, 1994).

29.6 Uses

Styrene is widely used to make polymerization products like polystyrene, polyesters, and synthetic rubbers. Other applications include latex paints and styrene-alkyd coatings (Collins & Richey, 1992; Miller, 1994).

This paint is known for its ease in handling and application and it also dries up very rapidly. Styrene is also used in the preparation of synthetic rubber which is used in paper coating, carpet backing, and molded articles such as thin gloves.

It is employed as a monomer in preparing polyester, a synthetic fiber, which is having a wide demand in the manufacturing of conveyor belts, safety belts, and plastic reinforcements with high energy absorption. Out of the abovementioned applications, polystyrene is the most important due to its wide spectrum of applicability. A few examples of polystyrene applications are being discussed as follows:

- Packaging material: Food packaging, meat/poultry trays, and egg cartons, CD, and DVD cases are made of polystyrene to avoid damage.
- Insulation materials: Lightweight polystyrene foam is an excellent thermal insulation material and finds its use in roofing, refrigerators, freezers, and industrial cold storage facilities. This insulation is inert, long-lasting, and waterproof.
- Food services: Polystyrene offers a convenient and cost-effective way of supplying food items to customers as hot stays hot, cold stays cold, and fresh stays fresh in these packaging.
- Automotive industry: Polystyrene (solid and foam) is used to prepare various car parts, instrumental panels, energy absorbing door panels, and sound dampening foam. Polystyrene (foam) is a major component in child protective seats.
- Electronics: Polystyrene is used for televisions, computers, and IT equipment.
- Medical: Polystyrene is used in various medical applications due to its clarity and ease of sterilization. It is used in culture trays, Petri dishes, test tubes, diagnostic equipment, and medical devices.

29.7 Source of contaminants

It can be deduced that humans may get exposed to styrene by breathing air contaminated with it due to any of the following reasons:

- Industrial units consuming or manufacturing styrene
- Automobile exhaust
- Photocopiers
- Disposal of styrene-based products
- Cigarette smoke
- Incineration of styrene polymers (Agency for Toxic Substances and Disease Registry, 2012)

29.8 Risk assessment

Risk assessment is an important procedure to be followed by every industrial unit processing styrene. Risk assessment involves identification of hazards, people at risk, extent of the risk involved, and periodic assessment of the safety measures taken in order to keep the danger at bay. Styrene is a chemical belonging to the hazardous substance category so a proper validation of safety measures is a must for safe production and usage. It is a flammable liquid, skin and eye irritant, harmful to the unborn child, and can cause organ damage.

The major class at risk is workers employed in industrial setup involved in the production and use of styrene. Automobile exhaust, cigarette smoke, and photocopiers also contribute to styrene concentration in the air but that does not usually cross a safe exposure limit of styrene.

The usual health concern in workers exposed to styrene is found to be related to the nervous system and respiratory system. The health issues involve tiredness, drowsiness, slowed reaction time, concentration problems, balance problems, and color vision. If it is swallowed, aspiration into the lungs may lead to chemical pneumonitis. Hearing loss is observed in animals also if they are exposed to a high concentration of styrene. Exposure at high levels could cause unconsciousness also. At high levels of concentration, styrene defats the skin, which may lead to dryness or cracking. Exposure to styrene may increase noise-induced hearing loss. This is possibly carcinogenic to humans.

As it can be visualized that styrene does not pose a serious threat at low concentration but at about 1000 times more concentration than usually found in air, it turns out to be a serious health issue. Therefore, units processing styrene must keep its concentration level under control and exercise all safety measures to ensure zero risks to people and the environment.

29.9 Environmental effects

Styrene is found to be an active component of photochemical smog. This happens when styrene present in the air reacts readily with ozone to give benzaldehyde and peroxides giving rise to a major class of irritants; peroxy bezoyl nitrate is regarded to be a potent eye irritant which is one of the peroxides yielded during the above-mentioned process (World Health Organization, 2000).

Styrene present in topsoil and surface water evaporates and that permeates into the lower layers of soil or water may be further decayed with the help of microorganisms (Agency for Toxic Substances and Disease

Registry, 2012). Its evaporation half-time in the water at a depth of about 1 m is approximately 6 h and thus, it is presumed that styrene does not bioaccumulate or bioconcentrate in food chains or organisms in any detectable magnitude (Air Quality Guidelines for Europe, 1987). In soil, bacteria and other microorganisms play a vital role in the removal of styrene.

Although there are no conclusive data as far as the effect of styrene on the environment is concerned as it readily breaks down in the air in 1–2 days after entering the atmosphere by any means but at the same time, its role as an active member of photochemical smog cannot be neglected.

29.10 Hazardousness

Styrene is associated with a broad spectrum of applications ranging from packaging, paints, synthetic rubbers, automotive industries, medical devices, etc. Due to its extensive usage, its presence can be well perceived everywhere in one form or the other. Though it is harmless at a low concentration which is manifested by its presence naturally in coffee beans, peanuts, balsamic trees, cinnamon, etc., styrene has been found to be present in unpolluted rural areas but in nominal concentrations. The concentration of styrene in the urban atmosphere is found to be around 0.3 $\mu g/m^3$, which accounts for an estimated daily intake of approximately 6 µg/person (Air Quality Guidelines for Europe, 1987). Within a 1-km range of styrene polymerization unit in the urban area, the value may rise to 20–30 $\mu g/m^3$ which leads to an estimated daily intake of about 400–600 µg/person living in that area (Air Quality Guidelines for Europe, 1987). The permissible exposure limit as per OSHA (Occupational Safety and Health Administration) is 100 ppm (426 mg/m^3 or 4.26×10^5 $\mu g/m^3$) averaged over an 8-h work shift; 200 ppm not to be exceeded at any time and 600 ppm as the 5-min maximum peak time should never be exceeding a 3-h shift in a day. Workers involved in the styrene production and polymerization units are exposed to an average limit of 20 ppm (85 $\mu g/m^3$). This figure considers even the occasional outbursts and leakage of reactors or tubing and other equipment (Some Industrial Chemicals, 1994).

Polystyrene and its copolymers find use in the food packaging materials which tend to contribute only a traceable amount of styrene in the food to an extent of 2.5–80 µg/kg as found in yogurt and other milk products stowed in polystyrene-based canisters (Air Quality Guidelines for Europe, 1987).

Although low concentrations of styrene do not pose much threat, a long-term exposure to styrene might result in blistering of the skin and consequently developing dermatitis (Some Industrial Chemicals, 1994).

Liquid styrene on direct contact can lead to erythema. Styrene has been a causative factor for irritation in eyes, throat, and respiratory tract at concentrations ranging between 10 and 100 ppm (43–426 mg/m^3) or higher (Environment Canada and Health Canada, 1993; Some Industrial Chemicals, 1994; World Health Organization, 1983).

Exposure to styrene beyond 100 ppm (426 mg/m^3) can lead to severe irritation of mucous membranes (the most vulnerable part of the human body to be affected by styrene) in the eyes and upper part of the respiratory tract (Some Industrial Chemicals, 1994). The odor threshold value for pure styrene is about 0.016 (70 μg/m^3). The characteristic pungent odor can be recognized at 3–4 times higher concentrations than the threshold value. As per the air quality guidelines, a peak concentration of styrene in the air should be maintained under the threshold level of 70 μg/m^3 as a 30-min average.

Sudden exposure to large amounts can result in the onset of "styrene sickness," which could prove lethal. This sickness is associated with nausea, headache, dizziness, and confusion leading to central nervous system depression. High concentration exposure could prove fatal as happened in the "Vizag gas leak" incident. A few inconclusive studies also associate styrene with an increased risk of leukemia and lymphoma which might happen due to long-term exposure to a high concentration of styrene.

29.11 Risk management

Following recommendations regarding the workplace of styrene, the processing is made to keep the concentration of styrene in check.

OSHA: Permissible exposure limit for two 8–h work shift is 100 ppm and for 5 min in any 3-h work shift is 600 ppm.

NIOSH: Recommended exposure limit for 10-h work shift is 50 ppm and 15 min short-term exposure limit is 100 ppm.

For transportation, only those persons/transports who are eligible and licensed to accept hazardous material should be authorized for transportation. The styrene should be properly classed, described, packaged, marked, and labeled, before shipment as needed or authorized by regulations pertaining to the hazardous materials guidelines.

29.12 Styrene toxicity: Case study

The deleterious effects of styrene come into existence as soon as it crosses its safe exposure limit which is certainly the outcome of either some serious negligence at human hands or malfunctioning of some major industrial equipment as it happened on May 7, 2020, at LG Polymers India

Pvt. Ltd. in Vishakhapatnam, India. It reportedly took 11 lives and affecting as much as 1000 more within hours of styrene leakage from the plant (Gettleman, Suhasini, Kai, & Sameer, 2020). This plant was reopened after a nationwide lockdown which was implemented in response to the COVID-19 pandemic. This plant reportedly stored 2000 metric tons of styrene in storage tanks maintained at 68–72°F (20–22°C) above which it vaporizes rapidly into the air (Gettleman et al., 2020). It is believed that the temperature of storage tanks exceeded safe levels due to some computer glitch resulting in the vaporization of styrene at about 3:00am, spreading styrene to nearby villages in a radius of 3km (2 mi.) (Vizag LG Polymers, 2020). Hundreds of people fell unconscious as soon as they came in contact with air laden with a high concentration of styrene as per the media reports. Many people were hastened to hospitals on complaining of breathing difficulties in addition to irritation and burning sensation in the eyes (Vizag LG Polymers, 2020). This happened due to inhalation of the styrene which affected the mucous membrane leading to damage in the lungs. Some casualties were reported with acute breathlessness that asphyxiated to death.

29.13 Safety measures

Styrene is harmful when its concentration in the air reaches beyond the safe exposure limit. The odor threshold limit plays a vital role as the pungent smell of styrene is perceptible at this limit which alarms the people around regarding the presence of unwanted gas in the air. If any such situation arises, one must check the source of such smell and if there is any chance of a styrene source nearby, a steady action must be taken by the concerned authorities in order to save people from an excessively higher concentration of styrene, which could prove fatal if left unchecked.

In case of an accidental leakage, following precautions must be taken for the persons affected by its exposure:

- In case of inhalation of styrene, the person must be carried away from the styrene-laden air toward an area where fresh air is available. Artificial respiration should be provided in case of breathlessness. In the case of acute breathing troubles, a physician is needed to be consulted.
- In case of dermal contact of styrene, wash the affected area with soap and plenty of water and consult a physician immediately.
- In case of eye contact, rinse the eyes with plenty of water for about 15min and thereafter consult a physician.
- If somehow styrene got swallowed, do not induce vomiting. Rinse the mouth with lots of water and hospitalize the victim immediately without a delay. Avoid giving anything orally to an unconscious person.

The following measures should be followed at site for controlling the leakage:

- Personal protective equipment (PPE) should be used by the people who are going to inspect the suspected area of leakage.
- All affected people should be evacuated to safe areas. Adequate ventilation of the area has to be ensured.
- All sources of ignition should be removed to avoid any mishap.
- If possible, further spillage should be prevented, and avoid discharge of the product into drains.
- An electrically shielded vacuum cleaner or wet-brushing can be handy in collecting liquid styrene.

In case of an accident of styrene leading to an explosion, any one of the following extinguishers can be used:

- Water sprayer
- Alcohol resistant foam
- Dry chemical or carbon dioxide

Firefighters must be well equipped with proper breathing apparatus and they should use water sprayers to cool unopened containers.

29.14 Handling and storage

Handling

- The person handling styrene should be well aware of the hazardous nature of styrene.
- Appropriate engineering controls should be in place.
- Styrene should be handled while following complete safety norms and good industrial hygiene.
- Equipment must be grounded properly before handling the product.
- Proper gloves face shield and safety glasses should be worn in order to avoid any contact with skin and eyes before handling the container of styrene.
- Inhalation of vapor or mist should be completely avoided.
- Gloves should be checked before use. A proper glove removal technique should be followed to avoid skin contact with styrene. Contaminated gloves must be disposed of properly after use as per the standard operating procedure of good laboratory practices.
- A person handling broken packages should be well equipped with proper personal protective equipment.

Storage

- Storage should be done in a cool and dark place.

- Store only if stabilized.
- Containers should be tightly closed in a dry and properly ventilated place.
- Styrene containers should be kept strictly in a no smoking zone.
- Opened containers should be resealed when not in use and kept upright to prevent any spillage.
- It is recommended to store styrene containers at a temperature of 2–8°C.
- Store in an area having no access to drain or sewer.

29.15 Precautions

- Do not eat, drink, or smoke during work.
- The use of contact lenses by industrial workers should be strictly avoided during work shifts.
- In case of fire broke out, the container may explode. Keep styrene away from sources of ignition as much as possible because its vapors form an explosive mixture in air.
- Electrostatic charges if developed can turn the situation serious, so measures must be in place to not let that happen.
- To avoid flammability hazards, clothing wetted by styrene should be removed with immediate effect.
- *Accidental release measures*: Personal protective equipment should be used. Vapors, mist, or gas must not be inhaled. Proper ventilation measures should be taken. All sources of ignition must be immediately removed. Evacuation of people to safe areas must be exercised. Vapors can pose a major threat by segregating in low areas and forming explosive mixture in the air; therefore precaution must be taken. Further leakage or spillage should be prevented if safe to do so. Avoid discharge of the product in drains and into the environment.
- *In case of fire*: Sparks, flames, and other ignition sources should be kept away if the material is not on fire. Avoid drainage of styrene in water sources and sewers. Dikes should be made to contain the flow if necessary. Attempt to stop leakage should be done avoiding undue personnel hazards. Use water spray to knock-down vapors.
- *Evacuation*: If leakage of styrene takes place, the evacuation process should be made in a one-half mile radius keeping amount of material spilled, location, and weather conditions under consideration.

References

Agency for Toxic Substances and Disease Registry (2012). *Public health statement styrene. retrieved from the agency for toxic substances and disease registry. (2012).* www.atsdr.cdc.gov/toxprofiles. (Accessed 07 August 2020).

Air Quality Guidelines for Europe (1987). *Copenhagen*: (pp. 118–126). World Health Organization Regional Office for Europe.
Berthelot, M. (1866). Sur les caractères de la benzine et dustyrolene, compares avec ceux des autres carbures d'hydrogene (On the characters of benzene and styrene, compared with those of other hydrocarbons). *Bulletin de la Societe Chimique de Paris (2nd Series), 6*, 289–298.
Berthelot, M. (1867). Sur les etats isomeriques du styrolene [On the isomeric states of styrene]. *Annales de Chimie et de Physique 4th Series, 12*, 159–161. (in French).
Blyth, J., & Hofmann, W. A. (1845). On styrole, and some of the products of its decomposition. *Memoirs and Proceedings of the Chemical Society of London, 2*, 334–358.
Collins, D. E., & Richey, F. A. (1992). Synthetic organic chemicals. In J. A. Kent (Ed.), *Riegel's handbook of industrial chemistry* (pp. 800–862). (9th ed.). New York: Van Nostrand Reinhold.
Duke, J. A. (1985). *CRC handbook of medical herbs*: (p. 323). Boca Raton, FL: CRC Press.
Environment Canada and Health Canada (1993). *Styrene*. (Priority Substances List Assessment Report PSL-24E), Ottawa.
Erlenmeyer, E. (1865). Ueber Distyrol, ein neues Polymere des Styrols (On distyrol, a new polymer of styrol). *Annalen der Chemiet, 135*, 122–123.
Erlenmeyer, E. (1866). Studien uber die s.g. aromatischen Sauren (Studies of the so-called aromatic acids). *Annalen der Chemiet, 137*, 327–359.
Gettleman, J., Suhasini, R., Kai, S., & Sameer, Y. (2020). *Gas leak in India at LG factory kills 11 and sickens hundreds*. The New York Times. ISSN 0362-4331.
Hoff van't, J. H. (1876). Die Identitat von Styrol und Cinnamol, einneuer Körper aus Styrax (The identity of styrol and cinnamol, a new substance from styrax). *Berichte der Deutschen Chemischen Gesellschaft, 9*, 5–6.
Marine Accident Investigation Branch (2019). *Interim report on the investigation of the explosion and fire on board the chemical tanker Stolt Groenland on 28 September 2019, Ulsan, Republic of Korea. Retrieved from the government of United Kingdom official (2019).* https://assests.publishing.service.gov.uk. (Accessed 28 July 2020).
Miller, R. R. (1994). Styrene production, use and human exposure. *Critical Reviews of Toxicology, 24*, S1–S11.
Semmler, F. W. (1906). *The essential oils according to their chemical components taking into account the historical development*. (Vol. 1)). Veit & Co..
Shimada, K. , et al. (1992). Styrene formation by the decomposition by *Pichia carsonii* of trans-cinnamic acid added to a ground fish product. *Applied Environmental Microbiology, 58*, 1577–1582.
Shirai, K., & Hisatsuka, K. (1979). Isolation and identification of styrene assimilating bacteria. *Agricultural and Biological Chemistry, 43*, 1595–1596.
Simon, E. (1839). Ueber den flüssigen Storax (Styrax liquidus). *Annalen der Pharmacie, 31*(3), 265–277.
Some Industrial Chemicals (1994). *Lyon, International Agency for Research on Cancer*: (pp. 233–320). (IARC Monographs on the Evaluation of the Carcinogenic Risk of Chemicals to Humans, Vol. 60).
Steele, D. H., Thornburg, M. J., Stanley, J. S., Miller, R. R., Brooke, R., Cushman, J. R. , et al. (1994). Determination of styrene in selected foods. *Journal of Agricultural and Food Chemistry, 42*(8), 1661–1665.
Vizag LG Polymers (2020). *Polymers Gas leak: 11 dead, over 5,000 fall sick after gas leak from LG Polymers. Siva G.* The Times of India. (Retrieved 07 May 2020).
World Health Organization (1983). *Styrene*. (Environmental Health Criteria, No. 26), Geneva.
World Health Organization (2000). *Styrene. Air quality guidelines* (2nd ed.). Europe.

CHAPTER

30

Sulfur dioxide: Risk assessment, environmental, and health hazard

Mayank Kumar Malik[a,b] *and Sonu Kumar*[b]

[a]Department of Pharmaceutical Sciences, Gurukula Kangri (Deemed to be University), Haridwar, Uttarakhand, India

[b]Department of Chemistry, Gurukula Kangri (Deemed to be University), Haridwar, Uttarakhand, India

Abbreviations

AIHA	American Industrial Hygiene Association
AP	action plan
CC	compound of concern
EPA	Environmental Protection Agency
ESP	emergency shutdown procedure
IDLH	immediately dangerous to life or death
IGF	industrial grate furnaces
NAAQS	National Ambient Air Quality Standards
NIOSH	National Institute for Occupational Safety and Health
OSHA PEL	Occupational Safety and Health Administration Permissible Exposure Limit
PCFB	pulverized coal boilers
RPE	respiratory protective equipment

30.1 Introduction

Sulfur dioxide is a colorless gas having the chemical formula SO_2. It is twice as heavy as air. It has a strong suffocating odor, which is noticeable when sulfur is burned. Under standard conditions, one volume of water dissolves 79.8 volumes of the SO_2 gas. It is certainly condensed to a colorless liquid at −10°C and freezes to a snow-like solid at −72°C. In moist air or fog, it combines with water to form a bisulfite anion (Alarie, Wakisaka, &

https://doi.org/10.1016/B978-0-323-89857-7.00011-6

Oka, 1973). It has many industrial and agricultural applications. Some foods and drinks are preserved with small quantities of SO_2 that are safe for human consumption. Commonly, it is derived from the combustion of sulfur-containing fossil fuels, which are not commonly used indoors. SO_2 gas is produced as a by-product during the smelting and converting processes. Domestic/Indoor air levels of SO_2 rarely exceed 30% of outdoor/industrial levels unless an unvented kerosene burner is used with extremely low-grade fuel. Most indoor exposure in the workplace results from the misuse of equipment including inadequate venting of oil-burning combustion appliances. When released into the environment, sulfur dioxide moves into the air. In the air, it reacts with the moisture and gets converted to sulfuric acid. Due to its high susceptibility to moisture, SO_2 can react with moisture available on the skin and other moist surfaces of the human body to form sulfuric acid.

30.2 Occurrence

Sulfur dioxide sinks have been extensively acknowledged by the soils and it could be springs of atmospheric SO_2 (Macdonald, Denmead, White, & Melville, 2004). Electrical power plants are the principal source of sulfur dioxide emissions (American Chemical Society Subcommittee on Environmental Improvement, 1969). Sulfur dioxide is released into the environment during the burning of coal and oil. Power plants, oil refineries, some motor vehicles, and domestic boilers and fires are all sources of sulfur dioxide. Active volcanoes also produce it naturally. To some extent, the colorless gas SO_2 is a natural constituent of the air and is, for example, produced by the decay of vegetation. However, in industrialized countries, only about one-tenth of total emissions are estimated to be from natural sources (Health and Safety Commission, 1978). Emissions of SO_2 from manmade sources are associated with the combustion of fossil fuels.

30.3 Historical background

Sulfur dioxide was first purified and recognized as a definite compound by Joseph Priestly in 1775. Romans revealed that sulfur dioxide could preserve the wine fresh. It also precludes the vacant wine barrels from vinegar smell characteristic of spoilt wine when sulfur candles burn inside them. Thus, sulfur dioxide is used in winemaking as a preserver. Consequently, SO_2 is the utmost significant preservative in wine production since the Roman era.

In ancient China and Greece, it was used to "drive out evil spirits," such as insects, and as a disinfectant. It features in myths, when Odysseus burns sulfur to fumigate his house after slaughtering Penelope's suitors; for at the time the Ancient Greeks believed that its choking smell was cleansing. In the Greece-Peloponnesian War around 429 BC, SO_2 was used as the first chemical weapon by the Syrians against the Romans. In warfare, it was manufactured by burning sulfur and pitch near an enemy city's walls, creating a poisonous cloud which would hang over the city.

30.4 Physicochemical properties

Sulfur dioxide consolidates at room temperature under a pressure of 2 atm. Hence, it is one and only the utmost straightforward gas that liquefy. It is an oxide of sulfur and is freely dissolved in an aqueous medium (Table 30.1).

Sulfur dioxide gives sulfurous acid after reacting with water (Eq. 30.1):

$$SO_2 + H_2O \rightarrow H_2SO_3 \quad (30.1)$$

It forms sodium sulfite when vigorously reacts with sodium hydroxide (NaOH), Eq. (30.2):

$$SO_2 + 2NaOH \rightarrow Na_2SO_3 + H_2O \quad (30.2)$$

If an excess amount of sulfur dioxide passed in this way, then it provides sodium hydrogen sulfite (Eq. 30.3):

$$SO_2 + Na_2SO_3 \rightarrow H_2O + 2NaHSO_3 \quad (30.3)$$

TABLE 30.1 Properties of sulfur dioxide.

Property	Value
Formula	SO_2
Molecular weight (lb/mol)	64.06
Critical temperature (°F)	315.5
Critical pressure (psia)	1142.0
Boiling point (°F)	14.3
Melting point (°F)	−103.9
Psat @ 70°F (psia)	49.3
Liquid density @ 70°F (lb/ft^3)	85.98
Gas density @ 70°F 1 atm (lb/ft^3)	0.1682
Specific volume @ 70°F 1 atm (ft^3/lb)	5.94
Specific gravity	2.285
Specific heat @ 70°F (Btu/lbmol °F)	9.58

In sulfur dioxide, sulfur atom is tetravalent. Subsequently, it could react with elements like C_{12} and O_2 to shape the comparing addition compounds and could increase its coordination number up to six. For example, when SO_2 reacts with chlorine gas, it gives sulfuryl chloride (SO_2Cl_2) under the impact of charcoal used as a catalyst (Eq. 30.4):

$$SO_2 + Cl_2 \rightarrow SO_2Cl_2 \tag{30.4}$$

Sulfur dioxide oxidized with vanadium pentoxide (V_2O_5) gives sulfur trioxide (SO_3), Eq. (30.5):

$$2SO_2 + O_2 \rightarrow SO_3 \tag{30.5}$$

It could start giving nascent oxygen in the existence of moisture and work as reducing agent. For example, it reduces halogens to halogen acids and ferric salts to ferrous salt, as shown in Eqs. (30.6) and (30.7), respectively:

$$2Fe^{3+} + SO_2 + 2H_2O \rightarrow 2Fe^{2+} + SO_4{}^{2-} + 4H^+ \tag{30.6}$$

$$X_2 + SO_2 + 2H_2O \rightarrow SO_4{}^{2-} + 2X^- + 4H^+ \tag{30.7}$$

30.5 Physiological role

SO_2 and its derivatives at physiological (1–10 mmoL/L) concentration cause an increase in cytosolic calcium concentration, which result in myocardial contractility due to the excitation of myocardial membrane calcium channels (Nie & Meng, 2005a; Nie & Meng, 2005b; Nie & Meng, 2006). Research studies have shown that level of lipid metabolism could be controlled by the usage of SO_2 (Lovati, Manzoni, Daldossi, Spolti, & Sirtori, 1994; Haider, 1985). SO_2 and more or less of its derivatives play biphasic roles in blood vessel tension.

Sulfur dioxide is an extremely water-soluble gas and is quickly absorbed through the mucosa of the nose and upper respirational zone; it has been specified by various studies in humans (Kleinman, 1984; Speizer & Frank, 1966). World Health Organization also stated that the upper respirational tract of the human being can be significantly affected by the absorption of sulfur dioxide through nose breathing followed by mouth breathing (WHO, 1979). The lower respiratory tract of the mucosa in humans absorbs a minimal amount of SO_2. Consequently, minimal delivery of sulfur dioxide to lower respirational tract occurs (EPA, 1986).

S-sulfonate (R-, S-, SO_3) was traced in the human plasma subjects succeeding the constant exposure through various concentrations of sulfur dioxide such as 0.3, 1.0, 3.0, 4.2, and 6.0 ppm in a compartment of phases up to 120 h (Gunnison & Palmes, 1974), which initiated the reaction of sulfur dioxide with metabolite, sulfite, and proteins. A progressive connection with appearance concentrations of sulfur dioxide in the air was revealed by the plasma levels of S-sulfonate.

A mixture of sulfite, hydrogen ions, and bisulfites ions generate due to inhalation of sulfur dioxide by humans. Once sulfite is formed, a reaction of oxidation of sulfite to sulfate is detected predominantly in the liver (Ellenhorn & Barceloux, 1988; Gunnison, Sellakumar, Currie, & Snyder, 1987), which is catalyzed through the sulfite oxidase. Sulfite oxidase has been identified in the utmost tissues and mitochondria (Cabré, Marín, Cascante, & Canela, 1990). The literature revealed that the brain, spleen, lungs, and testis have minor sulfite oxidase activity in comparison to the liver, heart, and kidneys. The activity of the low pulmonary sulfite oxidase could elucidate the conclusions of a study in which the sulfite dispersal through inhaled sulfur dioxide was controlled mostly to the main airways of the lung (Gunnison et al., 1987).

30.6 Uses

- The role of sulfur dioxide as a food additive can be classified as follows: (1) inhibition of nonenzymic browning; (2) inhibition of various enzyme-catalyzed reactions, notably enzymic browning; (3) use as an antioxidant and as a reducing agent; (4) inhibition and control of microorganisms (Roberts & McWeeny, 1972).
- Sulfur dioxide is sometimes used as a preservative (E number: E220) (Macdonald et al., 2004) in alcoholic drinks, or dried fruits and other dried apricots as it possesses antimicrobial properties. In order to preserve the existence of fruit, rather than precluding their decomposition, sulfur dioxide is used as a preservative. Fruits could be given a distinctive chemical taste due to the presence of preservative (sulfur dioxide).
- Sulfur dioxide is capable to decolorize the substances in the existence of water; subsequently, it has excellent reducing properties. Due to the reducing properties of sulfur dioxide, it is used as a convenient reducing bleach for papers and mild materials such as clothes. The bleaching effect through reducing bleach of SO_2 generally does not precede very long. The atmospheric oxygen reoxidizes the reduced dyes and reestablishes the color.
- Sulfuric acid is produced by sulfur dioxide. In this synthesis, sulfur dioxide is transformed into sulfur trioxide, which forms oleum with oxygen, and finally, produce the sulfuric acid. When sulfur combines with oxygen, it produces sulfur dioxide for the synthesis of sulfuric acid. This process of manufacturing sulfuric acid is called the contact process.
- In the early 19th century, the French emperor used sulfur dioxide as an execution poison to suppress a slave revolt in Haiti, which was specified by Claude Ribbe in *The Crime of Napoleon.*

- Sulfur dioxide blocks the nerve signals from the pulmonary stretch receptors and eliminates the Hering-Breuer inflation reflex (Najafi, 2017).
- Sulfur dioxide gas was utilized as a refrigerant in household refrigerators for the development of freons which are responsible for lowering the temperature of the refrigerator.
- In the synthesis of wine, SO_2 is an important constituent where it is applied in parts per million. SO_2 works as an antibiotic, antioxidant, and protects wine from spoilage organisms and bacteria, oxidation, and advantages to preserving volatile acidity at appropriate levels (Ferrer-Gallego et al., 2018).
- SO_2 is liable for the words that comprise sulfites, which originate on wine labels. The concentration of sulfur dioxide below 10 ppm in wines does not necessitate enclosed sulfites on the level through EU laws and the United States. The tolerable concentration of sulfur dioxide in wine is 350 ppm in the United States, 160 ppm for red wines, and 210 ppm in white and rose wines in the EU. It is the highest limit of sulfur dioxide in wines. Sulfur dioxide is frequently unobserved in wine at low concentration but above 50 ppm, it comes to be apparent in the nose and perception of wine.
- It is applied in the paper industry to eliminate extra chlorine from the bleaching application. Sulfur dioxide is useful to washout color-causing materials from the pulp easily (Bajpai, 2015).
- It is used in lithium batteries as an active cathode (Levy & Bro, 1994).

30.7 Sources of contaminants

Some major sources for the contamination of SO_2 are as follows:

1. Contamination through fossil fuels burning in power plants
2. By extraction of metals from ore through the industrial process
3. Natural sources such as volcanoes
4. By burning process of fuel with a high content of sulfur through heavy apparatus

30.8 Risk assessment

The moderate to a strong concentration of SO_2 penetrates the nose and throat with minimum concentration attained to the lungs. Sensitivity fluctuates from person to person but the short-term exposure (from 1 to 6 h) with SO_2 concentration as low such as 1.0 ppm could be a cause for reversible reduction in the functioning of lungs of the human system. The exposure with a low concentration (5 ppm) of SO_2 for 10–30 min originates the constraint of the bronchiole tubes (Badenhorst, 2007). The objectionable concentration of SO_2 is about 20 ppm, while 6–12 ppm instantaneously irritates the throat and nose; however, 0.3–1.0 ppm is the average concentration of it, which could be perceived individually via taste.

If a very high concentration of SO_2 is released in a closed or confined area, its origin for hypoxemia, pulmonary edema, and finally death within a minute of vulnerability. Pulmonary edema consists of shortness in breathing and coughing, which could last for hours or days after the exposure. These symptoms are subjected to intensify by physical effort. Severe exposure to SO_2 causes permanent injury of the lung. The respirational effects through exposure with low sulfur dioxide concentration revealed the symptoms of asthma (Badenhorst, 2007). In residential locations with elevated levels of SO_2 (13.3–592.7 $\mu g/m^3$), the danger of asthma occurrence was fundamentally found higher than that in residential locations with intermediate levels (6.7–13.3 $\mu g/m^3$) of SO_2 contaminant (Greenberg et al., 2016).

High concentration (100 ppm) exposure of sulfur dioxide is considered instantaneously hazardous to health and life. Fifty to hundred parts per million is considered the maximum exposure concentration in 0.5–1.0 h, which is stated by the American Industrial Hygiene Association (AIHA). Through concern to the atmospheric concentration, instantly dangerous to life AIHA (1955) stated that 400–500 ppm concentration of SO_2 is considered hazardous for even short time of contact and that experience to painful concentrations is not necessarily hazardous if escape is made within a few minutes (Henderson & Haggard, 1943).

30.9 Environmental effects

Sulfur dioxide in high concentrations in the environment could impede the growth of plants and damage the leaves through acid rain. Consequently, a higher concentration of SO_2 is harmful to living systems such as plants and trees. Oxides of sulfur existing in the atmosphere are one of the leading sources of acid rain. Sulfur dioxide forms sulfuric and sulfurous acid with the reaction of rainwater possessing a threat to a sensitive ecosystem. SO_2 and its other oxides could respond to other components in the environment to form very fine particles that diminished eye function.

Sulfur dioxide is measured individually as a major acidifying and atrophying air contaminant.

Sulfur dioxide mainly in the gaseous form enters into plant leaves through the stomata, while here is an indication for an inadequate pathway via the cuticle. The aperture of the stomata is organized largely by normal environmental situations, such as humidity, temperature, and light intensity. Thus, the rate of endorsement of sulfur dioxide is influenced by these exterior aspects and hence the degree of injury. Acute injury in the form of foliar necrosis may also occur through the high concentration of sulfur dioxide, even after a comparatively short period of exposure.

30.10 Hazardousness

Human-made secretions are considerably more important in terms of human health. They are manufactured predominantly through the combustion of sulfur-containing fossil fuels, coal, and oil (Collins, 1971). The concentration of SO_2 in the ambient air, even at low altitude, reasons the destruction of genetic material at the chromosomal level. SO_2 is consequently a clastogenic and genotoxic agent (Yadav & Kaushik, 1996).

Short-term experience (about 10 min) of the SO_2 causes changes in the respiratory symptoms and function of pulmonary. Consequently, the recommended concentration of SO_2 should not be more than 500 $\mu g/m^3$ in this period (10-min exposure). Long-term exposure of SO_2 revealed a threshold for health effects after 24 h on exposure of about the concentration of 5–40 $\mu g/m^3$.

SO_2 exposure causes the irritation of the throat, eyes, nose, and reflex bronchoconstriction. Liquid SO_2 could result in frostbite. SO_2 inhaled by the human can enter only the throat and nose if a negligible amount is inhaled. The person experiences a high concentration of SO_2 by breathing simply via the mouth. Long-term experience of the SO_2 causes various respiratory effects, which are given as follows:

- Reformed sense of smelling
- Permanent impairment of the pulmonary function due to the bronchoconstriction
- Cause for asthma and emphysema
- The improved vulnerability of breathing infection

The leading metabolite synthesized in the blood subsequent respirational experience to SO_2 is sodium sulfite. This metabolite origins for depression of DNA production and encourages chromosomal irregularities in human lymphocytes (Beckman & Nordenson, 1986; Nordenson,

Beckman, Beckman, Rosenhall, & Stjernberg, 1980; Schneider & Calkins, 1970; Mandal, 1990; Meng & Zhang, 1990).

Plasma proteins in the human system transform into S-sulfonates through chronic exposure to sulfur dioxide. These effects are also perceived by the external part of the respirational region at a high concentration of sulfur dioxide, for example, enlargement of the gall bladder and stomach in guinea pigs, hemorrhages in the stomach and paralysis of the hindquarters (U.S. Department of Health, Education, and Welfare, 1969).

30.11 Treatment

SO_2 is a highly toxic gas and released mostly during the combustion process of fossil fuel. So, it is essential to treat it before releasing it into the environment. To serve this purpose, various techniques have been discovered till now, such as wet scrubbing (Ding, Zhong, Zhang, Song, & Bu, 2014), adsorption (Atanes, Nieto-Márquez, Cambra, Ruiz-Pérez, & Fernández-Martínez, 2012), electron beam (Namba et al., 1995), and corona discharge process (Yan, Ninghui, Yimin, & Yanbin, 1998). In the 1990s, Lisovskii, Semiat, and Aharoni (1997) tried to adsorb SO_2 by active carbon treated by concentrated HNO_3. Adsorption is still in use as a successful method to remove harmful gases from the environment. Activated carbon is suitable for the removal of pollutants from flue gas or sulfur compounds from natural gas fuels before combustion. Activated carbon from petcoke proved to be outstanding materials for recovering sulfur compounds from flue gas with the benefit of making use of a petroleum solid waste (Jacobs et al., 2019). There are other technologies for removal of SO_2 (Cheng et al., 2003) that is, by industrial grate furnaces (IGF) and pulverized coal boilers (PCFB) with high flame temperature (1200–1600°C). However, the desulfurization activity of sorbents synthesized from CaO/coal ash/$CaSO_4$ is impacted by several factors. On increasing specific surface area desulfurization activity increases exponentially. Desulfurization activity increases with reaction temperature and also with decreased SO_2 concentration (Lee, Mohamed, Bhatia, & Chu, 2005). Microbial removal was also carried out (Gasiorek, 1994) to remove SO_2 from flue gases by utilizing *Thiobacillus ferooxidans* or *Desulfovibrio desulfuricans* bacterium, which results in a reduction of SO_2 to H_2S on contact with sulfate-reducing microorganism. This H_2S then oxidizes to sulfur by the ferric sulfate in a second stage where ferrous ion generates.

30.12 Risk management

Ventilation abstraction methods with source extraction hoods could prevent excessive exposure to SO_2. Extraction systems should be fitted with a cleaning device having exhaust air function like electrostatic precipitators for precluding the restoration of removed SO_2 at the workplace. All the probable users should be carefully trained for its use, maintenance, and restrictions and undergo retraining due to dependency upon the proper use of equipment for respiratory protection. Respiratory protective equipment (RPE) essentially sees government requirements.

Concerning the delivery of RPE, the succeeding must also be confirmed:

- The delivering equipment must be accurately used and selected.
- The equipment should be retained in virtuous circumstance and effectual work order.
- Detailed information, instruction, supervision, and training need to be compulsorily acknowledged by the employee for the use of equipment.

The following respiratory defense equipment for protecting employees against the extreme experience of SO_2 was suggested by the National Institute for Occupational Safety and Health (NIOSH) in the USA.

- Up to 20 ppm—Any chemical cartridge respirator with cartridge providing protection against the CC.
- Up to 50 ppm—Any supplied-air respirator operated in a continuous flow mode. Any powered, air-purifying respirator with cartridge providing protection against the CC.
- Up to 100 ppm—Any air-purifying, full-facepiece respirator (gas mask) with a chin-style, front- or back-mounted canister providing protection against the CC.
- Emergency or planned entry into unknown concentrations or IDLH conditions—Any self-contained breathing apparatus that has a full-facepiece and is operated in a pressure-demand or other positive pressure mode. Any supplied-air respirator that has a full-facepiece and is operated in a pressure-demand or other positive-pressure mode in combination with an auxiliary self-contained positive-pressure breathing apparatus (Badenhorst, 2007).

As it is well said that prevention is better than cure, it is important to take necessary precautionary measures in and around the industries dealing with sulfur and sulfur dioxide.

- RPE such as gas masks must be accessible near the production site for the workers to attend the massive gas leakage in case of emergency.

- Sensing equipment to sense minor leakages of sulfur dioxide must be installed near the possible leakage sites.
- ESP and AP should be displayed in the SO_2 production area in the local as well as most accepted language.
- Pressure and thickness investigation of all the piping and equipment associated with sulfur dioxide must be checked biannually and before the plant startup.

30.13 Current research

Recently, a gas therapy approach which is based on near-infrared light-triggered SO_2 generation has been developed in which a SO_2 prodrug-loaded rattle-structured upconversion@silica nanoparticles were built to enable light to convert near-infrared light to UV light to activate prodrug for SO_2 generation. SO_2 prodrug-loaded rattle-structured upconversion@silica nanoparticles show intracellular tracking ability, good compatibility, high cell uptake, and high near-infrared light-triggered cytotoxicity. Moreover, efficient inhibition of tumor growth was achieved along with prolonged survival of mice (Li et al., 2019). In recent years, the removal of SO_2 by the utilization of copper tailing has become a matter of research interest. In this process, a high-efficiency complex absorbent is prepared which is derived from copper tailing and $MnSO_4 \cdot H_2O$ and utilized for flue gas desulfurization (Tao et al., 2020). Zhou et al. (2019) used NiO–ZnO nanodisks as a highly sensitive and low-concentration SO_2 gas sensor.

30.14 Sulfur dioxide toxicity

On September 2, 1998, a sulfur dioxide related incident happened in the United Kingdom. The utmost 15-min mean concentration of SO_2 was 653 ppb; this was recorded in Nottingham, and where the 15-min mean standard of 100 ppb set by the Professional Board on Air Quality Standards was exceeded by 19 times (American Chemical Society Subcommittee on Environmental Improvement, 1969).

Various investigations show that introduction of SO_2 to surrounding air contribute to the seriousness and recurrence of asthma intensifications. To determine whether exposure to SO_2 may initiate long haul impact on the prevalence and seriousness of asthma in young adults, some clinical studies have been conducted. For this study, clinical records of 137,040 guys, 17 years of age, who underwent standard premilitary service health examinations during 1999–2008 were inspected.

In a case study done by researchers, the accumulative levels of SO_2 in leaves of plants and soil around a petroleum treatment facility were observed. For this study, four different sites around the treatment facility region were picked. Along with soil samples, plant samples of the dominant species *Myoporum pictum* were gathered from all the chosen sites. Noteworthy degrees of sulfate were found in soil and plant's leaves at all sites compared to the control. The most elevated level of sulfate in soil ($9000 \pm 1200\,\mu g/g$) and plant's leaves ($65,774 \pm 320\,\mu g/g$) were found on the southeast side. This high content of sulfate shows elevated levels of air tainting with SO_2 around the processing plant, which adversely impacted the climate and general well-being in this populated area (Al-Jahdali & Bisher, 2008).

30.15 Safety measures

In 2008, the American Conference of Governmental Industrial Hygienists reduced the short-term exposure limit to 0.25 ppm. The OSHA PEL is presently set at 5 ppm ($13\,mg/m^3$) time-weighted average. NIOSH has set the IDLH at 100 ppm. In 2010, the EPA revised the primary SO_2 NAAQS by inaugurating a different 1-h standard at a level of 75 ppb. EPA withdrew the two obtainable major standards because they would not deliver supplementary public health protection given a 1-h standard at 75 ppb. A 2011 organized assessment established that exposure to sulfur dioxide may be accompanied by preterm birth (Shah et al., 2011). The safety training programs and environmental awareness should be intensified.

30.16 Handling and storage

- Sulfur dioxide must be handled in a well-ventilated area, rather than in a hood through mandatory ventilation.
- During the handling of sulfur dioxide, it is mandatory to wear chemical protection goggles, face shields, appropriate shelter shoes, and neoprene gloves.
- Self-controlled gadgets, supplementary gas masks, and airline gas masks should be appropriately positioned for consumption in emergencies. Instant-acting protection cascades must be accessible in inappropriate situations. Cylinders must always be transported in the standing position, with the controller upmost, and be confidently protected.
- Use the "first in-first out" inventory arrangement to preclude occupied cylinders from being stored for extreme periods.
- Keep away from the reach of children.

30.17 Precautions

- The precautions to be taken by an employee to protect himself or herself against the health risks associated with the exposure include use of protective clothing and respiratory protective equipment.
- Increase ventilation to the area or move leaking containers to a well-ventilated and secure area.
- Stop or reduce leak if safe to do so. Knockdown gas with fog or fine water spray. The resulting sulfuric acid solutions are very corrosive and very toxic. Dike and recover contaminated water for appropriate disposal.

References

AIHA. (1955). Sulfur dioxide. In *Hygienic guide series* (Vol.16, pp.332–333). American Industrial Hygiene Association. quarterly.

Alarie, Y., Wakisaka, I., & Oka, S. (1973). Sensory irritation by sulfite aerosols. *Environmental Physiology & Biochemistry*, *3*(182), 1.

Al-Jahdali, M. O., & Bisher, A. B. (2008). Sulfur dioxide (SO_2) accumulation in soil and plant's leaves around an oil refinery: A case study from Saudi Arabia. *American Journal of Environmental Sciences*, *4*(1), 84–88.

American Chemical Society Subcommittee on Environmental Improvement (1969). *Committee on chemistry and public affairs, cleaning our environment, the chemical basis for action.* Washington, DC: American Chemical Society.

Atanes, E., Nieto-Márquez, A., Cambra, A., Ruiz-Pérez, M. C., & Fernández-Martínez, F. (2012). Adsorption of SO_2 onto waste cork powder-derived activated carbons. *Chemical Engineering Journal*, *211*, 60–67.

Badenhorst, C. J. (2007). Occupational health and safety risks associated with Sulphur dioxide. *The Journal of The Southern African Institute of Mining and Metallurgy*, *107*, 299–303.

Bajpai, P. (2015). Pulp and paper chemicals. Some excerpts taken from Bajpai, P., 2004. Emerging Technologies in Sizing. PIRA technology report, Smithers PIRA, the worldwide authority on the packaging, print and paper supply chains. In: *Pulp and Paper Industry, 25–273.*

Beckman, L., & Nordenson, I. (1986). Interaction between some common genotoxic agents. *Human Heredity*, *36*(6), 397–401.

Cabré, F., Marín, C., Cascante, M., & Canela, E. I. (1990). Occurrence and comparison of sulfite oxidase activity in mammalian tissues. *Biochemical Medicine and Metabolic Biology*, *43*(2), 159–162.

Cheng, J., Zhou, J., Liu, J., Zhou, Z., Huang, Z., Cao, X. , et al. (2003). Sulfur removal at high temperature during coal combustion in furnaces: A review. *Progress in Energy and Combustion Science*, *29*(5), 381–405.

Collins, A. C. (1971). Atmospheric pollution by sulfur dioxide. *Reports on Progress in Applied Chemistry*, *55*, 643–663.

Ding, J., Zhong, Q., Zhang, S., Song, F., & Bu, Y. (2014). Simultaneous removal of NO_X and SO_2 from coal-fired flue gas by catalytic oxidation-removal process with H_2O_2. *Chemical Engineering Journal*, *243*, 176–182.

Ellenhorn, M. J., & Barceloux, D. G. (1988). *Medical toxicology: Diagnosis and treatment of human poisoning*. New York, NY(USA): Elsevier.

EPA (1986). *Second addendum to air quality criteria for particulate matter and sulfur oxides (1982): Assessment of newly available health effects information*. Research Triangle Park, NC: Environmental Criteria and Assessment (MD-22), Office of Health and Environmental Assessment (ORD), U.S. Environmental Protection Agency. December 1986. EPA document number PB87-176574.

Ferrer-Gallego, R., Puxeu, M., Martín, L., Nart, E., Hidalgo, C., & Andorrà, I. (2018). Microbiological, physical, and chemical procedures to elaborate high-quality SO2-free wines. In *Grapes and wines—Advances in production, processing, analysis and valorization* (pp. 171–193).

Gasiorek, J. (1994). Microbial removal of sulfur dioxide from a gas stream. *Fuel Processing Technology*, *40*(2–3), 129–138.

Greenberg, N., Carel, R. S., Derazne, E., Bibi, H., Shpriz, M., Tzur, D. , et al. (2016). Different effects of long-term exposures to SO_2 and NO_2 air pollutants on asthma severity in young adults. *Journal of Toxicology and Environmental Health Part A*, *79*(8), 342–351.

Gunnison, A. F., & Palmes, E. D. (1974). S-sulfonates in human plasma following inhalation of sulfur dioxide. *American Industrial Hygiene Association Journal*, *35*, 288–291.

Gunnison, A. F., Sellakumar, A., Currie, D., & Snyder, E. A. (1987). Distribution, metabolism and toxicity of inhaled sulfur dioxide and endogenously generated sulfite in the respiratory tract of normal and sulfite oxidase-deficient rats. *Journal of Toxicology and Environmental Health Part A Current Issues*, *21*(1–2), 141–162.

Haider, S. S. (1985). Effects of exhaust pollutant sulfur dioxide on lipid metabolism of Guinea pig organs. *Industrial Health*, *23*(2), 81–87.

Health and Safety Commission (1978). *The hazards of conventional sources of energy*: (p. 13). London: HMSO.

Henderson, Y., & Haggard, H. W. (1943). *Noxious gases* (2nd ed.). New York, NY: Reinhold Publishing Corporation.

Jacobs, J. H., Wynnyk, K. G., Lalani, R., Sui, R., Wu, J., Montes, V. , et al. (2019). Removal of sulfur compounds from industrial emission using activated carbon derived from petroleum coke. *Industrial and Engineering Chemistry Research*, *58*(40), 18896–18900.

Kleinman, M. T. (1984). Sulfur dioxide and exercise: Relationships between response and absorption in upper airways. *Journal of the Air Pollution Control Association*, *34*(1), 32–37.

Lee, K. T., Mohamed, A. R., Bhatia, S., & Chu, K. H. (2005). Removal of sulfur dioxide by fly ash/CaO/CaSO4 sorbents. *Chemical Engineering Journal*, *114*(1–3), 171–177.

Levy, S. C., & Bro, P. (1994). Lithium/sulfur dioxide batteries. In *Battery hazards and accident prevention*. Boston, MA: Springer.

Li, S., Liu, R., Jiang, X., Qiu, Y., Song, X., Huang, G. , et al. (2019). Near-infrared light-triggered sulfur dioxide gas therapy of cancer. *ACS Nano*, *13*(2), 2103–2113.

Lisovskii, A., Semiat, R., & Aharoni, C. (1997). Adsorption of sulfur dioxide by active carbon treated by nitric acid: I. Effect of the treatment on adsorption of SO2 and extractability of the acid formed. *Carbon*, *35*(10 – 11), 1639–1643.

Lovati, M. R., Manzoni, C., Daldossi, M., Spolti, S., & Sirtori, C. R. (1994). Effects of subchronic exposure to SO2 on lipid and carbohydrate metabolism in rats. *Atherosclerosis*, *109*(1), 328.

Macdonald, B. C., Denmead, O. T., White, I., & Melville, M. D. (2004). Natural sulfur dioxide emissions from sulfuric soils. *Atmospheric Environment*, *38*(10), 1473–1480.

Mandal, P. K. (1990). Monitoring of cytogenetic effects in occupational workers of a fertilizer factory. XV. In: *Annual conference of indian society of human genetics and national symposium on chromosomes in health and disease*54.

Meng, Z., & Zhang, L. (1990). Chromosomal aberrations and sister-chromatid exchanges in lymphocytes of workers exposed to Sulphur dioxide. *Mutation Research/Genetic Toxicology, 241*(1), 15–20.

Najafi, M. (2017). The sulfur dioxide (SO_2) detection by O functionalized carbon and silicon Nanocones: A theoretical study. *Journal of Cluster Science, 28*, 1345–1353.

Namba, H., Tokunaga, O., Hashimoto, S., Tanaka, T., Ogura, Y., Doi, Y. , et al. (1995). Pilot-scale test for electron beam purification of flue gas from coal-combustion boiler. *Radiation Physics and Chemistry, 46*(4–6), 1103–1106.

Nie, A., & Meng, Z. (2005a). Study of the interaction of sulfur dioxide derivative with cardiac sodium channel. *Biochimica et Biophysica Acta (BBA) - Biomembranes, 1718*(1–2), 67–73.

Nie, A., & Meng, Z. (2005b). Sulfur dioxide derivative modulation of potassium channels in rat ventricular myocytes. *Archives of Biochemistry and Biophysics, 442*(2), 187–195.

Nie, A., & Meng, Z. (2006). Modulation of L-type calcium current in rat cardiac myocytes by sulfur dioxide derivatives. *Food and Chemical Toxicology, 44*(3), 355–363.

Nordenson, I., Beckman, G., Beckman, L., Rosenhall, L., & Stjernberg, N. (1980). Is exposure to Sulphur dioxide clastogenic? Chromosomal aberrations among workers at a sulphite pulp factory. *Hereditas, 93*(1), 161–164.

Roberts, A. C., & McWeeny, D. J. (1972). The uses of Sulphur dioxide in the food industry: A review. *International Journal of Food Science, 7*(3), 221–238.

Schneider, L. K., & Calkins, C. A. (1970). Sulfur dioxide-induced lymphocyte defects in human peripheral blood cultures. *Environmental Research, 3*(5–6), 473–483.

Shah, P. S., Balkhair, T., & Knowledge Synthesis Group on Determinants of Preterm/LBW births (2011). Air pollution and birth outcomes: A systematic review. *Environment International, 37*(2), 498–516.

Speizer, F. E., & Frank, N. R. (1966). The uptake and release of SO_2 by the human nose. *Archives of Environmental Health, 12*(6), 725–728.

Tao, L., Wang, X., Wang, L., Ma, Y., Zhang, Y., & Ning, P. (2020). Efficient removal of sulfur dioxide from flue gas through liquid catalytic oxidation using copper tailing as the in situ Iron ion donator. *Energy & Fuels, 34*(3), 3513–3521.

U.S. Department of Health, Education, and Welfare (1969). *Air quality criteria for sulfur oxides.* Washington, DC: National Air Pollution Control Administration Publication No. AP-50.

WHO (1979). *Environmental health criteria 8: Sulfur oxides and suspended particulate matter.* Geneva: World Health Organization.

Yadav, J. S., & Kaushik, V. K. (1996). Effect of Sulphur dioxide exposure on human chromosomes. *Mutation Research/Environmental Mutagenesis and Related Subjects, 359*(1), 25–29.

Yan, W., Ninghui, W., Yimin, Z., & Yanbin, Z. (1998). SO_2 removal from industrial flue gases using pulsed corona discharge. *Journal of Electrostatics, 44*(1–2), 11–16.

Zhou, Q., Zeng, W., Chen, W., Xu, L., Kumar, R., & Umar, A. (2019). High sensitive and low-concentration sulfur dioxide (SO2) gas sensor application of heterostructure NiO-ZnO nanodisks. *Sensors and Actuators B: Chemical, 298*, 126870.

Index

Note: Page numbers followed by *f* indicate figures, *t* indicate tables, and *s* indicate schemes.

C

I

J

K

L

M

FSC
www.fsc.org
MIX
Papier | Fördert gute Waldnutzung
FSC® C083411

Kolibri 360 GmbH
Ferdinand-Jühlke-Straße 7
99095 Erfurt, Deutschland
produktsicherheit@kolibri360.de